ADVANCES IN THERMAL MODELING OF ELECTRONIC COMPONENTS AND SYSTEMS

VOLUME 3

ASME PRESS SERIES ON ADVANCES IN THERMAL MODELING OF ELECTRONIC COMPONENTS AND SYSTEMS

Series Editors: Avram Bar-Cohen and Allan D. Kraus

Volume 1 Hemisphere Publishing Corp., New York, 1988

Volume 2 ASME Press, New York, 1990

Volume 3 ASME Press, New York/IEEE Press, New York, 1993

ASME PRESS SERIES

ADVANCES IN THERMAL MODELING OF ELECTRONIC COMPONENTS AND SYSTEMS

VOLUME 3

Editors:

Avram Bar-Cohen
Department of Mechanical Engineering
University of Minnesota
Minneapolis, Minnesota

Allan D. Kraus
Department of Electrical Engineering
Naval Postgraduate School
Monterey, California

ASME PRESS
The American Society of Mechanical Engineers
New York

IEEE PRESS
The Institute of Electrical and Electronics Engineers, Inc.
New York

1993

Shrinking product life-cycles and intense global competition in the high-technology marketplace have fundamentally altered the physical design process used in the electronic industry. Commercial success in the 1990s requires that packaging failures, which have come to dominate reliability statistics, be eradicated through the systematic application of a science-based physical design methodology.

This book, Volume III of the biannual series *Advances in Thermal Modeling of Electronic Components and Systems,* opens with a sweeping overview of the physical design of electronic systems—methodology, technology, and future challenges—establishing the need for detailed thermomechanical modeling of thermally-induced failures in electronic systems. Subsequent chapters examine the causes for thermally-induced failures of electronic components and the techniques used to analyze and prevent such failures. It gives a comprehensive bibliography of the literature of the field and will prove especially valuable to project managers and lead engineers, packaging engineers and mechanical analysts, consultants, and academic, industrial, and government laboratory researchers.

345 East 47th Street, New York, NY 10017

The cover figure was created by Gordon Ellison,
courtesy of Tektronix, Inc.

ISBN 0-7918-0018-0
ISSN 0897-7453

CONTENTS

PREFACE

Shrinking product life-cycles and intense global competition in the high-technology marketplace have fundamentally altered the physical design process used in the electronic industry. Commercial success in the 1990s requires that the packaging failures, which have come to dominate the reliability statistics, be eradicated through the systematic application of a science-based physical design methodology. Past experience suggests that the thermal management approach has a disproportionate impact on the life-cycle cost and performance envelope of an electronic product. Consequently, the development of *correct-by-design* and *failure-free* electronic systems is critically dependent on successful thermal modeling, at a level of detail unknown in past decades.

Against this background, ASME Press is pleased to offer *Advances in Thermal Modeling of Electronic Components and Systems — Volume III.* This latest edition to the biannual publication provides an extensive bibliography of the 1989 electronics thermal management literature, an in-depth exploration of future challenges in thermal packaging, and a comprehensive review of thermally-induced failures (three chapters). The present volume opens with a sweeping overview of "Physical Design of Electronic Systems — Methodology, Technology Trends, and Future Challenges," by Avram Bar-Cohen (University of Minnesota), which establishes the need, indeed the exigency, for detailed thermomechanical modeling of thermally-induced failures in electronic systems. The thorny issues associated with the "The Influence of Temperature on Integrated Circuit Failure Mechanisms," are meticulously explored by Michael Pecht, Pradeep Lall (University of Maryland), and Ed Hakim (US Army-Ft. Monmouth) in Chapter 2. The judicious selection and application of "Methodologies for Thermal Analysis of Electronic Components and Systems," at different packaging levels, are systematically discussed and evaluated by Gordon Ellison (Tektronix, Inc.) in Chapter 3. The fourth chapter, by David Steinberg (Steinberg & Associates) illustrates the procedures to be used in preventing "Thermal Stress Failures of Surface Mounted Components." The volume closes with Robert Simons' (IBM-Poughkeepsie) "Bibliography of Heat Transfer in Electronic

Equipment — 1989," which classifies and highlights the 200 papers, articles, and patents published during this year.

The editors are grateful to the authors for providing ASME with this opportunity to meet the information needs of the engineers, scientists, and managers in the electronic packaging community. The production support of Jim and Pat Allen (Allen Computype) and our dedicated editor, Caryl Dreiblatt (ASME Press), are similarly acknowledged.

It is the editors' hope and expectation that this third volume of *Advances in Thermal Modeling of Electronic Components and Systems* will find the widespread acceptance in the electronic packaging community, attained by the 1988 and 1990 volumes (contents at back). Moreover, we anticipate that Volume III will prove especially valuable to:

- project managers and lead engineers in need of a conceptual understanding of thermal packaging trends in high-performance computational platforms and thermally-induced failures in electronic systems;

- packaging engineers and mechanical analysts in need of an update on thermal and thermal-stress analysis techniques for 1st and 2nd level packaging;

- the in-house and independent thermal consultant in need of a critical review of the recent literature and overview of industry trends; and

- academic, industrial, and government laboratory researchers in need of a clear understanding of the state of the art in thermal management and thermally-induced failure of electronic systems, prior to commencing a new research effort.

Avram Bar-Cohen
Allan D. Kraus

Chapter 1

PHYSICAL DESIGN OF ELECTRONIC SYSTEMS — METHODOLOGY, TECHNOLOGY TRENDS, AND FUTURE CHALLENGES

Avram Bar-Cohen
Department of Mechanical Engineering
University of Minnesota

1 INTRODUCTION

1.1 Device Trends

As the 1990s begin, the electronic industry is maturing and coming under the influence of traditional economic forces. As in other economic sectors, high-technology electronic products must now provide a benefit that is commensurate with the "cost-of-ownership." A major commitment to reducing the "time-to-market" characterizes start-up electronics companies and restructured conglomerates alike. Control of manufacturing, assembly processes and costs has come to dominate the attention of company executives but failure to develop a pervasive "Quality Culture" can quickly marginalize even the best staffed technology organizations.

Concurrently, the cadence established by the incessant "Moore's Law" [Moore (1975)] improvements in solid-state microelectronic devices continues to drive a crescendo of performance/price expectations. The remarkable technological achievements of the past decade have failed to quench mankind's thirst for computing power. By the end of this century, it may be expected that super-workstations, delivering performance approaching 2000 MIPS (million instructions per second) and/or 10 TERAFLOPS (trillion floating point operations per second) in far smaller "packages" and lower prices than are common today, may well be commonplace. It remains unclear whether such computing platforms will evolve from the workstation up, driven by companies such as Intel, Stardent and Silicon Graphics, from the supercomputer down, propelled by Cray Research Corporation, Supercomputer Systems Inc., Cray Computer Corporation and NEC, or by the traditional U.S. and Japanese mainframe companies such as IBM, Fujitsu and Hitachi developing modular systems capable of spanning the entire range of computing performance. It has become apparent, however, that microprocessors will be at the heart of nearly all such systems.

A computer central processing unit (CPU) has traditionally required approximately 1 million circuits and in the past has been composed of many hundreds of circuit packages interfacing through tens of printed circuit boards and several common back-planes. By the mid-1990s, increasing integrated circuit densities can be expected to allow a complete CPU to be implemented in a single CMOS chip (or 5–10 bipolar chips) and a 16 to 64 CPU parallel processor computer could then be housed in a single multichip module.

In recognition of this trend, leading US mainframe computer companies, including IBM and DEC, as well as Cray Research Corporation, have recently made a major commitment to incorporate microprocessor technology in future high-end machines. Two such prototype systems, Intel's Touchstone Delta and Thinking Machine's CM-200, have already used large numbers of microprocessors in a "massively-parallel" array to attain a commanding lead in some categories of supercomputer performance [Markhoff (1991)]. Interestingly, longer-term projections suggest that, by the year 2000, it may be possible to combine four, 64-bit CPUs in a single microprocessor on a 2.5 cm, 50 to 100 million transistor chip [Gelsinger et al. (1989)] and to package an entire massively-parallel computer in a single multichip module.

Thus, it is clear that, by the end of this decade, it will be possible to obtain nearly unimaginable scalar, vector and parallel processor capability in small, accessible and relatively cheap computing platforms. The development of such "information appliances" will, however, place extraordinary demands on packaging and physical design. To keep pace with the electronic design of such advanced systems (e.g., signal routing and timing), the packaging approach will, of necessity, need to be "correct by design" and "failure-free" in the operating range of the product. The development of the science-based physical design methodology required to attain this level of sophistication constitutes the primary packaging challenge of the 1990s.

1.2 Packaging Hierarchy

The physical design of high-performance electronic systems such as digital computers and airborne navigation/communication systems, as well as sophisticated medical, laboratory and manufacturing equipment, requires engineering design and development at many distinct architectural levels. As shown in Fig. 1.1, the "packaging" of an electronic system often commences with the design or selection of the structure that houses the silicon integrated circuit or chip (level 1 – the IC package), proceeds to the chip-to-chip interconnect (level 2 – typified by the printed circuit board), continues to the board-to-board interconnect structure (level 3) conveniently represented by the back-plane or "mother" board and concludes at the box, rack or cabinet (level 4), which houses the complete system. In some configurations, such as those relying on direct attachment of the chips to the printed circuit board, the so-called "chip-on-board" technology, the first level of packaging may be absent. Alternatively, the use of very high gate-count ICs may allow the entire electronic system to be placed on a single printed circuit board and truncate the packaging hierachy at that level. The use of high-performance multichip modules, containing as many as 100 bare chips bonded to, and communicating through, a common substrate may eliminate both the first and the fourth level of packaging.

1.3 Aims of Electronic Packaging

In any and all of the packaging levels appropriate to a particular system, the packaging design must, first and foremost, provide the signal distribution needed to support the "architecture," that is, the electronic implementation of the logic design of the system, while introducing the minimum delay in the propagation of signals throughout the network linking logic and memory components. In addition, it must include appropriate means for delivering properly conditioned power and removing dissipated heat, at the prescribed component and/or device temperature, as well as providing mechanical support and environmental protection for both the functional and intercon-

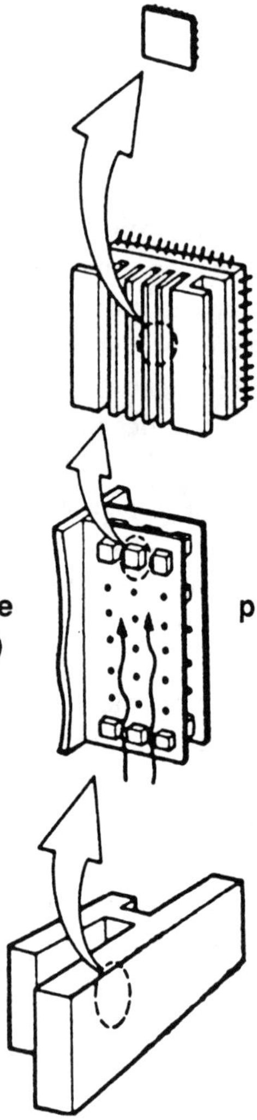

Figure 1.1 Architectural levels in electronic systems [Nakayama (1988)].

nect circuitry. In nearly all of today's electronic systems, the cost, performance (as measured in time-delays) and failure rate are dominated by packaging at one or several of the four levels. The incessant drive toward higher quality, functional density and speed, in nearly all categories of electronic equipment, is thus severely constrained by the available packaging technology. Consequently, the use of conventional packaging and current design methods for the advanced IC chips now becoming available will significantly dilute their potential contribution to system-level gains.

2 THE PACKAGING DEVELOPMENT PROCESS

The selection and/or development of a packaging strategy is one of the critical steps in the evolution of a microelectronic product, and yet, it is just one part of a larger product development process. To assure success, the physical design process must be pursued iteratively, within both the realm of packaging technology and the overall product development process. As shown schematically in Fig. 1.2, packaging decisions cannot be reached in isolation but rather must be tightly coupled to the selection of enabling component and manufacturing technologies, as well as the definition of the market or market segment for the final product.

It is convenient to divide the packaging process into several stages: I – understanding and interpreting the packaging environment; II – generating candidate design concepts based on packaging "building blocks;" III – identifying and selecting among the relevant enabling technologies; IV – hardware modeling and simulation; V – comparison to specifications and decision making; and VI – preparation for fabrication.

The physical design of the electronic system commences with an understanding of the "environment" which has been defined during the initial consultations within the product development team. The integrated circuit technologies selected for use, the system architecture (or partitioning) and performance specifications, including the anticipated operating environment, as well as manufacturing know-how and field-support capability, delineate the boundaries of the "playing field" for the development of the physical design. Thus, the packaging approach appropriate to a "smart" typewriter, to be used in an office or home environment, may be totally inappropriate for an airborne navigation computer. Similarly, the device and chip technologies chosen for a supercomputer may dictate substantially different packaging approaches than those used for a computer-aided-design workstation. Or, as a further example, access to an in-house, sophisticated lamination facility may allow extension of epoxy-glass printed circuit board technology to specifications far beyond the perceived "state of the art" at commercial board fabricators.

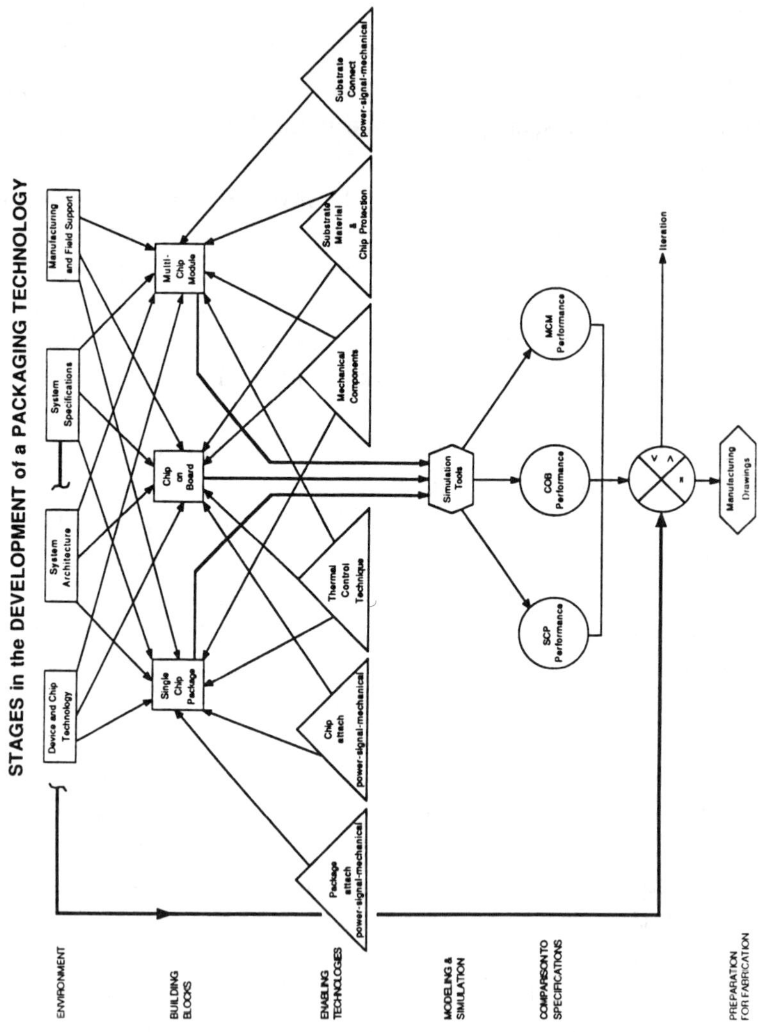

Figure 1.2 Stages in the development of a packaging technology.

The articulation of a candidate physical design of an electronic system often begins with the selection of single-chip packages, multichip modules, or fully-populated (with unpackaged chips) printed circuit boards, as the packaging "building blocks" or field-replaceable-units (FRUs). In developing or purchasing single chip packages (SCPs), it is necessary to choose among the many means available for electrically, structurally and thermally "connecting" the chip to the package, alternative encapsulation or housing materials, and the various pin-insertion and surface-mount techniques by which the SCPs can be connected to the next level of packaging. The elimination of a packaging level in the "chip-on-board" strategy reduces the complexity of the physical design but places greater demands on the signal distribution capability and structural integrity of the printed circuit board. When designing multichip modules, the technology selection process must be extended to the choice of substrate material, as well as the substrate-to-board interconnect.

In configuring an electronic system, the physical design engineer must acquire or develop specific packaging techniques in some or all categories of the enabling technologies, including power supplies, chip attach, chip housing, thermal control technique, substrate material, the method of substrate attachment and mechanical subsystems such as pumps, blowers, heat exchangers, and racks. These must then be integrated into a viable entity, capable of meeting the performance and reliability specifications, as well as price, weight/volume and delivery date targets established by the product development team. During the final stages of product development, detailed thermal, fluid and structural analyses or physical tests will be performed to verify and/or quantify the physical behavior of the complete electronic system. However, the complexity of modern systems and the many alternatives which exist for providing a specific capability or system attribute, require that quantitative evaluations and comparisons, based on a judicious combination of analysis and test results, also be made throughout the packaging development process.

The quantification of packaging capability and performance is greatly aided by the use of "metrics" or technology figures-of-merit (FOMs). Unfortunately, there is no single set of such packaging metrics, rooted in the governing mechanistic relations, that has gained universal acceptance in the industry. FOMs based on the salient attributes of the packaging strategy can, nevertheless, be used to great benefit during the conceptual and preliminary design stages of product development. The efficacy and relative merits of competing choices in each enabling technology category and/or packaging subsystem can be determined by expressing their capability in such familiar parameters as: input/output (I/O) terminal density, propagation delay, temperature or power cycles-to-failure and chip-to-case thermal resistance. Comparison of these values to the system requirements (expressed in similar form) serves to identify the alternatives which meet the "threshold" expectations in each technology category and the most promising candidates among them. The generation of an appropriately-weighted matrix of such FOMs, spanning the entire range of disciplines involved in electronic packaging, can also facilitate the first-level comparison of competing packaging strategies.

Completion of the candidate "building-block" design, by selection of a synergetic set of enabling technologies, sets the stage for the evaluation of system performance and comparison to the initial specifications. The electronic performance of the system, i.e., signal and current flow at all architectural levels and/or the time required to perform a specified set of computations, is typically evaluated with the aid of powerful computer-based simulation tools. In recent years, considerable progress has been made in applying equally powerful thermal and mechanical software to many of the common packaging configurations and efforts are underway to develop integrated packaging simulation tools for use on advanced graphic workstations.

Unfortunately, the understanding of the thermal and structural behavior of packaging configurations is largely inadequate for detailed prediction of performance and susceptability to failure. Furthermore, there are only very few software packages for simulating the manufacturing and assembly processes used in the electronic industry and detailed simulation of complete electronic systems is, even on current computing platforms, very time-consuming. The determination of the costs involved in fabricating an electronic system, i.e., capital costs, yields, work-hours and the failure rate of the resulting system can only be estimated from past experience, with the consequent increase in uncertainty and risk for innovative approaches. This limitation creates a serious barrier to the implementation of advanced packaging techniques, because it is often the price/performance ratio, measured in \$/MIP or \$/MFLOP, or the projected cost of ownership of a specified computing platform, i.e., life-cycle-cost, that is of greatest interest to the customer.

In consideration of all these factors and despite the growing availability of integrated computer-aided-design software for electronic hardware, the use of existing physical design tools is generally limited to the generation and evaluation of generic design alternatives, at the first and second levels of packaging. Comprehensive system-level numerical simulation can be used to validate the conceptual design and technology choices, but testing of working prototypes, under conditions mandated by the development contract or company procedures, is still needed to verify system performance.

When the simulation of performance (physically and/or numerically) and estimation of costs is complete, it is possible to compare the projected performance of the candidate electronic system with the initial requirements. Failure to meet the target values of essential parameters may, at this stage, terminate the development process or lead to a further design iteration. Close agreement between the specifications and the projected performance and costs would, of course, allow the product development process to continue to the final stage, i.e., preparation of detailed engineering specifications for purchase, fabrication and assembly of the appropriate hardware.

3 THE PACKAGING STATE OF THE ART

Although the state of the art in electronic packaging often appears to be the proverbial "moving target," a review of the strategy implemented in several recently released computers can, nevertheless, help to portray the capability and degree of sophistication achieved in packaging technology at the beginning of this decade. It is to be noted that the rapid advances in the gate and I/O density of VLSI chips, as well as the need to provide significantly longer system mean time between failures (MTBF) at substantially lower cost, have focused the efforts of many of the leading manufacturers of electronic equipment on the development of multichip modules.

3.1 UNISYS A-16

Every processor of the Unisys A-16 high-end mainframe is implemented on a single printed circuit board, populated with 24 individual application-specific integrated circuits (ASICs) and 27 multichip modules (MCM), each dissipating 28 W [Fulcher (1990)]. The 46 × 46 × 15 mm module, shown in Fig. 1.3, is based on a 14-layer

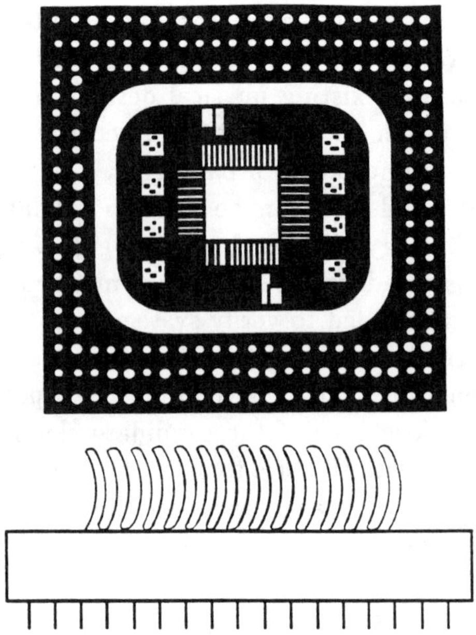

Figure 1.3 Schematic of Unisys A-16 multichip module [Fulcher (1990)].

co-fired alumina substrate with 155 pins, which provides a down-facing cavity for a 10,000-gate, 14 W ECL ASIC surrounded by eight, 1 k $\times$ 4, 1.75 W (average) ECL SRAMs. A copper/tungsten die-attach "slug" serves to conduct the heat dissipated by the ASIC (through a silver-filled glass bond) to an air-cooled heat sink soldered to the slug on the back of the pin grid module. Both the copper/tungsten slug and the copper-clad-invar heat sink were chosen to provide high thermal conductance, as well as thermal expansion coefficients nearly identical to that of the alumina. The convoluted surfaces of the heat sink were designed to operate efficiently in both impinging and axial air flow.

Experimental results for an impinging air flow of 2.9 CFM (approximately 1 m/s across the projected area of the fins), with an ASIC dissipation of 14.1 W and total SRAM dissipation of 12.5 W, revealed the ASIC chip-to-ambient temperature difference to equal 39.5 K, while the SRAMs experienced a 43.5 K temperature difference.

3.2 DEC VAX-9000

Digital Equipment Corporation's (DEC) latest large-scale computer, the VAX-9000, is reported to achieve scalar computing performance 117 times that of the VAX 11/780, with a 16 ns machine cycle-time, and offers a vector processing capability of 125–500 MFLOPS [Pei et al. (1990)]. In order to achieve this performance DEC has, for the first time, adopted multichip modules in the central processing unit and the system control unit of the VAX-9000. Although improvement in chip density was modest, the reduction in interconnect delay was found to be very significant.

DEC's air-cooled multichip module, or Multichip Unit (MCU), shown in Fig. 1.4, is approximately 140 × 178 × 51 mm high and includes a 102 × 108 mm, 9-layer high-density signal carrier (HDSC) laminated to a 9 mm thick, nickel-plated, chrome-copper baseplate. An aluminum heat sink, containing 600 staggered pin fins, is bolted to the baseplate. Copper-polyimide thin-film wiring technology is used in the HDSC to reduce signal delay and to achieve high-density packaging. The HDSC, beneath each chip, is laser cut to allow direct mounting of the chips to the chrome-copper base plate [Godsell et al. (1990)].

At the semiconductor level, the VAX-9000 contains 77 different gate-array chips, 5 custom chips, and 2 densities of a self-timed RAM (STRAM). The MCU can accommodate the face-up, 360 I/O tape automated bonding of up to 8 Motorola, MCA III, gate-array chips or up to 48 STRAM ECL chips with a 5 ns access time. The MCA III, used in the MCU, is 9.8 mm square and 0.5 mm thick, dissipates up to 30 W and provides 10,000 ECL gates with a 200 ps switching-time. The RAM chips are either 3.6 × 4.9 mm or 4.2 × 6.3 mm and dissipate 2 W. At each MCU location, a 50 mm diameter nozzle directs air perpendicularly to the aluminum heat sink. The air flow is reported to range from 0.8 to 3 m^3/min, facilitating the removal of as much as 300 W, but more typically 170 W, from each multichip module. For a power dissipation of 250 W, the MCU is capable of limiting the chip temperature rise to 60° C above the ambient.

3.3 NEC SX-3

Nippon Electric Corporation's (NEC) latest supercomputer is reported [Dohya et al. (1990)] to achieve a peak performance of 5.5 GigaFLOPS in a single processor and 2.9 ns machine-cycle-time,

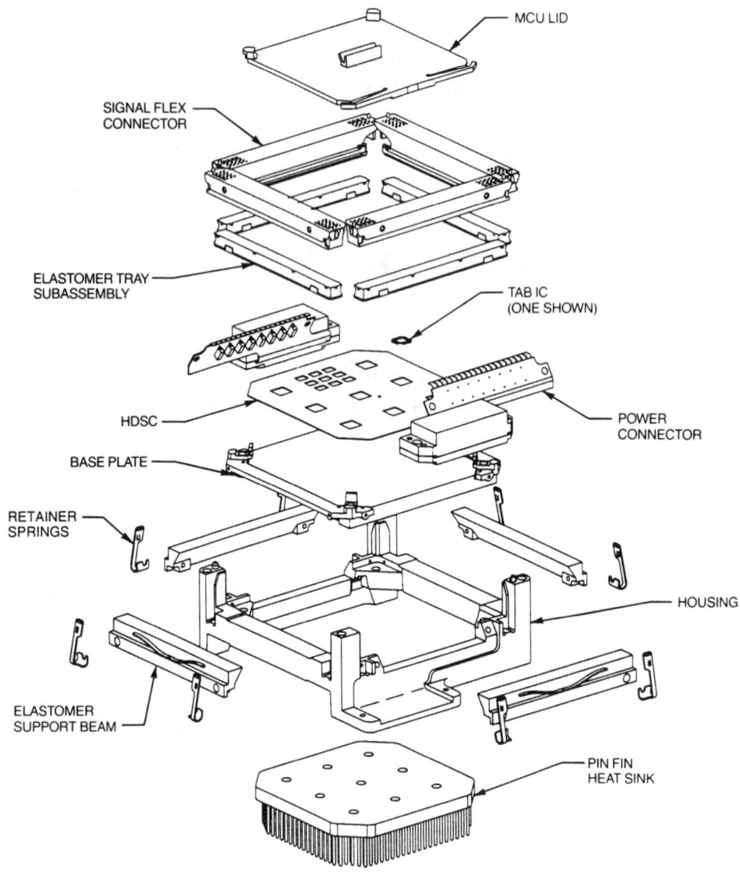

Figure 1.4 Digital Equipment Corporation's VAX-9000 multichip unit—exploded view [Pei et al. (1990)].

roughly half of the previous NEC supercomputer, the SX-2. The packaging strategy employed by NEC relies heavily on the use of a very large size, 4000 W, multichip module (MCP), shown in Fig. 1.5. Six MCPs constitute a single processor in the SX-3.

The packaging technology used by NEC to achieve these extremely high data processing capabilities, features a 3-level packaging hierarchy: flipped-TAB carriers (FTC), a liquid-cooled module (LCM), and a high-density, multilayer printed circuit board. The 18.5 × 18.5 mm FTC contains a 20,000 gate, 70 ps, bipolar low current mode logic (LCML) chip. Each chip is 14 × 14 mm, with 485 TAB leads, and dissipates 33 W. The cache memory FTC contains a 40 Kbit SRAM (1.6 ns access time), implemented with 7000 gates and 487 TAB leads, and dissipating 38 W. In the FTC, a heat spreader, which is currently made of a copper-tungsten alloy, is attached to the top of the LSI chip. This heat spreader transfers the chip power dissipation, through a layer of thermal grease, to a monolithic cold-plate, within which impinging water jets provide efficient heat removal. The chip is electrically mated on its lower surface, via TAB leads, to the glass-ceramic substrate.

To create the multichip module, a 10 × 10 array of FTCs is solder-bumped onto a 225 × 225 × 5.5 mm substrate, consisting of 15 layers of glass-ceramic, with tungsten-metallized inner layers for the power and ground planes, 11,540 pins brazed on its bottom surface, and eight thin-film signal layers separated by polyimide insulating layers, as well as a top layer for the FTC attachment. The overall dimensions of the module are approximately 300 × 300 × 60 mm. At a total water flow rate of 10 liters/min, the SX-3 appears to be able to maintain chip temperatures in the range of 50-55° C, with a water inlet temperature of 25° C.

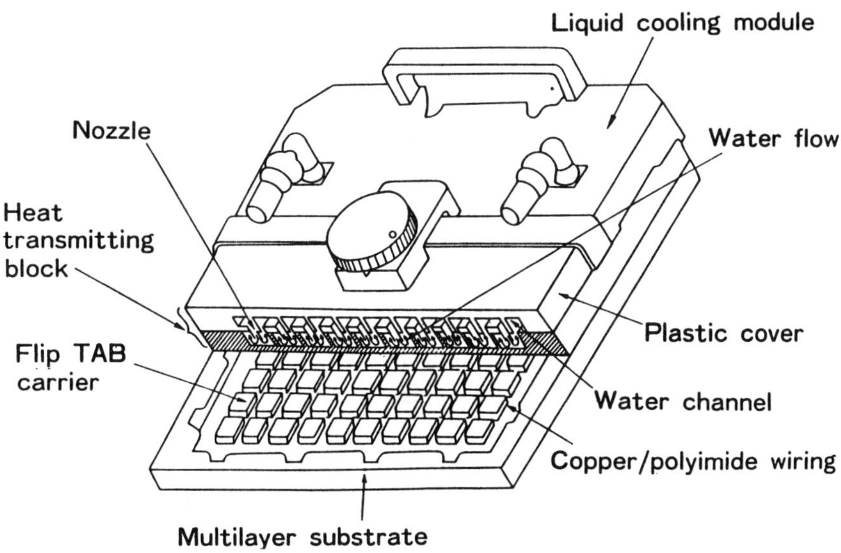

Figure 1.5 Nippon Electric Corporation's SX-3 multichip module (MCP) [Dohya et al. (1990)].

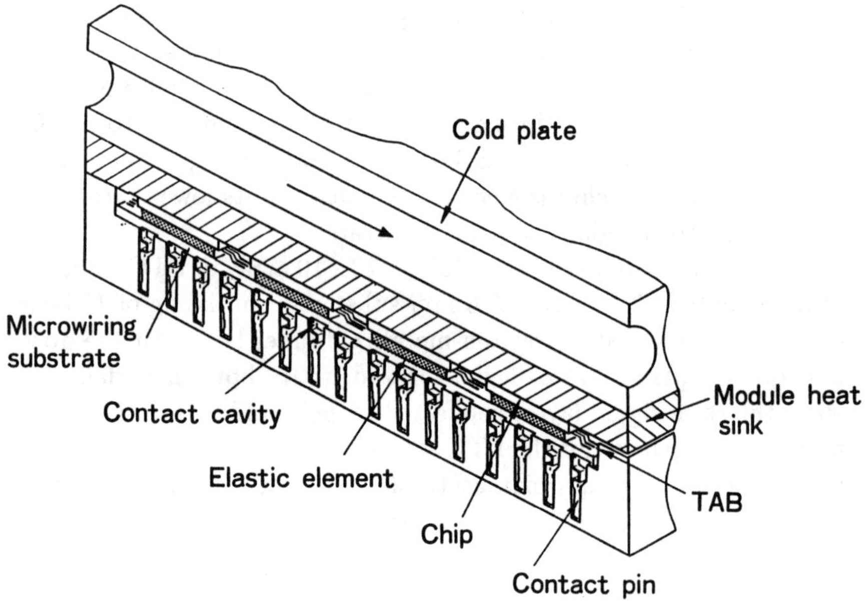

Figure 1.6 Siemens H-90 multichip module (MCM) [Wessley et al. (1990)].

3.4 SIEMENS H-90

Siemens' new H-90 mainframe computer is based on water-cooled multichip modules (MCMs), which are approximately 115 × 115 × 52 mm high and include a high-performance substrate, a module heat sink, and a cold-plate, as shown in Fig. 1.6 [Wessley et al. (1988)]. Each MCM allows the placement of up to 16, 11.5 × 12.7 mm LSI chips (320 I/O, 200 ps switching delay, 5850 gates), dissipating as much as 20 W, or up to 144 memory devices, 3.7 × 3.7 mm in size, with 52 I/Os. The components are mounted on a 16-layer, 1850 I/O, 98 × 98 × 2.1 mm thick microwiring substrate, consisting of fiberglass reinforced, acrylic resin and polyimide layers and capable of electrically supporting as many as 50,000 gates. This substrate is inserted into the connector housing, fitted with press-down contact springs. Face-down TAB is used to attach the components to the substrate. A single micropackaging board, 460 × 478 × 4.2 mm thick, can accomodate up to 9 MCMs.

Heat generated in the chips is transferred, across a dry interface, to an anodized aluminum heat sink, pressed firmly against the chips. Water cooling allows a 20 W LSI chip, in a 280 W H-90 MCM with the specified water inlet temperature of 17° C, to operate at approximately 50° C.

4 THE PACKAGING ENVIRONMENT

4.1 Chip Technology

Extrapolation of current trends suggests that in the mid-1990s, it will be possible to fabricate silicon chips with as many as 10 million CMOS logic circuits and nearly 300,000 bipolar logic circuits per cm^2 of silicon (and considerable on-chip memory) [Cavin et al. (1989)]. The reduction in chip feature size to 0.25 micron and a lowering of the operating voltage to 2.5 V can be expected to lead to dramatic reductions in the silicon transistor switching energy, which is forecasted to reach 5 and 10 fJ/circuit for CMOS and bipolar circuits, respectively [Cavin et al. (1989)]. In this same time frame, it is anticipated that transistor rise times will decrease to less than 200 ps for CMOS and approach 50 ps for commercial bipolar technology.

If all these developments are fully utilized, an unlikely 1 kW ultra large scale integration (ULSI) CMOS chip, requiring delivery of 400 Amps and thermal management of a prodigious 250 W/cm^2, and a 250 W bipolar chip, dissipating approximately 65 W/cm^2 and requiring 100 Amps of 2.5 V power, could become available in 1995. The formidable challenge posed by the power delivery requirements of these chips, as well as advances in BiCMOS devices, can be expected to moderate these trends. Nevertheless, many packaging experts foresee 100 to 200 W chips, with cooling requirements above 50 W/cm^2, in use by the mid-1990s.

The number of leads needed to provide signal flow to and from a chip generally depends on the number of active gates on the chip raised to an exponent less than unity, typically between 0.5 and 0.7 [Vilkelis (1982)]. The exponent in this so-called Rent's Rule [Landman and Russo (1971)] relationship has been found to decrease as chip complexity increases. The large number of gates available on the mid-1990s chips can be expected to substantially increase the required number of chip input/output (I/O) leads. However, the addition of on-chip cache memory and more effective partitioning of logical functions, made possible by higher gate densities, will reduce the Rent's Rule exponent for these advanced chips. Consequently, the mid-1990s leading-edge chips can be expected to require approximately 1000 I/Os [Cavin et al. (1989)] and may allow current inner lead spacings of 3 to 4 mils (75 to 100 microns) to be maintained.

While larger chip sizes, thus, provide greater functionality and ease the restrictions on lead-to-lead spacings, they are unfortunately more likely to experience and/or induce thermal-stress failures. This is especially true for plastic-encapsulated single chips, as for example 4 Mbit dynamic random access memory (DRAM) chips which have grown to cover 70% of the cross-sectional area of the standard dual-in-line package [Nakayama (1988)]. Reliable packaging of these and similar chips requires careful selection, filling and curing of the molding compound and of the die-bond material.

Despite worldwide efforts to accelerate the development of gallium arsenide (GaAs) semiconductor technology, silicon (Si) chips continue to dominate the microelectronic industry and to establish the pace of technology introduction and replacement. As seen in Fig 1.7, successive improvements in both crystal growth processes and photolithography have increased typical chip dimensions from 0.5 cm in 1985 to 1.0 cm in 1990 and can be expected to yield 2.0 cm chips by the mid-1990s [Cavin et al. (1989)]. In parallel with these developments, chip feature sizes have decreased to today's submicron dimension and are forecasted to reach 0.25 microns by 1995.

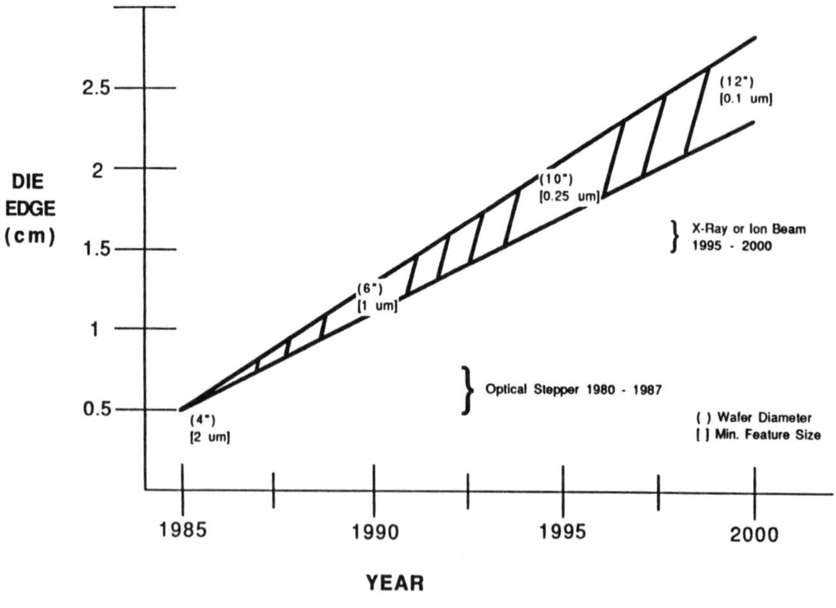

Figure 1.7 Integrated circuit chip size projections.

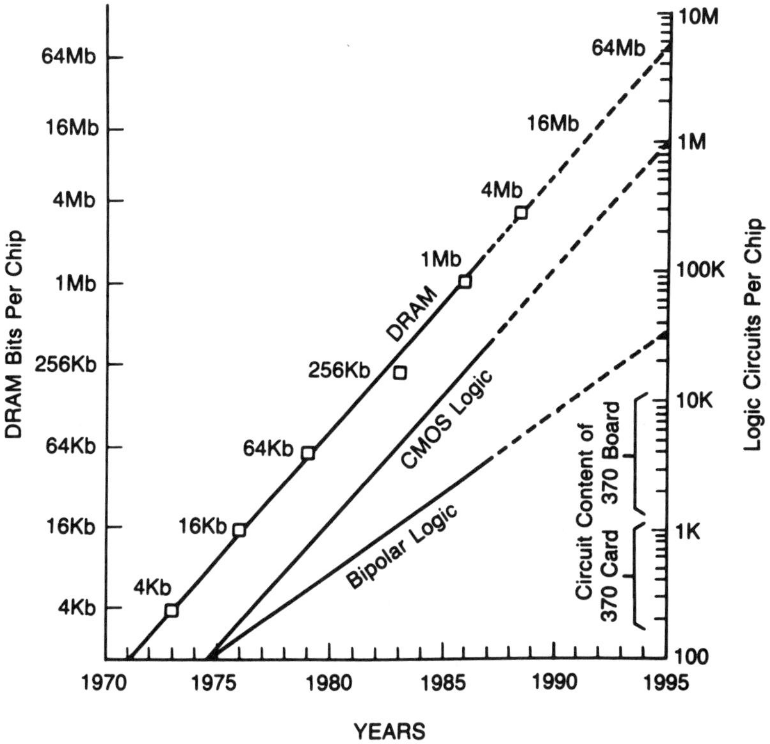

Figure 1.8 Trends in integrated circuit density [Rymaszewski and Tummala (1989)].

The rapid, "Moore's Law" [Moore (1975)], improvements in these two parameters have made it possible to dramatically increase the number of circuits in logic chips and the number of bits in memory chips. The most advanced microprocessor currently available, Intel's 486 chip, provides 1.2 million transistors, while leading-edge components offer in excess of 150,000 circuits for CMOS logic, 20,000 circuits for ECL logic, and 4 megabytes of DRAM per chip (Fig 1.8). As a measure of the rapid pace of innovation in the electronics industry, it is to be noted that by March of 1991, Hitachi Ltd, Fujitsu, Matsushita and Mitsubishi had already announced laboratory prototypes of a 64 MB memory chip, using 0.3–0.4 micron features to place approximately 140 million transistors on a 1 × 2 cm chip [Rosenblatt (1991)].

4.2 System Architecture

The architecture of an electronic system establishes the signal bandwidth, the number of cycles needed to complete a set of instructions (calculations, transactions), and the number of circuits and I/O channels needed at each packaging level. In conventional "CISC" digital computers, implementing so-called "complex instruction sets," 10–20 machine cycles may be needed to complete a single instruction. Alternatively, when relying on "reduced instruction-sets," computers "RISC" can be designed to perform specific instructions in as little as one cycle. As shown in Fig 1.9, the cycle-time required to achieve a specified system performance goal, typically stated in terms of millions-of-instructions-per-second (MIPS), is inversely proportional to the number of cycles-per-instruction. Because the cycle-time is determined by the switching time of the logic transistors, the access time of the memory circuits, and the time required for the signal to flow through the network linking the chips (i.e., the packaging structure), short cycle-times require fast, closely-spaced chips and advanced packaging. Thus, the architecture of an electronic system, whether CISC, RISC or a hybrid, has a profound impact on the packaging technologies considered in the development of the physical design of the system.

4.3 System Specification and Operating Environment

While distinctions in the computer market may vanish in the future, it is still common today to classify computers in categories of increasing computational capacity; starting with personal computers, workstations and minicomputers, continuing on to mainframes, and ending with supercomputers. A similar situation exists for special-purpose electronic systems such as navigation computers, automatic tellers and tomographers. Traditionally, each category was associated with a distinct operating environment, varying from the appropriate military specification on temperature, humidity and vibration for avionics, to carefully controlled conditions in supercomputer and/or mainframe computer centers. Moreover, these categories also have an impact on the office environment for workstations, a factory floor environment for a minicomputer or small mainframe and almost any possible environment for personal computers. Performance in each of these categories is increasing geometrically and by the mid-1990s workstation performance can be expected to exceed

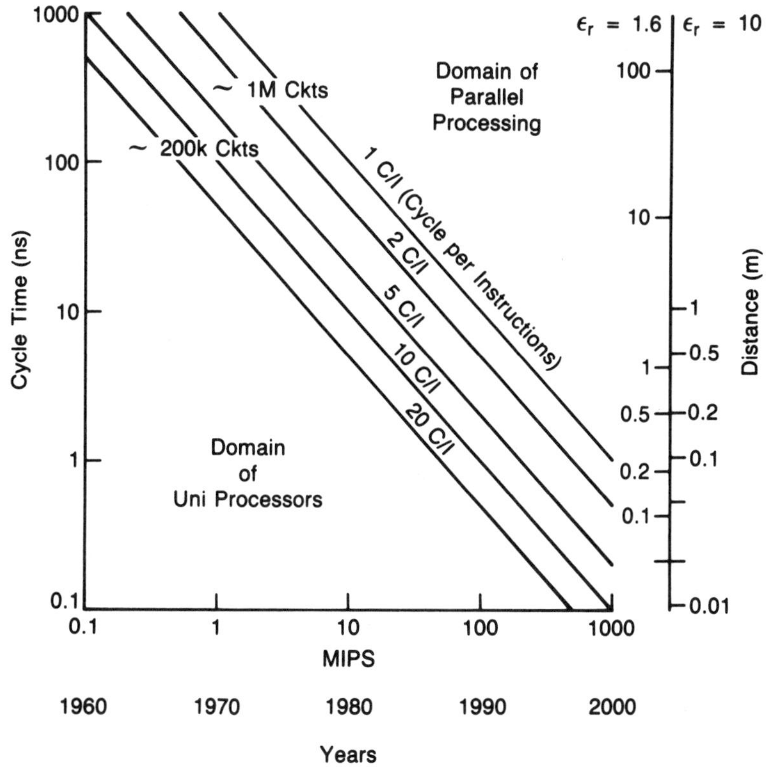

Figure 1.9 Cycle time dependence on computer architecture [Rymaszewski and Tummala (1989)].

100 MIPS, mainframes to reach, perhaps, 500 MIPS, and supercomputers to exceed 100 GFLOPs.

The decision to develop a product in one of these categories imposes a relatively narrow and distinct price/performance ratio target on the product design team. For the larger systems, this is often expressed in terms of the life-cycle-cost or ownership cost of the system. The shipping, delivery and installation requirements, unique to each category, frequently dictate a maximum weight, volume, electrical power demand and cooling load for the system. Furthermore, particular customers may impose requirements on the amplitude and frequency of the acoustic noise generated by the system and the heat sink to be used, as well as a preferred maintenance schedule and maximum allowable system down-time during a year. Clearly, any and all of these factors can substantially affect the choice of packaging technology for the electronic system under development.

4.4 Manufacturing Capability and Field Support

The worldwide competition in electronic systems and the consequent need to reduce the "time-to-market" for any product under development dictates close cooperation between the design and manufacturing engineers or vendors [Smith and Reinertsen (1991)]. To successfully complete a product development cycle in a relatively short time (often just 12 months), it is essential that the necessary manufacturing/assembly techniques and equipment be acquired or developed concurrently with the product design. The capability, reliability, and cost of the available chip encapsulation techniques, inner- and outer-lead bonding, chip bonding, heat sink attachment, printed circuit board fabrication, multilayer ceramic substrate processing, multichip module assembly and the monitoring/distribution of coolants in a complex geometry can all affect the packaging strategy adopted for a particular product.

The reliability of large electronic systems has been improving at a rapid pace and it is to be expected that further attention to reliability in the design process will yield mean-times-between-failures (MTBFs) for 1995 commercial workstations and mainframe computers approaching 50,000 hours. Nevertheless, maintaining customer satisfaction will continue to necessitate rapid and effective correction of in-field malfunctions. To eliminate or reduce machine downtime associated with necessary maintenance, the use of accepted field-support procedures should be addressed, both in the conceptual design, and in the detailed design of electronic products. Alternatively, when new technology is to be used in a product, field maintenance procedures should be developed concurrently with the product design.

5 PACKAGING BUILDING BLOCKS

The physical design of an electronic system generally begins with the selection of the first-level packaging structure which must house and protect the integrated circuit chip. Recent technical developments have considerably broadened the options in first-level packaging, which today include plastic and ceramic single chip packages, the traditional choice; multichip modules, the "high technology" choice; and bare chip-on-board, that is, elimination of the chip package, a prime contender in some application categories.

5.1 Single Chip Packages

Single chip packages can be attached to the second packaging level, usually a printed circuit board, with leads that penetrate into solder plated holes in the board (the "pin-insertion" type), or with leads that are soldered to the surface of the board (the "surface mounted" type). A representative sample of these two single chip package (SCP) types is shown in Table 1.1.

In the dawn of semiconductor applications, chips were housed in metallic transistor packages, or so-called "TO-cans," with 8 to 12 "pin-insertion" leads. For digital IC devices, these were quickly replaced by the dual-in-line packages (DIP), with 14 to 16 leads on 2.54 mm (100 mil) spacings, fabricated of alumina or plastic. The ease of fabrication and assembly has led to the remarkable longevity of the plastic DIPs and the evolution, since their introduction in the mid-1960s, of various derivative packages with as many as 144 leads, including the shrink DIP, with 1.8 mm (70 mil) lead spacings; the skinny-DIP, with a narrow body; the zigzag package (ZIP), with staggered leads; and the single-in-line package (SIP) for narrow profile, low lead count applications. Alternatively, the pin grid array (PGA), providing a uniform distribution of 2.54 mm centered pins across the bottom of a ceramic capsule, has been used extensively for the high lead-count applications. PGA lead spacings are expected to decrease to 1.27 mm by the mid-1990s and to provide in excess of 400 first-to-second level package interconnections for future applications [Rymaszewski and Tummala (1989)].

Surface-mounted packages offer significant advantages over the pin-insertion type and, by the mid-1990s, are expected to account for more than 65% of the first-level packages in assembled systems. Due to the smaller component "footprint" and the ability to attach components on both sides of the printed circuit board, surface-mounted packages have become the technology of choice for high packaging density applications. This technology has also spawned a family of derivative packages, including the small outline package (SOP), with gull-wing or J-bend leads on two sides and on 1.27 mm spacings; the quad flat pack (QFP), with leads on all four sides spaced 1 mm to 0.65 mm apart; the leadless ceramic chip carrier (LCCC), with leads spaced 0.75 mm to 1.27 mm apart on all four sides and the plastic leaded chip carrier (PLCC), providing J-bend leads spaced 1.27 mm apart on all four sides of the package.

Table 1.1 Characteristics of Single Chip Packages

I. SURFACE MOUNTED	SHAPE	TYPICAL FEATURES		
		MATERIAL	LEAD PITCH	# OF I/O PINS
SOP SMALL OUTLINE PACKAGE	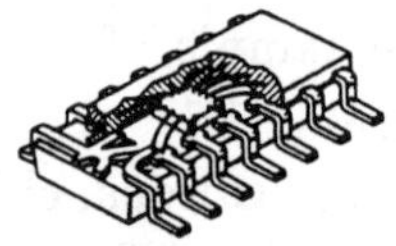	PLASTIC	•1.27mm (50MIL) •2 direction lead	8-40
QFP QUAD FLAT PACKAGE	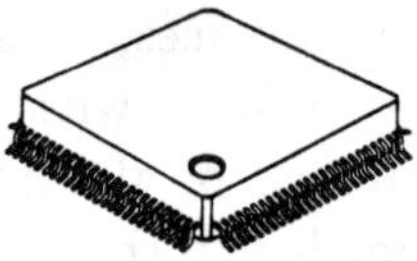	PLASTIC	• 1.0mm •0.8mm •0.65mm •4 direction lead	88-200
FPG FLAT PACKAGE OF GLASS	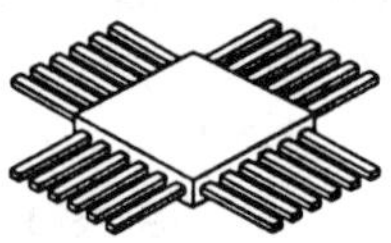	CERAMIC	•1.27mm (50MIL) •.762mm (30MIL) •2 direction lead •4 direction lead	20-80
LCC LEADLESS CHIP CARRIER	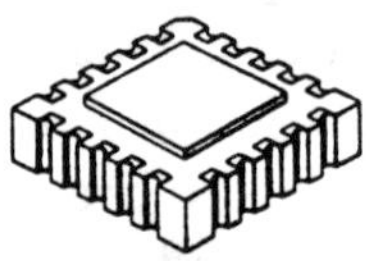	CERAMIC	•1.27mm (50MIL) •1.016mm(40MIL) •.762mm (30MIL)	20-40
PLCC PLASTIC LEADED CHIP CARRIER		PLASTIC	•1.27mm (50MIL) • J-shaped bend • 4 direction lead	18-124
VSQF VERY SMALL QUAD FLAT PACKAGE	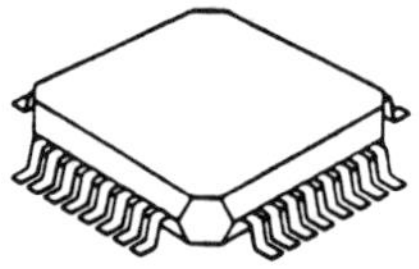	PLASTIC	•0.5mm	32-100

Due to the ease of assembly and attachment, the TAB process, originally developed as a chip-attach technology and typically providing copper leads on 0.5 mm (20 mil) centers in a four-sided plastic (Kapton) sheath, is one of the leading surface-mounting techniques in large volume, low lead-count applications. More recently, TAB has been applied in the packaging of high I/O, high performance, chips in

Table 1.1 Continued

II. THROUGH HOLE MOUNTED		SHAPE	TYPICAL FEATURES		
			MATERIAL	LEAD PITCH	# OF I/O PINS
DIP	DUAL INLINE PACKAGE	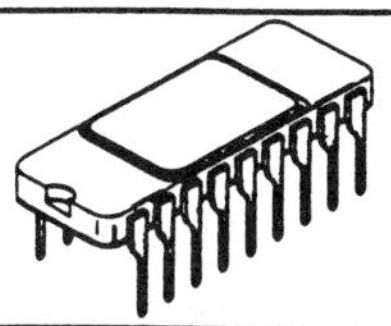	CERAMIC PLASTIC	• 2.54mm(100MIL)	8-64
SIP	SINGLE INLINE PACKAGE		PLASTIC	• 2.54mm(100MIL) • 1 direction lead	3-25
ZIP	ZIGZAG INLINE PACKAGE		PLASTIC	• 2.54mm(100MIL) • 1 direction lead	16-24
S-DIP	SHRINK DIP		PLASTIC	• 1.778mm(70MIL)	20-64
SK-DIP	SKINNY DIP	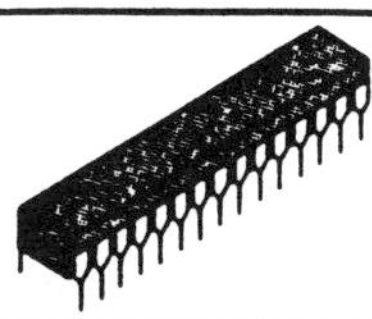	CERAMIC PLASTIC	• 2.54mm • Half-size pitch in the width direction	24-32
PGA	PIN GRID ARRAY	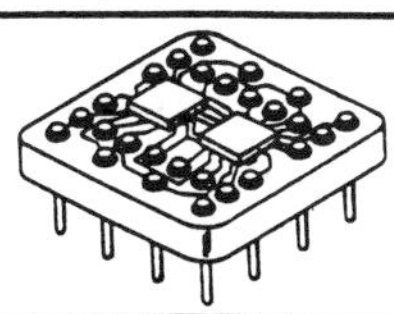	CERAMIC PLASTIC	• 2.54mm(100MIL)	

Data obtained from [Seraphim et al. (1989)] and [Tummala and Rymaszewski (1989)].

mainframes and supercomputers. Plastic encapsulated or bare chips on TAB frames, with lead spacings reduced to between 0.25 mm and 0.125 mm, appear capable of seriously challenging PGA packages for dominance in future high lead-count packaging applications.

5.2 Printed Circuit Boards (and COB/COS)

Despite the many changes in single chip package requirements, the packaging engineer has continued to rely extensively on laminated printed circuit boards (PCBs) to interconnect the many chips required in a functional unit. The copper-clad epoxy glass laminate (better known as an FR-4 board), first used in the mid-1950s, has been the most popular choice, but PCB technology has expanded to include many other options. Designers have chosen PCBs with polyimide, teflon and benzocyclobutene for their reduced dielectric constants and the resulting higher signal propagation rates, and they have relied on quartz glass and carbon fibers to strengthen the laminate. They have also added copper-clad Invar, copper-clad-molybdenum, Kovar and Alloy-42 to reduce the (in-plane) PCB thermal expansion coefficient and, thereby, lower thermal strain/stress on the solder connections and leads. During the past 30 years, PCBs have gone from 4 to as many as 50 layers, line widths have been reduced from 0.25 mm (10 mils) to 0.05 mm (2 mils), via density has increased from 9 to 64 per cm^2 and maximum board size has grown from approximately 30 cm to more than 65 cm on each side [Seraphim et al. (1989)].

These many advances have made it possible for PCBs not only to support high performance packages but also to accept direct attachment of chips in so-called "chip-on-board" (COB) packaging. The direct chip attachment, with wire-bonds or TAB, eliminates one level of packaging and provides a smaller, simpler and cheaper chip packaging approach. A thin layer of epoxy, covering the chip and leads, is generally used to provide some protection against physical contact and moisture penetration, although silicon nitride coatings used alone or interwoven with epoxy or polyimide are of growing interest. The difficulty of testing bare chips prior to assembling the product, the chip I/O density limitations imposed by wire-bonding and TAB and the difficulty in removing heat from the chip have limited the applications of COB in high-performance systems. On the other hand, COB technology has found wide acceptance in consumer products, such as calculators and digital watches, where only a few components need to be interconnected, power dissipation is low and cost considerations are paramount.

Many of the performance limitations encountered in the use of chip-on-board technology can be overcome by replacing the printed circuit board with a ceramic substrate, silicon substrate, glass-ceramic substrate or a composite substrate, consisting of thin film dielectrics/ conductors on the ceramic. Thin film copper/polyimide layers, as well as silicon substrates, are expected to yield 10 micron lines on 25 micron spacings by the mid-1990s. The rigidity, low thermal expansion coefficients and high line density of such substrates facilitate reliance on the "flip-chip," solder-bump chip-attach technique pioneered by IBM. When using the solder-bump chip attachment, heat generated in the chip can flow through the solder connections to the substrate and/or be removed by a fluid or conductor from the unencumbered back side of the chip. Thus, the chip-on-substrate (COS) approach allows a significant increase in chip I/O density and in the chip (or gate) density on the substrate as well as in the allowable chip power dissipation. As a consequence, the COS approach is at the heart of many of the multichip modules in use and under development for high performance applications.

5.3 Multichip Modules

Although hybrid packages with multiple and often diverse chips, attached to a common substrate, have been used for many years in military applications, sophisticated multichip modules (MCMs) only first began to appear during the 1980s. The high packaging density and reliability achievable in co-fired ceramic substrates, silicon substrates and in thin conductor/insulator films placed on ceramic substrates, led companies such as IBM, Honeywell, Hitachi, Mitsubishi, and NEC to offer products incorporating COS in multichip modules [Bar-Cohen (1987, 1988)]. The commercial success of these multichip modules, ranging in size from 27.5 mm on a side to 150 mm on a side and containing between 6 and 130 VLSI chips, motivated a large number of commercial computer companies, as well as military electronic system developers, to explore multichip module approaches. Thus, Rockwell International has recently completed development of a dual-1750A processor multichip module [Johnson (1990)], Boeing and Hughes are reported to have developed in-house multichip module programs and Fujitsu, Hitachi and NEC, as well as Unisys, DEC, IBM and Siemens, have all recently released new, high-performance systems, based on multichip modules [Ohki et al.

(1989), Nikkei (1990), Dohya et al. (1990), Fulcher (1990), Pei et al. (1990), IBM (1987), and Wessely et al. (1988)].

As previously noted, a single, high-performance central processing unit has traditionally required approximately 1 million circuits and in the past has been composed of many hundreds of circuit packages interfacing through tens of printed-circuit-boards and several common back-planes. The early, or first-generation, multichip modules substantially reduced the number of packaging "building blocks" used in a system. However, 10–20 multichip modules were still often needed for each CPU. Using current, or second-generation, multichip module technology, NEC and Fujitsu [Ohki et al. (1989) and Dohya et al. (1990)] have already found it possible to provide in excess of 2 million circuits in a single multichip module. By the mid 1990s, increasing integrated circuit densities can be expected to allow a complete CPU to be implemented in a single CMOS chip (or 5–10 bipolar chips) and an 8-16 parallel processor computer could then be housed in a single multichip module.

Many in the packaging world expect multichip modules to serve as the primary building blocks for the advanced electronic systems of the late 1990s [Ohsaki (1990), Wessely et al. (1990), and Markstein (1991)].

6 ENABLING TECHNOLOGIES

The preceding sections reveal the packaging "environment" in the 1990s to be highly demanding and extremely competitive. Chip gate density and speed continue to increase at a prodigious rate and consumer demands for quality and performance at an affordable price show no sign of abating. Moreover, the steady stream of innovative packaging concepts and technologies emerging from research laboratories threatens to overwhelm product design procedures based on antiquated "build and test" techniques. Yet, while semiconductor failures have decreased precipitously over the past two decades, packaging failures, and especially those associated with manufacturing defects, have come to dominate the reliability statistics. Thus, there is growing concern that the use of conventional packaging and design methods will significantly dilute the system-level gains promised by the highly advanced chips scheduled for release in the near future. To facilitate cost-effective utilization of these microprocessors, gate-arrays, ASICs and DRAMs, the design, manufacturing and performance of first- and second-level packaging will need to improve dramatically relative to the current state of the art.

The successful design of a packaging "building block" involves the selection of a synergetic set of enabling technologies from among numerous alternatives and sub-options and exploitation of opportunities for component and system optimization. Unfortunately, packaging science is still in its infancy and well-defined procedures for packaging design and optimization have yet to emerge. Nevertheless, use can be made of technology figures-of-merit (FOMs), based on one or more significant attributes of the technology, to evaluate and identify the most promising alternatives in each category of enabling technology.

This section will be devoted to the articulation of just such an approach involving die-attach, package attach, thermal management, and substrate material. Because power supplies, fluid handling subsystems and cabinetry are often supplied by external vendors (or other divisions of an electronic systems company), these important technologies will not be examined.

6.1 Die-Attach

The technique and/or material used to connect the chip to a substrate (printed circuit board or ceramic) must provide for electrical function, signal distribution and power delivery; a structural function, permanent physical attachment to the substrate; and a thermal function, heat transfer with low thermal resistance. Certain chip attachment techniques, notably elastomeric pads and solder bumps, can provide all three functions simultaneously. More often, these functions are performed by distinct attachment techniques.

Wire bonding is the oldest and most common technique for attaching signal and power leads to silicon chips. First introduced by Bell Laboratories in the 1950s with thermocompression-bonded gold wires, wire bonding has grown to encompass ultrasonic and thermosonic bonding with gold, aluminum-alloy, silver and copper wires, typically 25 microns in diameter. The lead frame used in tape automated bonding (TAB) replaces the discrete wires with an etched foil (typically copper), coated by a polymer film (often polyimide), and provides 20 to 40 micron thick leads on all four sides. TAB lead spacings range from approximately 0.25 and 0.5 mm for the inner and outer leads, respectively, in low end applications to 0.15 mm and 0.3

mm for implementation in supercomputers and mainframes. In the common TAB implementations, the leads on one side or on all four sides are bonded simultaneously. It is anticipated that during the next few years, TAB technology will evolve to less than 50 micron inner-lead spacings and less than 100 micron outer-lead spacings and provide improved electrical and thermal performance [Koopman et al. (1989)]. At this inner-lead spacing, a 12.5 mm chip could accommodate the 1000 I/O requirement anticipated for the mid-1990s.

While wire bonding and TAB are "peripheral" interconnect techniques, which offer closely spaced attachment pads along the perimeter of the chip, the solder-bump technique, which provides uniformly distributed I/O pads across the base area of the chip, is the most successful representative of the "areal" attachment category. Solder bump interconnects were first introduced in the mid-1960s by IBM, where they are known as controlled collapse chip connections (C4), and have found growing acceptance in the packaging of high-performance chips. In the IBM implementation, solder bumps are deposited on wettable metal terminals on the chip and a matching footprint of solder wettable terminals on the substrate. The bottom surface of the chip is then aligned with the terminals on the substrate, held in place with a weak adhesive, and then heated in an oven to simultaneously reflow the solder and establish the interconnect at all of the terminals [Koopman et al. (1989)]. Although in typical solder-bump applications, 0.1 mm bumps are placed on a square grid with 0.25 mm centers, the most dense array used to date has involved 25 micron solder bumps on 60 micron centers to provide 16,000 input/output terminals for a single chip [Weston (1985)].

Other "areal" interconnect techniques presently under development include the use of fine gold bumps (10 micron bumps on 50 micron centers) pressed together by the shrinkage of a cured epoxy adhesive [Hatada et al. (1987)] and reliance on electrical conduction from an array of chip pads to substrate pads through a thin layer of electrically conducting epoxy.

Elastomeric pads with embedded electrically-conducting rubber contacts have also been used to electrically mate chips and substrates and to provide mechanical cushioning against vibration in digital watches. The relatively large pads and substantial loading pressures needed to minimize contact and conduction resistances limit the application of this technique.

As the density and speed of silicon devices increase, the electrical performance of the chip may well be compromised by shortcomings of the first-to-second level interconnect. Although it is recognized that several factors, including I/O density, lead capacitance and lead inductance, play an important role in the performance of the interconnect structure, no composite figure-of-merit, which captures and quantifies the interplay of these factors, has found widespread acceptance. In the absence of such a composite FOM, use can be made of several FOMs, each based on a specific attribute, to evaluate and compare the electrical merits of the available chip interconnect techniques.

Figure 1.10 displays a comparison of wire bond, TAB and solder-bump technologies using the I/O density as an FOM. A high value of this FOM is desirable, as it will maximize the number of I/Os per chip or minimize the chip area needed to provide the necessary number of input/output terminals. A lead spacing of 100 to 150 microns has been assumed for wire bonding, 50 to 100 microns for the inner-lead bond in TAB, and 60 to 250 microns for solder bumps. Clearly, the smaller the lead spacing the higher the I/O density provided by any of these techniques. More significantly, the density of the "peripheral" attachment methods (TAB and wire bond) is seen to decrease with increasing I/O count. Consequently, reliance on peripheral attachment will require a progressive decrease in lead spacing to keep pace with the increasing I/O counts of more complex chips. Alternatively, the I/O density for the areal solder bump technique does not vary with the number of I/Os and, once developed, can be extended to more advanced chip technology.

A similar comparison of TAB, wire bonding, and solder bumping on the basis of the lead inductance as an FOM, was reported by Koopman et al. (1989). Due to significant variations in lead length within each technology category, this FOM was found to display a broad range of values and substantial overlap among the three techniques. Thus, the potential exists for using any of these techniques to meet a specified inductance or switching noise requirement.

The first-level packaging structure must not only provide the electrical interconnect, but must also structurally support the chip and establish a path for effective heat removal. The physical attachment of wire-bonded silicon chips in ceramic chip carriers and PGAs is most often accomplished with a gold-eutectic bond, while epoxies are used in attaching wire-bonded chips to lead-frames and heat spread-

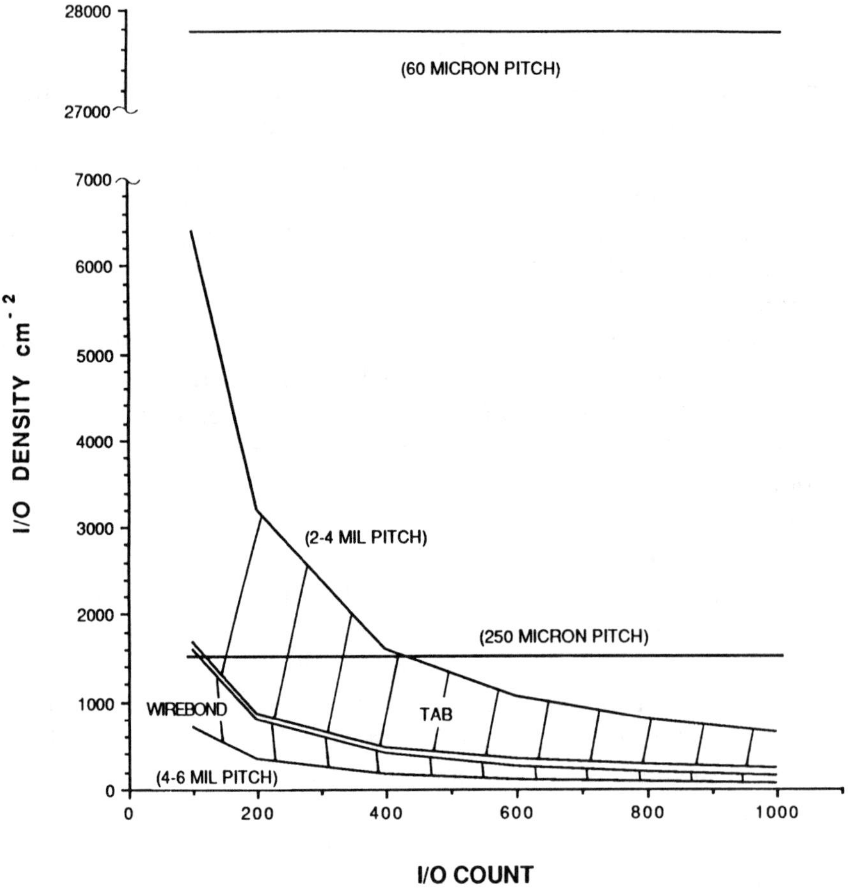

Figure 1.10 Limiting values for chip I/O lead density.

ers in plastic packages. Chips, which are electrically interconnected by TAB, are also attached, most often, with a layer of epoxy. When solder bumps, conductive epoxies or conductive elastomers are used for the signal, power and ground leads, they are also called upon to carry the structural load.

In all but the chip-on-board packaging configurations, where the flexing of the printed circuit board under static and dynamic loads can lead to chip-attach failures, the most severe mechanical stresses in chip packages result from differential thermal expansion/contraction between the chip and the package and/or substrate. This differential thermal expansion is encountered during chip and package temperature changes and results from differences among the thermal expansion coefficients of the constituent materials. The cooling of the chip package (or multichip module) following the assembly process, environmental temperature variations and component heating during chip operation all lead to thermal stress in the chip, package, leads and die-attach layer. While cool-down from the assembly process is generally a single occurrence, it is not uncommon for a chip package to experience more than 1000 fluctuations in environmental temperature and power dissipation during its operating life. These latter events produce a cyclic variation in the induced thermal stress and may result in a fatigue failure of the chip package. Consequently, it is common to impose 1000 temperature and/or power cycles on devices undergoing reliability testing.

When epoxies and solders are used for chip-attach, their relatively high compliance serves to structurally decouple the expansion-mismatched materials and substantially lowers the normal stresses in the chip and substrate. However, the resulting differences in displacement across the relatively thin die-bond layers makes these materials especially vulnerable to shear stress failures in the bulk of the material (cohesive failure) or at the interfaces (adhesive failure). Thermostructural analysis of epoxy and solder die-bonds, as well as solder-bump chip attachment, suggests that, for a given temperature excursion, the shear stress in the die-bond layer can be reduced by limiting the bond area to the center regions of the chip, increasing the thickness of the bond layer (or height of the solder bumps) and selecting a material with a low shear modulus [Suhir (1988)]. The ratio of the maximum shear stress to the adhesion or rupture strength of the die-bond material can serve as a thermostructural FOM for chip attach. Low values of this FOM are indicative of a wide safety margin and are to be preferred, but process-related variations in die-bond adhesion and rupture strength make it difficult to accurately quantify this FOM.

To facilitate removal of the heat generated in the chip, the first-level packaging technique must connect the chip to a low resistance thermal path. In wire-bonded and TAB-attached chips, heat generally flows through a gold-eutectic or epoxy layer into the substrate, heat spreader or, for COB packaging, into the printed circuit board. When flip-chip designs are used, with solder-bump or TAB interconnect, heat can be removed from the back of the chip by low-pressure contact with a thermal conductor or conduction through a thermal paste, epoxy or solder bond. Due to the relatively high heat flux at the chip surface, great care must be exercised in the design and implementation of this interface. Consequently, the thermal interface resistance constitutes a convenient thermal FOM for die-attach techniques.

The thermal efficacy of a variety of thermal interfaces is compared in Fig. 1.11, where a nominal 50% voidage has been assumed for the bonding layers. It may be noted that only very thin, void-free, epoxy or grease layers would be capable of transferring the projected heat flux of 50 W/cm^2 at a typical allowable conduction temperature drop of approximately 10° C across this interface.

6.2 Chip Housing

The structure housing a chip must protect the semiconductor device and the interconnect leads/wires against environmental degradation while facilitating signal, power and heat flow from the first-to-second-level packaging structure. In the past, applications involving advanced components with strict hermeticity requirements dictated the use of ceramic packages with glass seals, while plastic packages, formed by injection or transfer molding, were used primarily in low-cost consumer products. In the present decade, it is no longer possible to clearly separate the domains of plastic and ceramic packages. Intense price competition in the electronic industry, as well as improvements in chip passivation and plastic encapsulation, has induced many design teams to consider plastic chip housings for a wide range of high-performance applications.

In selecting the chip housing from among the many available options, attention must thus be devoted not only to the I/O density, thermal resistance and signal delay, but also to the cost per lead and documented failure rate in the environment of interest. These performance factors, taken individually or in appropriate combinations, can serve as chip housing FOMs. Regrettably, nearly all these attributes are highly sensitive to details of the fabrication process and, in the absence of process sequence and yield information, generalized conclusions are likely to be misleading. On the other hand, the standardization of lead spacings, across the industry, does make it possible to compare chip housing options on the basis of I/O density, or package area, versus I/O count.

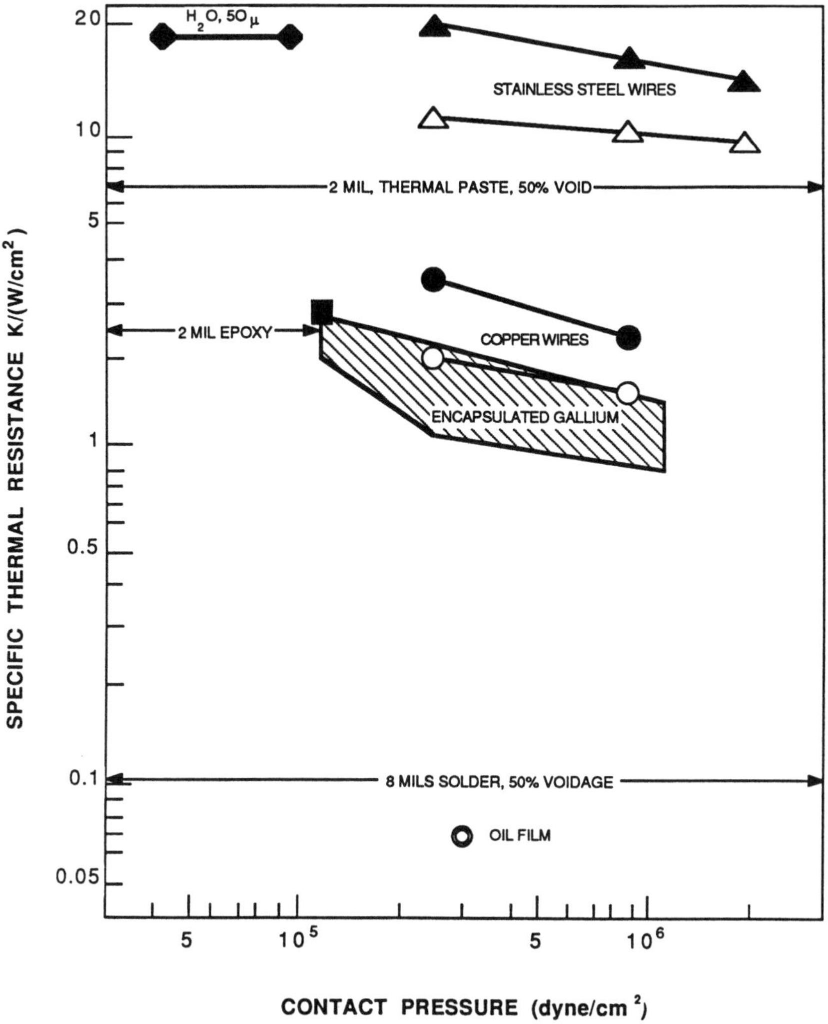

Figure 1.11 Specific thermal resistance at lightly-loaded interfaces [Nakayama (1988)].

Figure 1.12 compares several single chip packages on the basis of the package area needed to provide a specified number of leads. Examining this figure, it may be observed that, at the high lead counts needed for future applications, the areal lead arrays provided by PGA

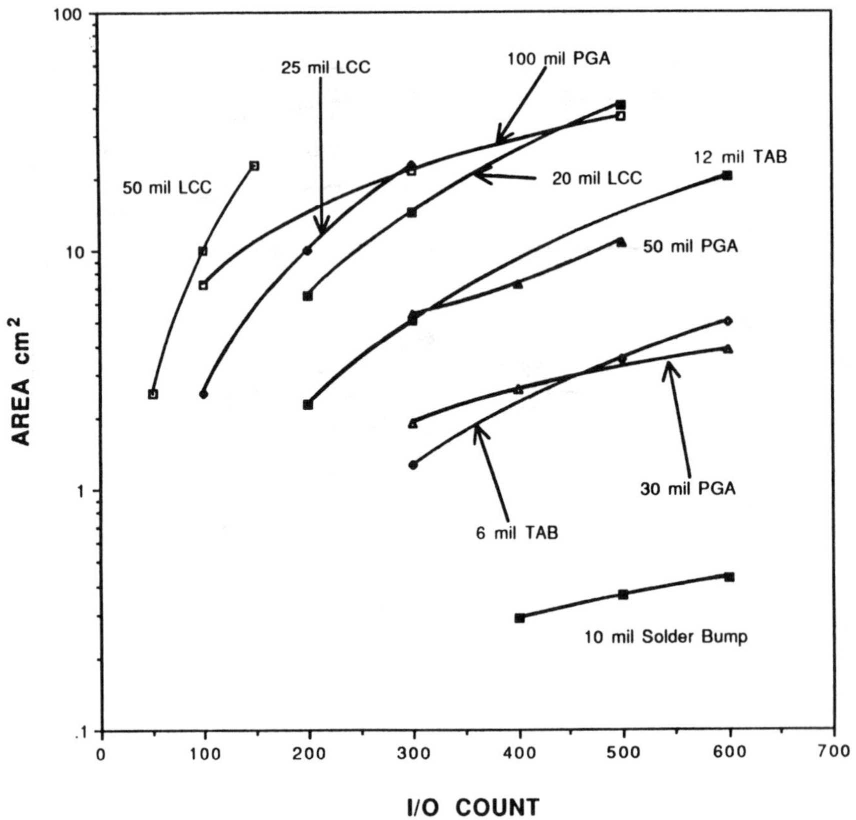

Figure 1.12 Variation of integrated circuit package area with I/O lead count.

and solder bumps offer a significant advantage over all the techniques relying on peripheral arrays. Similar conclusions can be drawn when examining the I/O density of the substrate-to-board interconnect in multichip modules. However, rapidly increasing chip sizes and lower "Rent's Rule" exponents (relating the number of needed I/Os to the number of gates) may make it possible to continue the reliance on peripheral lead arrays even for the most advanced chips. Recognizing that the per lead costs of ceramic and even plastic PGAs are typically 2 to 4 times higher than TAB and wire-bonded plastic leaded chip carriers (PLCC), and 4 to 8 times more expensive than plastic DIPs [Rymaszewski and Tummala (1989)], cost considerations could be expected to reinforce the "staying power" of the peripheral lead arrays. It is interesting to note that, while PGAs are common in today's large multichip modules, peripheral arrays are a frequent choice in the smaller multichip modules and may become even more popular as more effective functional partitioning reduces the I/O density required in future applications of COS and multichip modules.

6.3 Substrate Selection

The rapid advances in material science during the 1980s have considerably broadened the range of substrate options available to the packaging and product development team. Large silicon wafers, multilayer co-fired alumina, cast slabs of silicon carbide and aluminum carbide, as well as printed circuit boards with up to 50 laminated layers, can all be used to interconnect advanced VLSI chips. To select the most appropriate technology, it is necessary to evaluate the electrical, thermal and structural capability of these substrates. Electrical FOMs (dealing with I/O pad density, line density, impedance and signal delay) and mechanical FOMs (dealing with heat diffusion, thermal expansion, stiffness and yield stress) are among the parameters that can be used to identify those alternatives likely to meet the system specifications.

To determine the propagation rate of short rise-time signals in a substrate, it is generally possible to rely on "transmission line" analysis, which relates the propagation delay to the square-root of the dielectric constant of the insulating material. As may be seen in Fig. 1.13, while the delay is just 3.3 ps/mm in vacuum, it increases to approximately 6 ps/mm for polyimide, 7 ps/mm in an epoxy-glass FR-4 board, 11.5 ps/mm in silicon and a largely unacceptable value of 21 ps/mm for silicon carbide. Based on these values it may be surmised that the current enthusiasm for thin-film substrates, with copper conductors in polyimide insulators, is well warranted. By way of contrast, Fig. 1.13 also reveals that, due to their higher dielectric constants, component density on an alumina substrate and silicon substrate (so-called silicon-on-silicon packaging) would have to be nearly 2 and 2.7 times higher, respectively, than on an FR-4 board to generate a comparable propagation delay.

Comparisons such as these suggest that it may be premature to discount the possibility of using printed circuit boards as substrates in advanced applications. The large number of signal layers coupled with the relatively high via density achieved in the late

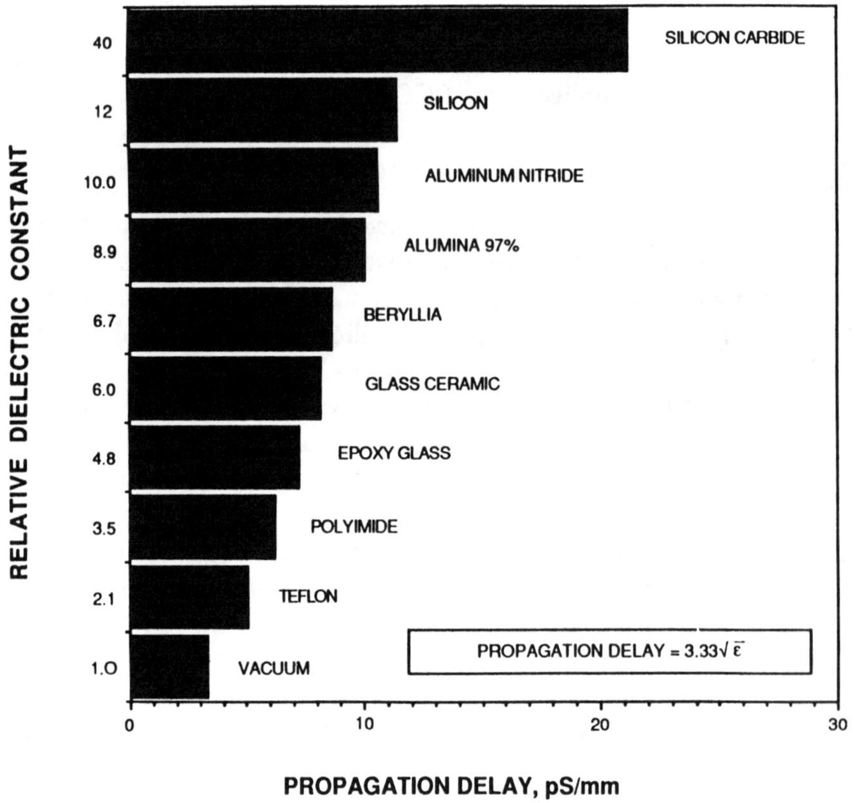

Figure 1.13 Signal propagation delay in electronic substrates.

1980s printed circuit board technology have made it possible to overcome the large line-spacing limitation of PCBs and permit the use of laminated boards to electrically interconnect high gate-density, high speed IC packages.

In layered composite structures, such as chip-on-board and chip-on-substrate configurations, as well as in populated printed circuit boards, spatial and temporal temperature variations, operating through the thermal expansion coefficient of each material, lead to thermally-induced strain and stress. These mechanical loads often threaten the integrity of chips, boards and packages but it is the bond layers and joints bridging the two packaging levels that appear most vulnerable to such thermally-induced failures [Engelmaier

(1984) and Suhir (1988)]. The mechanical criteria for the selection of a substrate material are, therefore, largely related to the thermal strain/stress behavior of these composite structures. However, in selecting a substrate material from among the ceramics, silicon and the organic-based printed circuit boards, the designer must consider not only the operating conditions of the populated substrate but also the acceptance and screening tests used by the customer of the product or electronic assembly.

It is often suggested that use of a substrate with the same thermal coefficient of expansion as the attached component (a so-called thermally-matched substrate) will prevent the development of thermal stress. Although this approach is valid for ideal, isothermal, structures, the components constituting a populated PCB or multichip substrate are unlikely to attain and maintain this isothermal condition. The temperature of chips and chip packages cooled by conduction to the substrate will exceed the substrate temperature throughout the operating envelope. Similarly, convectively cooled chips, without attached heat sinks, are generally found to operate above the printed circuit board temperature. The temperature of flip-chips and chip-packages cooled by contact with a conductive structure, as well as chip packages with highly effective heat sinks, may, on the other hand, lie below that of the substrate or board. In these latter, relatively high-power configurations, the large currents flowing through the power distribution layers result in significant heat generation in the board or substrate. This acts to further increase the substrate/board temperature relative to the chip or chip package.

Temperature cycling of electronic components and assemblies is a common technique for the evaluation of product integrity and reliability. In such a screening and/or acceptance test, the assembly is alternately heated and cooled in an oven or in dielectric liquid baths, typically in the range of -55 C to $+125°$ C. Neglecting the generally modest effects of differences in the thermal time constants of the various components, such tests induce nearly isothermal changes in the entire electronic assembly. Thus, temperature cycling provides a generic simulation of the fluctuations likely to be encountered during component operation and shut-down (or storage), but is incapable of replicating the internal temperature variations that can be expected to accompany these fluctuations. Alternatively, power cycling tests, in which the chips are periodically activated for a prescribed duration, are generally effective in creating the internal gradients expected to occur in the electronic assembly. However, they may not produce the desired temperature levels. A combination of temperature cycling and power cycling tests may, therefore, be needed to faithfully simulate the thermostructural behavior of an electronic assembly.

Due to the use of relatively thin layers and/or short leads in connecting surface-mounted components to the second packaging level, modest variations in thermal expansion between the two levels can produce enormous shear strain and unacceptably large shear stresses in these structures. Large variations in the thermal coefficients of expansion (TCE) among the common packaging materials, such as silicon (2.5 E-6), alumina (6.3 E-6) and an epoxy-glass laminate (12 to 16 E-6) draw attention to the difficulties encountered in attempting to design a stress-free module or assembly.

Elimination of shear stress across the bond layers and/or leads requires that the product of the thermal expansion coefficient of the substrate and its temperature rise during operation (or testing) equal the product of the component TCE and its temperature rise. Or,

$$\alpha_{\rm sub}(T_{\rm sub} - T_{\rm in}) = \alpha_{\rm comp}(T_{\rm comp} - T_{\rm in})$$

where $T_{\rm in}$ is the stress free initial temperature, $T_{\rm sub}$ is the temperature of the substrate, $T_{\rm comp}$ is the temperature of the component and $\alpha_{\rm sub}$ and $\alpha_{\rm comp}$ are, respectively, the thermal expansion coefficients of substrate and component.

The ratio of these two products can serve as a mechanical FOM for substrate evaluation and selection with values close to unity to be preferred. But, unfortunately, it is rarely possible to obtain, a priori, the necessary information on the temperature distribution internal to the electronic assembly and, by default, the designer is forced to limit his choices to substrate materials which closely match the TCE of the attached component. However, caution must be exercised in pursuing this approach, because, in addition to the thermal gradient effects, residual stress (which often plays a significant role in establishing the stress state of the structure) can vary widely with the substrate fabrication and component attachment techniques as well as with the storage temperature. Furthermore, due to the temperature variation of the thermal expansion coefficient, it is unlikely that "matching" can be maintained over the entire testing and operating temperature range. Consequently, even though thermally-matched substrates are well suited to the near-elimination of thermal stress in temperature cycling tests and can serve to reduce the magnitude of the stress and the associated component failure rate under more realistic conditions, they cannot be expected to provide the desired (thermal) stress-free state in any but the most fortuitous circumstances.

Applying TCE-matching mechanical selection criteria, silicon is seen to be the most appropriate substrate for bare chips. The more common choice of alumina is also generally acceptable, especially when the chip temperature is higher than the substrate. Similarly, printed circuit boards are inherently well suited to interconnecting plastic chip packages. Difficulties do arise in finding an appropriate substrate for chip-on-board and PCB surface-mounted ceramic components. These needs have stimulated the development of several materials with low and/or adjustable TCE values which can be used to restrain the natural thermal expansion of laminated PCBs and achieve effective TCEs compatible with silicon and alumina.

Prime among the PCB restraining materials is the "sandwich" structure of copper-clad-invar (made by Texas Instruments), available in both millimeter and tenths of millimeter thicknesses. The thermal expansion coefficient of copper-clad-invar, shown in Fig. 1.14, is reported to vary from 3 E-6 (close to that of silicon) with approximately 20% copper to nearly 7 E-6 (close to that of alumina) with approximately 50% copper. In thicker "slabs" this material can be used as a "metal core" to add rigidity and thermal conductance, along with a matched TCE, to the printed circuit board. In the thinner "sheets" copper-clad-invar can be used, as implemented by Schnorr and Lutz (1986), to replace the power and ground layers in the PCB and directly reduce the thermal expansion of the laminated structure. This latter approach may overcome the delamination and via cracking difficulties observed with the matched-thermal-expansion metal-core boards, using copper-clad-invar, copper-clad-molybdenum and Alloy-42.

6.4 Thermal Management

The previously discussed trends in the packaging environment appear to suggest that the successful thermal management of electronic systems under development in the latter part of the 1990s, will require the removal of as much as 500 W from a single chip, at a heat flux of

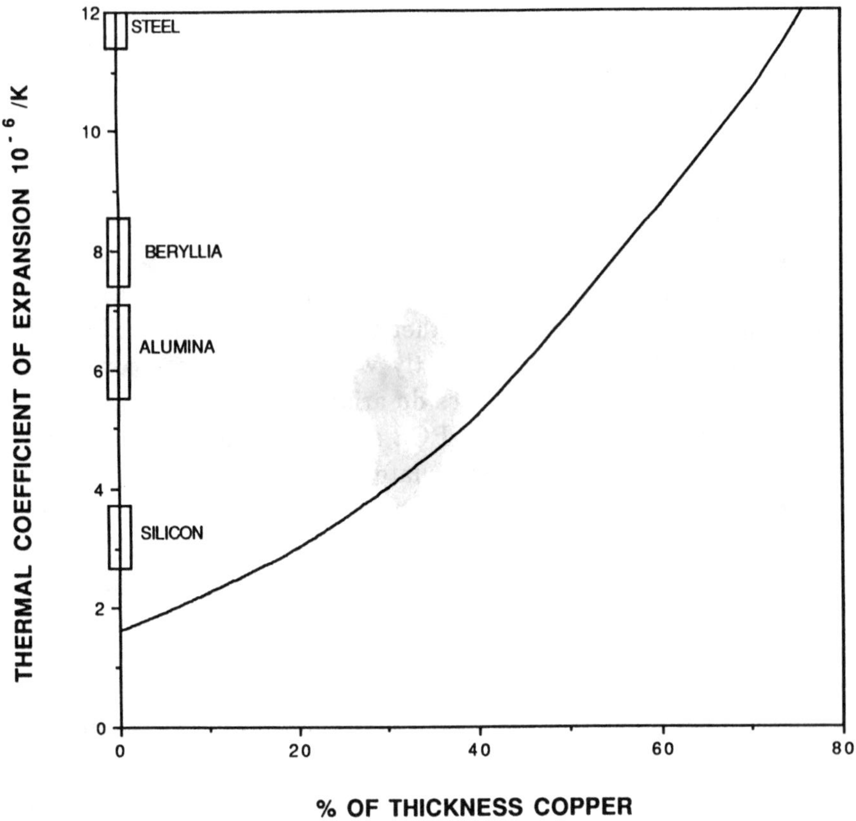

Figure 1.14 Thermal expansion coefficient of copper-clad-invar [courtesy of Texas Instruments].

50 to 100 W/cm^2. For a multichip module, the removal requirement may be in excess of 10 kW at a heat flux of, perhaps, 25 W/cm^2. Module-based, volumetric heat release rates can also be expected to increase dramatically and approach 10 W/cm^3. Silicon chips, with embedded bipolar circuits, have traditionally been maintained at temperatures ranging from 65° C, for commercial computers, to 110° C – 125° C, for military equipment. But while typical, average coolant temperatures have remained constant at 30° C, recent commercial multichip modules display a decreasing trend in chip (or junction) temperatures, to approximately 55° C from the more common 75° C to 85° C. Combining these various factors, it would appear that future electronic thermal control systems would require chip-to-coolant thermal resistances of less than 0.1 K/W and overall areal

heat transfer coefficients of approximately 10,000 W/m^2 K to 30,000 W/m^2 K (1–3 W/cm^2 K) with volumetric heat transfer coefficients greater than 200,000 W/m^3 K (0.2 W/cm^3 K).

It is to be noted that recent developments have introduced several new categories in the thermal management of electronic components. The performance of silicon-CMOS, superconductor interconnects and/or devices and the newly-developed GaAs high energy mobility transistor devices appears to benefit from operation at cryogenic temperatures [Vacca et al. (1987) and Fujitsu (1990)], while power transistors and the emerging Power Integrated Circuits (PICs) will be required to achieve reliable operation at temperatures approaching 200° C. Alternatively, the use of optoelectronic (hybrid optical/semiconductor) and/or optical components can be expected to require near-room-temperature operation and to impose far more severe restrictions on temporal and spatial temperature variations than are common today. These developments are further broadening the scope of thermal design in electronic systems.

Successful thermal control requires that an acceptable "microclimate" be provided at all four packaging levels of the electronic system. The prescribed microclimate must, first and foremost, assure the electronic function of the devices, components, interconnects and assemblies used to implement the architecture of the system. In early electronic systems, catastrophic integrated circuit failure at elevated temperatures was thought to result from changes in the bias voltage, thermal runaway (produced by regenerative heating) and accelerated dopant migration. While these failure modes may still occur during the device development process, improved silicon simulation tools and thermally-compensated integrated circuits have largely quieted these concerns and substantially broadened the operating temperature range of today's silicon-based logic and memory devices.

The use of many near-room-temperature materials, as well as the structural and metallurgical complexity of first- and second-level packaging, has shifted attention to thermally-induced catastrophic failure in the chip housing and interconnect structures. These failures include the melting and/or vaporization of organic materials, the melting of solders, and the thermal stress fracture of leads, joints, seals, die-bonds, packages and printed circuit boards. Unfortunately, thermal stress failures are, by their nature, dependent on the details of the local temperature fields, as well as the assembly, attachment and local operating history of the component. Furthermore, as previously noted, thermal stress generation in packaging materials and structures is exacerbated by power transients, as well as by periodically varying environmental temperatures experienced by most electronic systems during both qualification tests and actual operation. The development of an acceptable thermal packaging approach is especially difficult when different thermal requirements are imposed on adjacent components, as is often encountered in the design of avionics modules and hybrid or power-integrated circuits (PICs). Consequently, the optimal "microclimate" for the various categories of components and interconnect structures, as well as the most suitable combination of conductive, convective and radiative thermal management techniques, can only be defined in the context of specific applications.

In developing a thermal management strategy for an electronic product, it is, consequently, no longer sufficient to merely limit the temperature, as dictated by functional and reliability requirements, at each packaging level. Rather, today's packaging engineer must strive to understand and eliminate or, at least, control the thermally-induced failure mechanisms at work throughout the the electronic system. The attainment of this goal, in the presence of severe electrical, manufacturing cost and reliability constraints, poses a formidable challenge to the thermal packaging community and demands the selection and proper melding of the specific techniques which are best suited to the needs of each packaging level. Nevertheless, past success suggests that skillful crafting of the thermal control system can facilitate successful product development.

It must be noted that the choice of the thermal management strategy for an electronic product has a large impact on the cost, reliability, operating environment and performance of this system. Thus, while thermal control is just one of several enabling packaging technologies, it deserves and must receive special attention.

The precise thermal control that is required to insure the performance and reliability of advanced electronic products, necessitates detailed thermal analysis and empirical validation of the thermal management strategy. Nevertheless, at the conceptual design stage, use can be made of appropriate "metrics" to evaluate and rank the most promising choices for heat removal at the chip and package level. Single-chip packages can be characterized by their internal, or so-called junction-to-case, resistance. The convective heat removal techniques applied to the external surfaces of the package, including the effect of finned heat sinks and other thermal enhancements, can be compared on the basis of the external thermal resistance. The complexity of heat and coolant flow paths in a multichip module generally require that the thermal capability of these packaging configurations be examined on the basis of overall, or chip-to-coolant, thermal resistance. Clearly, low thermal resistances are to be preferred in nearly all categories of electronic packaging.

The internal thermal resistance in a single-chip package is generally defined, as

$$R_{jc} = (T_j - T_c)/q \quad \text{(K/W)}$$

Although two- and three-dimensional conduction effects, as well as non-uniformities in the applied external cooling, limit the accuracy of temperature predictions based on R_{jc}, this parameter does serve to characterize the thermal paths from the active chip volume, through the materials used to support and bond the chip, to the case of the integrated circuit package. Multiple interfaces between the constituent package materials and the use of low thermal conductivity bonding and encapsulation materials, as is often the case in plastic packages, can be expected to result in relatively high thermal resistances. Alternatively, encapsulation in ceramic chip packages or direct attachment of the chip to a thermally-conductive element ("heat spreader"), which extends to the case — or protrudes from the surface — of a plastic package, will generally yield relatively low values of this FOM. The internal thermal resistance of a chip package can be expected to vary from approximately 80 K/W for a plastic package with no heat spreader, to 15 to 20 K/W for a plastic package with heat spreader, and 5 to 10 K/W for a ceramic package or an especially designed plastic chip package. Table 1.2 displays R_{jc} values for some specific classes and sizes (lead counts) of plastic and ceramic single chip packages.

The external thermal resistance of a chip package can be defined as

$$R_{cf} = (T_c - T_f)/q \quad \text{(K/W)}$$

Table 1.2 Electrical and Thermal Characteristics of Some Plastic Packages [Manzione (1990)].

Module Type	No. of Pins	THERMAL CHARACTERISTICS Thermal Resistance Junction-to-Case (θ_{JC}) A42 L.F.	Cu L.F.	Thermal resistance Junction to-Ambient (θ_{JA}) A42 L.F.	Cu L.F.
DIP	8	79	35	184	110
	14	44	30	117	85
	16	47	29	122	80
	20	26	19	88	68
	24	34	20	76	66
	28	34	20	65	56
	40	36	18	60	48
	48	45	—	—	—
	64	46	—	—	—
SOP	8	—	45	236	159
	14	—	29	172	118
	16	—	27	156	110
	16w	—	21	119	97
	20	—	17	109	87
	24	—	15	94	75
	28	—	71	92	71
PLCC	20	—	56	—	20
	28	—	52	—	15
	44	—	45	—	16
	52	—	44	—	16
	68	—	43	—	13
	84	—	42	—	12
PQFP	84	—	14	—	47
	100	—	13	—	44
	132	—	12	—	40
	164	—	12	—	35
	196	—	11	—	30

R_{ext} TYPICAL VALUES FOR VARIOUS FLUIDS AND COOLING MODES

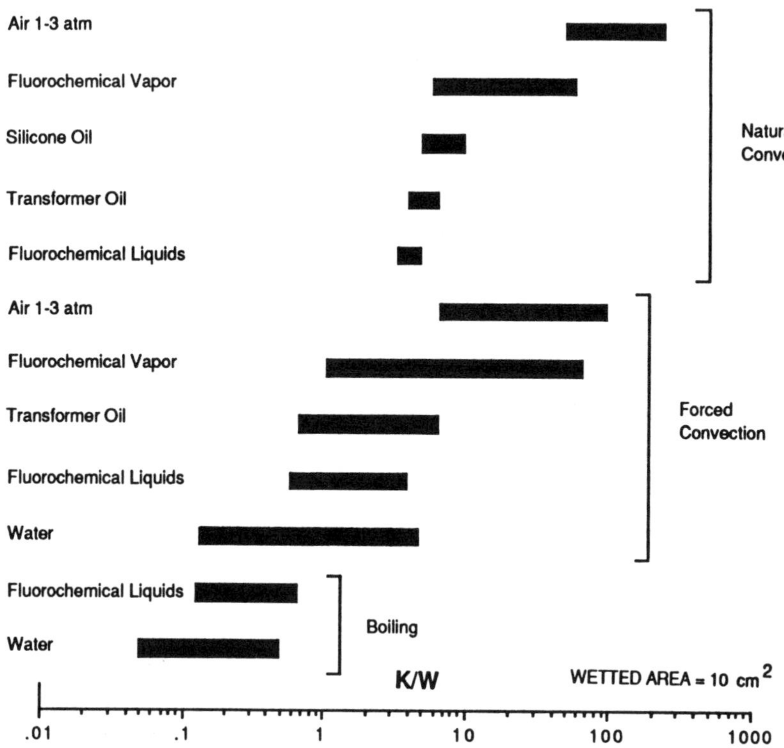

Figure 1.15 Typical external (convective) thermal resistances for various coolants and cooling modes.

When the component is cooled directly by contact with a gas or liquid, the resistance to convective heat removal from the wetted surface is inversely proportional to the product of the heat transfer coefficient and the wetted area, 1/(hA). Values of this external resistance, for a variety of coolants and heat transfer mechanisms, are shown in Fig. 1.15 for a typical component wetted area of 10 cm^2. They are seen to vary from a nominal 100 K/W for natural convection in air, to 33 K/W for forced convection in air, to 1 K/W in fluorocarbon liquid forced convection, and to less than 0.5 K/W for boiling in fluorocarbon liquids. Clearly, larger chip packages will experience proportionately lower external resistances than the tabulated values and conduction of heat through the leads and package base into the printed circuit board or substrate will serve to further reduce the effective thermal resistance.

When the direct cooling of the package surface is inadequate to maintain the desired chip temperature, it is common to attach finned heat sinks or compact heat exchangers to the chip package. These heat sinks can considerably increase the wetted area, but may act to reduce the convective heat transfer coefficient and most definitely introduce additional conductive resistances, through the adhesive used to bond the heat sink to the package and the body of the heat sink. Typical air-cooled heat sinks can reduce the external resistance to approximately 15 K/W in natural convection and as low as 5 K/W for moderate forced convection velocities. Several studies of finned structures for the enhancement of liquid (fluorocarbon) forced convection cooling of 1.27 cm chips are summarized in Fig. 1.16 and are seen to yield external resistances which range from 1 to 0.1 K/W for various geometries and liquid velocities.

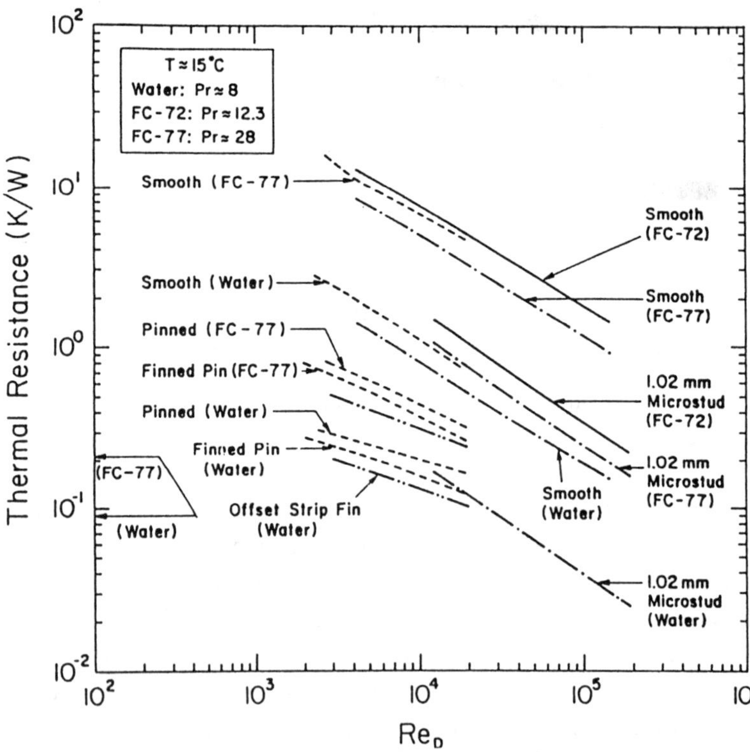

Figure 1.16 Thermal resistances for liquid-cooled heat sinks on 12.7 × 12.7 mm area [Incropera (1988)].

The thermostructural complexity of the multichip modules in current use hampers effective thermal characterization and introduces significant uncertainty into any attempt to compare the thermal performance of these packaging configurations. Variations in heat generation patterns across the active chips (e.g., devices versus drivers), as well as nonuniformities in heat dissipation among the chips assembled in a single module, further complicate this task. To establish a common, though approximate, basis for comparison of multichip modules, it is possible to neglect these variations and consider that the heat generated by each chip flows through a unit cell of the module structure to the external coolant [Bar-Cohen (1987, 1988)]. For a given structure, increasing the area of the unit cell, allows heat to spread from the chip to a larger cross-section, reducing the heat flux

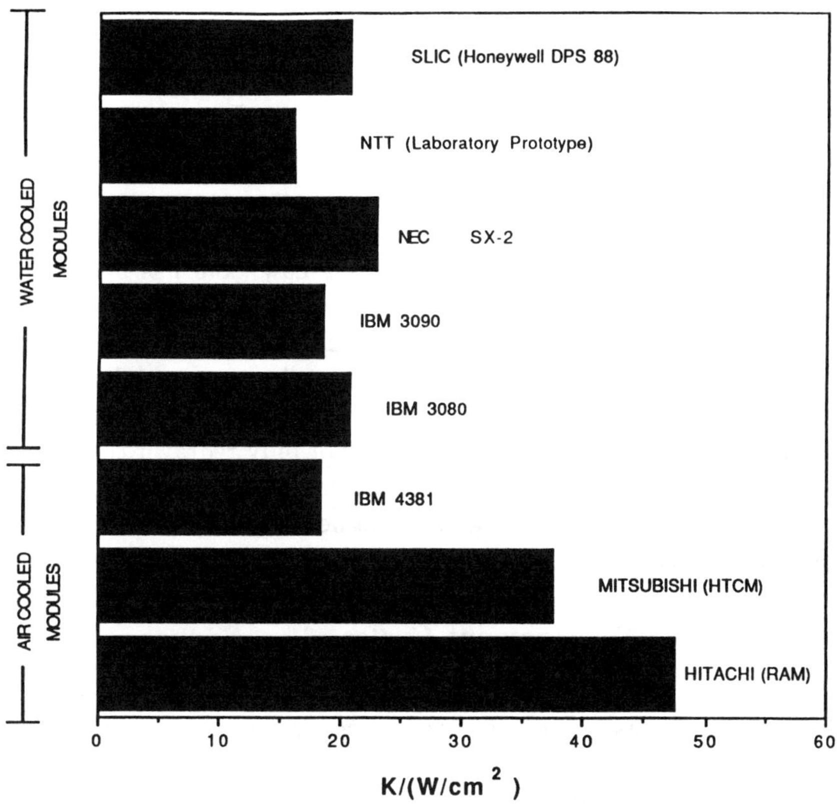

Figure 1.17 Specific thermal resistances for multichip modules [Bar-Cohen (1988)].

at some of the thermally-critical interfaces and at the convectively-cooled surfaces.

Consequently, the overall chip-to-coolant thermal resistance of a multichip module can be expected to vary directly with the packaging density, along a gradient that depends on the details of the thermostructural design, fabrication processes and the coolant. It is this gradient, equivalent to the inverse of the overall heat transfer coefficient in the compact heat exchanger literature, U^{-1}, which can best serve as a thermal figure-of-merit for multichip modules, with lower values to be preferred. Examination of Fig. 1.17 reveals that, in general, water-cooled modules offer superior thermal performance to air-cooled modules and that distinct combinations of materials and interfaces can result in very similar values of U^{-1}. Similar values of U^{-1} signify near-equivalence in thermal performance.

7 TRENDS IN ELECTRONIC PRODUCT DEVELOPMENT

In the mid-1970s, Moore observed that the complexity of integrated circuits, as measured by the number of components per chip, appeared to be doubling each year [Moore (1975)]. This empirical correlation soon became known as "Moore's Law" and has been found to apply to a surprisingly wide range of technologies within, as well as external to, the electronics industry. The integrated circuits, chip packages, printed circuit boards and magnetic storage media, as well as numerous derivative technologies developed under the cumulative influence of this "doubling rate," have not only redrawn the map of human expectations, but have also dramatically shortened the life cycle of electronic products and fundamentally altered the physical design process.

7.1 Product Development Cycles

The useful life of a high-technology product in the 1980s was typically five to seven years and commercial success required the introduction of a major new product every three years. In the early part of this decade, the useful life and product development periods in nearly all segments of the industry have continued to shrink. While supercomputer development appears to still require nearly five years, mainframe development periods have been reduced to approximately two years and in the fiercely competitive market for workstation and personal computers, new products are today created in less than one year. The reduction in product cycle times has had a direct impact on the engineering design and development process, necessitating more rapid evaluation and selection of competing enabling technologies, the expeditious migration of technology from the laboratory to the manufacturing line and substantially reduced "time-to-market" of new product concepts.

The accelerated, technology-driven growth, which had characterized the early years of the electronic industry, has given way to a greater role for market forces in determining success and profitability. The need to strike a balance between technology and market considerations underpins the broad corporate commitment to the achievement of so-called "world-class" quality in consumer products. Defined in the broadest terms, quality in the 1990s has come to include not only the performance and reliability of the product/system, but also a price commensurate with utility/benefit, minimal use of scarce resources, "fail-safe" operation and environmental safety, during both the use and disposal of the product. The achievement of world class quality thus requires a comprehensive understanding of the customer and market, the development of optimum and error-free designs and the prevention, as well as swift elimination, of manufacturing defects. In the past, many of these factors were addressed in new models, upgrades and "cost-reduced" versions of successful products, substantially lengthening the product development process. The success of many Japanese electronics companies in providing such quality immediately upon new product introduction, has permanently altered the relationship between engineering changes (in the product) and product delivery schedules.

In the past, engineering change orders (ECOs) continued to flow from the development engineers to the manufacturing teams well past the release, or first customer ship (FCS), of the product. Reference to Fig. 1.18 shows that this is no longer acceptable. Today's marketplace is intolerant of trial-and-error solutions, demanding instead that performance and quality be designed into the product and that the purchaser of the first, as well as the last, unit sold be able to "plug and play" even the most sophisticated computer system. The commitment to world-class quality, and the resulting requirement that the product design and manufacturing plan be completed prior to FCS, has further shortened the product development period.

A further consequence of the intense global competition for a major share of the high-technology market is a new emphasis on the design and manufacture of customized products for so-called "niche markets." These niche markets, characterized by high volatility, unique requirements and small product volumes, are no longer the exclusive province of start-up and low-volume producers, but rather, have become an important source of revenue for all serious competitors in the electronic industry. Extensive use of flexible manufacturing systems and just-in-time techniques can lead to success in this arena but often necessitates development of product families consisting of many individually-optimized or "personalized," designs sharing a common technology base. Expanded customer use of life cycle costing (LCC) and reliability/availability/maintainability analyses, as well as imposition of acoustic, electromagnetic and other environmental specifications, are also necessitating far greater precision and control in hardware development, than was common in past decades.

7.2 Physical Design Methodology

To deal with the new realities of shrinking product development cycles, demand for world-class quality, the growth of niche products, the expanded use of life cycle costing and the imposition of ever more stringent environmental specifications, leading companies in

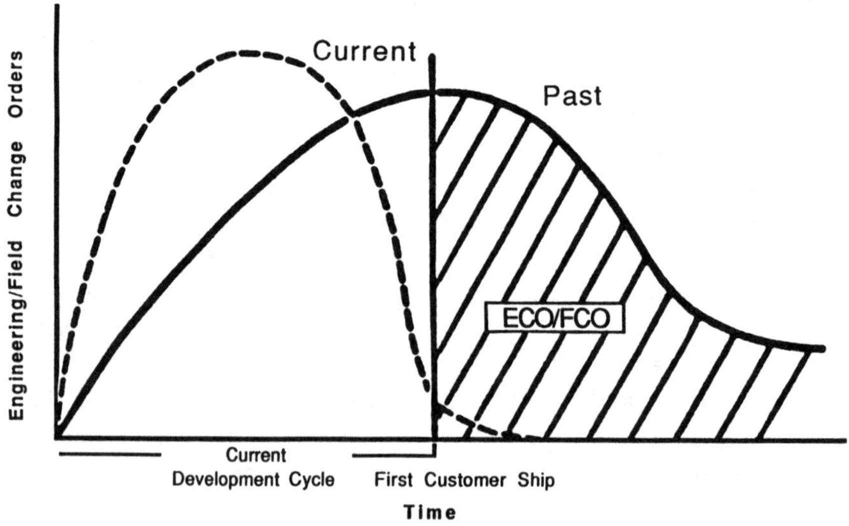

Figure 1.18 Time distribution of engineering change orders.

the electronic industry have substantially improved the methodology and tools used in the physical design of computer systems and have dramatically expanded the role of mechanical modeling and simulation. To rapidly implement the many intricate steps needed to successfully design, manufacture, and market a sophisticated electronic product, it is necessary to considerably shorten each stage and to perform many of the requisite tasks simultaneously. To accelerate the physical design process, mechanical engineers have been empowered with modeling and simulation tools, analogous to those presently available to electrical engineers, and have been challenged to develop product concepts that are "correct-by-design." However, due to physical limitations on the time required to carry out specific tasks, it appears that much of the time savings will need to come from the simultaneity made possible by concurrent, holistic design.

A concerted, effort during the 1980s resulted in the development of a mechanical computer-aided design (MCAD) infrastructure throughout the industry. The early MCAD systems were used primarily for drafting, layouts and parts lists – replacing the use of the drawing board, but only marginally affecting the realm of engineering analysis. In recent years, the well-publicized efforts of software vendors, such as Mentor Graphics, Cadence, SDRC, PCAD, as well as Swanson Associates (ANSYS), Fluid Dynamics International (FIDAP) and Nektronics, research centers such as CALCE at the University of Maryland and a growing number of ASME and IEEE technical publications and conference presentations, have spurred growing use of mechanical modeling and simulation.

The available modeling tools are typically used in three distinct ways: to provide first-order insight into the behavior of the system, to determine the parametric sensitivities and inflection points and to confirm that the final design can perform within the imposed specifications. Stress, fluid flow and heat flux patterns can be obtained by appropriate "post-processing" of the numerical results produced by the relatively large finite-element, finite volume and finite-difference codes. With these it is, thus, possible to determine air flow patterns through cabinets and printed circuit board arrays, regions of high mechanical stress and hot spots on printed circuit boards or substrates for geometrically coarse representations of the system. Unless carefully calibrated against experimental data, present codes can only be expected to provide qualitative results, as is in fact appropriate for the conceptualization stage of the design process, when trial combinations of technologies and packaging "building blocks" are being synthesized. It is in this mode, thus far, that modeling tools have had the greatest impact on the physical design process.

Following empirical determination of key parameters, such as interface thermal resistances, stiffness of elastomeric restraints and inlet velocity profiles, these same codes can be calibrated to yield results for the key variables that are within engineering accuracy (typically $\pm$ 10%). It must be noted that proper calibration can often require substantial, perhaps even massive, effort. But, when completed, the calibrated computer models can be used in place of extensive verification tests, to determine the success of the physical design in meeting the system specifications.

The modeling studies needed to support the detailed design of the candidate packaging configurations, including determination of the preferred materials, dimensions, velocities, conductances and strengths, can be performed numerically with these same codes, with analytical models or with empirical correlations. Despite some claims to the contrary, the mechanical complexity of advanced electronic systems and the throughput limitations of existing computing platforms preclude the use of numerical models for detailed design of all but the most profitable commercial systems. Unfortunately, the development of analytical models, which address the salient failure mechanisms in electronic packaging, has only just begun and empirical data and/or correlations for the relevant configurations are often available only in proprietary databases. Clearly, it is this aspect of electronic physical design that is in the greatest need of improvement.

The availability of computer-based tools, as well as the growing recognition of their importance and utility in tackling the multidisciplinary issues which arise in electronic product development, has not only improved product quality, but has profoundly altered the physical design process. In the traditional mode, the design of primary system elements is decoupled and much of the engineering effort is carried out by individual engineers working on isolated subsystems or components, such as chip packages, printed circuit boards, connectors and racks. The subsequent need to integrate these diverse elements into a single, functioning entity is often the "Achilles' heel" of the traditional design process, with much labor and managerial attention devoted to interfacing requirements and design features from the various packaging levels.

In such a design process, which requires that decoupled solutions be melded to synthesize a viable system, product optimization is nearly impossible to achieve. Instead, it has been observed [Simon (1981)] that the design process continues until all major requirements are met, thus producing a "sufficient" rather than an optimum system. Although, in the past, such a product may have found market acceptance and may have even generated a modest return on the investment, in the 1990s such a non-optimized product may well become a financial liability. As a consequence of this concern, a holistic, concurrent engineering design process is taking root in the electronic industry [Rosenblatt and Watson (1991)]. This change is manifested by the formation of multidisciplinary product development teams actively involving physical designers, electronic designers, software specialists, R&D scientists, manufacturing technologists, marketing experts, lawyers and, at times, even sociologists and psychologists, in the articulation, evaluation and prioritization of design concepts.

The efforts of these product development teams are aided immeasureably by the use of common design tools and databases operating on shared computer platforms. The successful exploitation of these packaging CAD tools and the associated electronic communication among team members serves, not only to bridge across disciplinary and geographical distances, but also to facilitate concurrent design and analysis of various subsystems and simultaneous progress in implementing distinct parts of the physical design process. The ability to observe, nearly instantaneously, the impact of a component change on the reliability or cost of the entire product, to learn that the via parameters in a proposed printed circuit board are beyond the capability of the organization's fabrication processes or that redistribution of high heat-flux components will allow effective thermal management of a module at an even higher clock-rate, provides the development team with "real-time" insights into the characteristics of the evolving product. This insight is essential for parallel development of components and subsystems and one of the essential means for shortening the design process of a new product.

Along with the expanded use of packaging CAD, product development teams in the electronic industry must master the use of computer-integrated-manufacturing (CIM) to reduce tooling costs and to shorten the time required to set up or modify manufacturing lines. CIM systems that can be accessed freely, or in an appropriately restricted manner, by part, component or subsystem vendors, can dramatically shorten the "time-to-market" of an electronic product. The computer-based management of engineering drawings, qualified parts, and material lists, manufacturing/assembly procedures and relevant standards provide significant secondary benefits to users of such systems.

8 FUTURE CHALLENGES

During the 1980s, the electronic packaging community was enriched by an influx of mechanical engineers and researchers with expertise in critically important disciplines, but with little experience in the development of electronic products. Initially, these engineers were motivated to generate failure-free mechanical designs, as is the common practice in the development of the more familiar mechanical systems, such as pressure vessels, automotive transmissions and heat exchangers. However, their encounter with established procedures in the electronic industry led them, all too often, to accept palliative, rather than curative, approaches to the attainment of reliability targets. The institutionalized use of "de-rating" to avoid electrical and thermal stress and "burn-in" to screen out defective components and assemblies projected an aura of scientific rigor and legitimacy that left scant room for the application of more traditional approaches to failure prevention. Consequently, although the availability of much-needed skills in thermal and structural design, as well as in manufacturing, has improved product quality, the reliability/availability/maintainability of many US-built electronic products has lagged behind the expectations of consumers and taxpayers, alike. Moreover, packaging failures have come to dominate the reliability statistics for electronic systems [Jeannotte et al. (1989)].

Future success in the electronic industry, thus, requires that the packaging strategy not only represent the optimum balance among the many imposed requirements and constraints, but that recognized packaging failure modes be eradicated to generate "failure-free" products which are "correct by design." While a better understanding of the packaging development process, recognition of the many alternative "enabling technologies" and effective use of the available CAD tools are all necessary to the achievement of these twin goals, little progress will be realized without a concerted effort to both broaden and deepen the science-base underpinning the packaging configurations anticipated for use in the advanced electronic systems of the 1990s.

The requisite in-depth, multi-disciplinary effort must begin by identifying the fundamental failure mechanisms of the advanced VLSI/ULSI chips, single-chip packages and multi-chip modules, 20–50-layer printed circuit boards and of all the intervening interconnect structures. Semiconductor failure mechanisms (such as dopant migration, piezoresistive effects and electromagnetic effects), metallurgical failures modes (such as corrosion, electromigration and intermetallic formation), structural failure mechanisms (such as fracture, creep and fatigue), and failures associated with thermofluid transitions (such as the onset of turbulence and the onset of boiling) must all be considered. Mechanistic formulations, describing the relationship between the failure characteristic (e.g., voltage, temperature, velocity) and the salient independent parameters, must then be formulated for each of the relevant failure mechanisms. Empirical studies can then be performed to establish the scientific validity of these relations as well as the severity and rank-order of each failure mechanism.

The availability of empirically validated mechanistic relations would make it possible to determine the scaling laws and parametric sensitivities of the governing phenomena. On this foundation, it would be possible to develop a comprehensive set of figures-of-merit to be used, as previously discussed, in the selection of candidate enabling technologies. Moreover, the understanding gained from the empirical studies and functional form of the governing equations would facilitate the development of computer subroutines and/or embedded algorithms that could be used to enhance existing packaging CAD software. As accuracy improves and confidence grows in the mechanistic relations for the primary failure mechanisms, use of this science-based physical design methodology, in appropriately configured software, can be expected to facilitate the demarcation of the parametric region within which component, assembly or system operation can proceed unimpeded for the useful life of the product.

It is the development of this science-based physical design methodology and its incorporation in packaging computer-aided-design tools which constitute the primary packaging challenges for the 1990s. Organizations which succeed in:

- gaining the understanding of packaging failure mechanisms,

- developing proven figures-of-merit for material and technology selection,
- generating and embedding the appropriate failure algorithms in software,
- promoting the use of integrated design tools (packaging CAD), and
- fostering the principles of concurrent engineering

will come to dominate the electronic industry with products that are "correct-by-design" and effectively "failure-free." The companies and organizations which prove unable to make the transition from "trial-and-error" packaging development to a science-based physical design methodology are unlikley to survive into the next century.

NOMENCLATURE

α_{sub}	Linear thermal expansion coefficient of the substrate, 1/K
α_{comp}	Linear thermal expansion coefficient of the component, 1/K
q	Component power dissipation, Watts
R_{cf}	External, or case-to-fluid, thermal resistance, K/W
Re_D	Reynolds number based on diameter, dimensionless
R_{jc}	Internal, or junction-to-case, thermal resistance, K/W
T_c	Case temperature, K or °C
T_{comp}	Temperature of the component, K or °C
T_f	Fluid temperature, K or °C
T_{in}	The initial, stress-free temperature, K or °C
T_j	Junction or chip temperature, K or °C
T_{sub}	Temperature of the substrate, K or °C
U	Overall heat transfer coefficient, W/m^2 K

REFERENCES

Bar-Cohen, A (1987). Thermal management of air- and liquid-cooled multichip modules, *IEEE CHMT Trans*, **CHMT-10** (2), 159–175.

Bar-Cohen, A (1988). Addendum and correction to thermal management of air- and liquid-cooled multichip modules, *IEEE CHMT Trans*, **CHMT-11** (3), 333–334.

Cavin, R K, Sumney, L W, and Burger, R (1989). The semiconductor research corporation: Cooperative research, *IEEE Trans*, **77** (9), 1327–1344.

Dohya, A, Watari, T, and Nishimori, H (1990). Packaging technology for the NEC SX-3/SX-X supercomputer, *Proc 40th IEEE ECTC Conf*, 525–533.

Engelmaier, W (1984). Functional cycling and surface mounting attachment reliability, *Surface Mount Technology*, Technical Monograph Series, 6984-02, ISHM.

FUJITSU (1990). Cryogenic cooling system for HEMT LSI, (translation via Tech Search).

Fulcher, E (1990). Multichip evolution — From desktops to mainframe computers, *Proc 1990 Intl Elec Packaging Soc Conf*, 34–45.

Gelsinger, P P, Gargini P A, Parker, G H, and Yu, A Y C (1989). Microprocessors circa 2000, *Spectrum*, Oct, 43–47.

Godsell, S, Richard, B, Dischler, J, and Westbrook, S (1990). Implementing a packaging strategy for high-performance computers, *High Performance Systems*, Jan, 28–31.

Hatada, K, Fujimoto, H, and Kawakita, T (1987). Insulation resin bonding: Chip on substrate assembly technology, *Proc ICICE Conf.*

IBM (1987). The IBM 9370 Information System: Technology changes the shape of computing, IBM Technology Marketing Support Center, Gen Tech Div, Essex Junction, VT.

Incropera, F P (1988). Convection heat transfer in electronic equipment cooling, *ASME J of Heat Transfer*, **110**, 1097–1111.

Jeannotte, D A, Goldman, L S, and Howard, R T (1989). Package reliability, Chap 5 in *Microelectronic Packaging Handbook*, E J Rymaszewski and R R Tummala, eds, Van Nostrand Reinhold, New York, NY.

Johnson, R R (1990). Multichip modules: Next generation packages, *IEEE Spectrum*, Mar, 34–36, 46–48.

Koopman, N G, Reiley, T C, and Totta P A (1989). Chip-to-package interconnections, Chap 6 in *Microelectronic Packaging Handbook*, Van Nostrand Reinhold, New York, NY.

Landman, B J and Russo, R L (1971). Pin versus block relationship for partitions of logic graphs, *IEEE Trans on Computers*, **C20** (12), 1469–1479.

Manzione, L T (1990). *Plastic Packaging of Microelectronic Devices*, Van Nostrand Reinhold, New York, NY, 14–15.

Markhoff, J (1991). Supercomputing's speed quest, *New York Times*, May 31.

Markstein, H W (1991). Multichip modules pursue wafer scale performance, *Elec Packaging and Prod*, Oct, 40–45.

Moore, G E (1975). Progress in digital integrated electronics, *Proc IEEE Intl Electron Devices Meeting*, Dec, 11–13.

Nakayama, W (1988). Thermal management of electronic equipment: A review of technology and research topics, Chap 1 in *Advances in Thermal Modeling of Electronic Components and Systems – Vol I*, A Bar-Cohen and A D Kraus, eds, Hemisphere Pub Corp, New York, NY.

NIKKEI, (1990). Hitachi announced large scale general purpose computer M-880, *Nikkei Electronics*, Jun 25, 104–106.

Ohki, N, Murase, Y, Nishihara, M, and Yamamoto, M. (1989). Packaging technology for the VP-2000 supercomputer series, *Nikkei Microdevices*, Jun, 50–54.

Ohsaki, T (1990). Electronic packaging in the 90s — A perspective from Asia, *Proc 40th IEEE Elec Comp & Tech Conf*, **1**, 1–8.

Pei, J, Heng, S, Charlantini, R, and Gildea, P (1990). Cooling componenets used in the VAX 9000 family of computers, *Proc IEPS*, 587–601.

Rosenblatt, A (1991). ISSCC hears of 100-MHz version of 486, *The Institute*, Mar.

Rosenblatt, A and Watson, G F (1991). Concurrent engineering, *IEEE Spectrum*, Jul, 21–37.

Rymaszewski, E J. and Tummala, R R (1989). Microelectronics packaging – An overview, Chap 1 in *Microelectronics Packaging Handbook*, E J Rymaszewski and R R Tummala, eds, Van Nostrand Reinhold, New York, NY.

Schnorr, D P and Lutz, R A (1986). TCE-tailored, high-density surface mounting printed wiring assembly development for military systems, *Proc 6th Intl Elec Packaging Conf*, San Diego, Nov, 140–154.

Seraphim, D P, Barr, D E, Chen, W T, Schmitt, G P, and Tummala, R R (1989). Printed circuit board packaging, Chap 12 in *Microelectronics Packaging Handbook*, E J Rymaszewski and R R Tummala, eds, Van Nostrand Reinhold, New York, NY.

Simon, H (1981). *The Science of the Artificial*, MIT Press, Cambridge, MA.

Smith, P G and Reinertsen, D G (1991). *Developing Products in Half the Time*, Van Nostrand Reinhold, New York, NY.

Suhir, E (1988). Thermal stress failures in microelectronic components – Review and extension, Chap 5 in *Advances in Thermal Modeling of Electronic Components and Systems – Vol I*, A Bar-Cohen and A D Kraus, eds, Hemisphere Pub Corp, New York, NY.

Tummala, R R and Rymaszewski, E S (1989). *Microelectronics Handbook*, Van Nostrand Reinhold, New York, NY.

Vacca, A, Resnick, D, Frankel, D, Bach, R, Kreilich, J, and Carlson, D (1987). A cryogenically-cooled CMOS VLSI supercomputer, *VLSI Systems Design*, 80–88.

Vilkelis, W (1982). Lead reduction among combinational logic circuits, *IBM J Research and Development*, **26** (3), May, 342–348.

Wessely, H, Turk, W, Schmidt, K H, and Nagel, G (1988). Computer packaging technology, *Siemens Research and Development Reports*, (5), 234–239.

Wessely, H, Fritz, O, Klimke, P, Koschnick, W, and Schmidt, K H (1990). Electronic packaging in the 90s — A perspective from Europe, *Proc 40th IEEE Elec Comp and Tech Conf*, **1**, 16–33.

Weston, W (1985). High density 128×128 area arrays of vertical electrical interconnections, *4th Ann Microelec Interconnect Conf*, Jul.

SOURCES OF FURTHER DATA

Chu, R C and Simons, R E (1988). Evolution of cooling technology in medium and large scale computers – An IBM perspective, *Proc Intl Symp on Heat Transfer in Elec and Microelec Equip*, Intl Center for Heat and Mass Transfer, Dubrovnik, Yugoslavia, Hemisphere Pub Co, Aug 29–Sep 2.

Hannemann, R (1990). Physical technology and the air cooling and interconnection limits for mini- and microcomputers, in *Advances in Thermal Modeling of Electronic Components and Systems, Vol. II*, A Bar-Cohen and A D Kraus, Eds, ASME Press, New York, NY.

Nakayama, W and Bergles, A E (1988). Cooling electronic equipment: Past, present, and future, *Proc Intl Symp on Heat Transfer in Elec and Microelec Equip*, Intl Center for Heat and Mass Transfer, Dubrovnik, Yugoslavia, Hemisphere Pub Co, Aug 29–Sep 2.

Chapter 2

THE INFLUENCE OF TEMPERATURE ON INTEGRATED CIRCUIT FAILURE MECHANISMS

Michael Pecht and Pradeep Lall
CALCE Center for Electronics Packaging
University of Maryland
College Park, MD 20742

Ed Hakim
U.S. Army Laboratory
Ft. Monmouth, New Jersey

1 INTRODUCTION

Temperature is generally considered to be a key parameter in the design of electronic equipment and cautions concerning temperature and its relationship to reliability are widely documented. While some studies suggest that temperature is the most critical stress influence on microelectronic device failures, the actual failure mechanisms have generally not been quantified in terms of whether a steady state temperature, temperature change, rate of temperature change, or spatial temperature gradient induced failure.

Presently, temperature dependent models such as the Eyring and Arrhenius models, originally proposed to model the effect of temperature on chemical reaction rates, are widely applied to microelectronic devices. These models predict the effect of temperature on electronic component failure rates under the assumption that the dominant component failure mechanisms depend on the steady state temperature. However the correlation between steady state temperature and failure mechanisms is one that is often difficult to confirm. The usual case is that failure occurrences are compared to some measurement or assumption of operating temperature with little in-depth failure diagnosis as to the root cause of failure. In fact, detailed assessment of the role of steady state temperature after failure analysis of field failure returns often does not provide clear evidence that a lowered temperature would have prevented the failure event, or that a higher temperature would have caused an earlier failure. Only too often the apparent agreement between elevated temperatures and high failure rates forms its own rationale for the conclusion that steady state temperature is the culprit causing the failures.

Distributions of microelectronic device failures collected over the last 20 years [Dantowitz et al. (1971), Manno (1983), Westinghouse (1987), DELCO (1988), Devanay (1989), Bloomer et al. (1989), and the *Semiconductor Reliability News* (1990)] show that with advances in technology, as problems are solved, new failure mechanisms become dominant, and, depending on manufacturing quality and defects, the actual failure mechanism experienced by a device can vary. Because of this dynamic and competitive nature of the electronic component industry, no failure mechanism can remain dominant, or even significant, for very long. Thus it is beyond the scope of this chapter to model all types of defects or to discuss the relative impact of the various temperature related failure mechanisms on overall device reliability.

In this chapter, only the influence of temperature on major integrated circuit failure mechanisms is discussed, with emphasis placed on those failure mechanisms which occur in the temperature range of −55° C to 125° C. The die or chip contains the primary circuit elements. Package elements include die, die attach, lead frame, leads, substrate, passive elements, substrate attach, wire bonds, interconnection wiring, tape automated bonds, flip-chip bonds, lid seals, lead seals, case, lead finish, and case finish. The die is attached to the substrate by a die attach usually composed of gold eutectic, epoxy, polyimide or solder. For hybrid packages, the substrate may also contain printed passive component and routing paths. The substrate provides the mechanical support to the die and acts as a thermal path for heat dissipation from the die to the die package case. The chip carrier connects the die to the second level of the package. This is accomplished by bonding external leads on the lead frame using wires or tape. The processes of interconnection are called wire bonding or tape automated bonding, respectively. An alternative method for interconnection is flip-chip bonding. The external leads are sealed to provide structural rigidity and to avoid any contaminant ingress into the package. The assembly consists of a die mounted on the substrate connected to the external leads by wires or tape and sealed in a case.

This chapter shows that no simple expression can adequately describe temperature as a failure accelerator for all integrated circuit failure mechanisms. In fact, a generic statement that can be attributed to temperature is lacking. This suggests that a much deeper level of insight into temperature dependencies is necessary to achieve reliable equipment and avoid unnecessary complexities. Thermal management in electronic equipment can involve additional costs and system complexities that can be of consequential importance, and temperature control should not be routinely employed without close study and justification.

Failures in microelectronic devices are often classified into electrical, mechanical and corrosion failures. Electrical failures occur predominantly at the device (i.e., die) level, and include slow trapping, hot electrons, electrical overstress, electrostatic discharge, dielectric failure, oxide breakdown, and electromigration. Some of the electrical failures are temperature dependent. The mechanical failure mechanisms are predominately package related. Mechanical failures arise from differential thermal expansion between bonded materials, large time dependent temperature changes, and large spatial temperature gradients, all of which can cause tensile, compressive, bending, fatigue and fracture failures. Corrosion induced failure mechanisms include electrolyte formation, and galvanic and ionic corrosion. Corrosion failures are complex functions of contamination, temperature, humidity and bias.

2 TEMPERATURE DEPENDENT MECHANICAL FAILURES

2.1 Wire Fatigue

The wires used to connect die bond pads to leads, or die bond pads to die bond pads in the case of hybrid packages, can fail due to cyclic temperature changes resulting from repeated flexing of the wire. Gaffeny (1968) points out that the most prevalent failure site is the heel of the wire and this has been confirmed by Villela and Nowakowski (1970), Ravi and Philosky (1972), and Phillips (1974), who report on extensive studies on wire bond failures. These show that failures are due to the differential coefficient of thermal expansion between the wire and the package as the device is heated and cooled down during temperature and power cycling. Failures in small diameter wires may be inhibited by high loops, though the small dimensions of present day packages impose stringent requirements on the loop dimensions making this solution largely impractical. However, such a solution is only helpful for large diameter wires because wire stiffness prevents easy wire flexing.

The cycles to failure N_f, estimate is [Pecht et al. (1989)], where

$$N_f = A(\epsilon_f)^n \tag{2.1}$$

where

$$\epsilon_f = \frac{r}{\rho_o}\left[\frac{\cos^{-1}\{\cos\lambda_o[1-(\alpha_w-\alpha_s)\Delta T]\}}{\lambda_o} - 1\right] \tag{2.2}$$

and

r	=	wire radius, m
λ_o	=	angle of the wire with the substrate, radians
α_w	=	coefficient of thermal expansion of the wire, 1/° C
α_s	=	coefficient of thermal expansion of the substrate, 1/° C
ΔT	=	temperature cycle encountered by the structure, 1/° C
ρ_o	=	initial radius of curvature of the wire, m
A	=	material constant
n	=	material constant

The time to failure of a wire undergoing a cyclic temperature variation is thus dependent on the magnitude of the temperature cycle, the geometry of the wire and the material properties, including the coefficient of thermal expansion. However, wire flexing is, for the most part, dependent directly on the magnitude of temperature change and is steady state temperature independent.

Temperature and power cycling can also contribute to wire failure, especially in plastic encapsulated packages in which the wire is encapsulated by a material with a high coefficient of thermal expansion compared to the wire. This can result in the partial neckdown and eventual breakage of the wire. However, wires commonly used (i.e., gold for smaller packages and aluminum for larger packages which require high current density) are ductile enough to absorb large deformations. The number of cycles to failure are typically over 200,000 power cycles for 0.002 inch diameter aluminum, 1% silicon, ultrasonic bonds with loop heights greater than 25% of the bond to bond spacing, for temperatures ranging from 38° C to 170° C [Harman (1974)]. The number of cycles to failure were found to be in the range of 18,000 cycles for 0.008 inch diameter 99.99% pure aluminum bonds on a 2N4863 power transistor subjected to temperatures from 25° C to 125° C [Harman (1974)].

2.2 Wire Bond Fatigue

A wire bond subjected to temperature change experiences shear stresses between the wire, the bond pad and the substrate as a result of differential thermal expansion [Philosky and Ravi (1973) and Pecht et al. (1989)]. The number of cycles to failure, N_f, as a result of shear between the bondpad and the substrate, due to a temperature change ΔT is typically given by the closed form fatigue relation

$$N_S = C_{p'}(\tau_{\max})^{-m_p} \tag{2.3}$$

where

$\tau_{\max}$	=	maximum shear stress, N/m^2
$C_{p'}$ and m_p	=	shear fatigue properties for the bond pad materials
N_S	=	number of cycles to failure

$$\tau_{\max} = Q\Delta T \tag{2.4}$$

where

$$Q = \left(\frac{G_p}{b_p Z}\right)\left((\alpha_w - \alpha_s) - \frac{(\alpha_s - \alpha_p)}{\left(1 + \dfrac{E_s A_s(1-\nu_s)}{E_p A_p}\right)}\right) \tag{2.5}$$

with Z defined by Eq. (2.10) and

G_p	=	shear modulus of the bond pad material, N/m^2
ΔT	=	temperature cycle magnitude, ° C
b_p	=	bond pad thickness, m

b_s	=	substrate thickness, m
α_s	=	coefficient of thermal expansion for the substrate, 1/° C
α_p	=	coefficient of thermal expansion for the pad, 1/° C
α_w	=	coefficient of thermal expansion for the wire, 1/° C
E_p	=	modulus of elasticity of the pad material, N/m^2
E_s	=	modulus of elasticity of the substrate material, N/m^2
ν_s	=	Poisson's ratio for the substrate material, dimensionless
A_p	=	cross-sectional area of the pad, m^2
A_s	=	effective cross-sectional area of the substrate, $b_s(W_p + W_s)/2$, m^2
W_p	=	width of the bond pad, m
W_s	=	width of the substrate, m
$C_{p'}$ and m_p	=	shear fatigue properties for the bond pad materials

The life prediction estimate for the number of cycles to failure as a result of shear between the wire and the bondpad under thermal cycling, N_S, can be empirically related to the shear stress amplitude by the following (Coffin-Manson relation) relations:

$$N_S = C_{w'}(\tau_{w,M})^{-M_{w'}} \tag{2.6}$$

Assuming that the wire and substrate near the bonding layer are subjected to longitudinal normal stress, the normal traction in the wire and substrate near the bondline can be determined by integrating the equations of equilibrium at the wire substrate interface. The shear stress under thermal expansion mismatch is then given by [Hu et al. (1991)]

$$N_S = C_{s'}(\tau_{s,M})^{-M_{s'}} \tag{2.7}$$

where

$$\tau_{w,M} = \left(\frac{r^2}{4Z^2 A_w^2} \left(\frac{\cosh(Zx_w)}{\cos(Zl_w)} - 1 \right)^2 + \frac{\sinh^2(Zx_w)}{\cosh^2(Zl_w)} \right)^{1/2} Q\Delta T \tag{2.8}$$

and

$$\tau_{s,M} = \left(\left(\frac{W_p Q}{2Z^2 A_s^2}\left(1 - \frac{\cosh(Zx_w)}{\cosh(Zl_w)}\right) + \frac{(\alpha_s - \alpha_p)}{\frac{(1-v_s)}{E_s A_s} + \frac{1}{E_p A_p}}\right)^2 + Q^2 \frac{\sinh^2(Zx_w)}{\cosh^2(Zl_w)}\right)^{1/2} \Delta T \tag{2.9}$$

where

$$Z^2 = \frac{G_p}{b_p}\left(\frac{r}{E_w A_w} + \frac{(1-\nu_s)W_p}{E_s A_s}\right) \tag{2.10}$$

where, in addition to the nomenclature defined previously, ℓ_w is the bonded length (m) of the wire, x_w gives the location of the maximum stress in the wire (m), Q is defined by Eq. (2.4) and $C_{w'}$, $m_{w'}$, $C_{s'}$ and $M_{s'}$ are the shear fatigue properties of the wire and substrate materials.

Intermetallic brittleness and growth is enhanced by temperature cycling. When bonded together, Au-Al present a problem when subjected to temperature cycling. Au-Al intermetallics are more susceptible to flexure damage than pure Au and Al wires. Kirkendall voids occur at the interface between the wire and the bond pad when the contacting materials are gold and aluminum. Kirkendall voids form when either the aluminum or gold diffuses out of one region faster than the other can diffuse from the other side of that region. Vacancies pile up and condense to form voids, normally on the gold rich side along the gold–to–aluminum interface.

The rates of diffusion vary with temperature and are dependent upon the adjacent phases as well as on the number of vacancies in the original metals. The compounds formed in the intermetallic are often called "purple plague" because of the purple appearance of the gold-aluminum compound. Intermetallic compounds are formed during thermocompression or thermosonic bonding. There are five intermetallic compounds which can arise: Au_5Al_2, Au_2Al, $AuAl_2$, $AuAl$, and Au_4Al. Prolonged exposure to high temperatures results in continued diffusion until gold or aluminum is consumed. Initial growth rate is indicated by a parabolic relationship,

$$X = Kt^{1/2} \tag{2.11}$$

where X is the intermetallic layer thickness, t is the time, and K is the rate constant, given by

$$K = Ce^{-[E_a/kT]}, \qquad T > 150^\circ \text{ C} \tag{2.12}$$

where C is a constant, E_a is the activation energy (eV) for layer growth, k is the Boltzmann constant (eV/K), and T is the steady state temperature (K). The value for K changes for each intermetallic phase. Figure 2.1 shows that the formation of significant amounts of purple plague can take years at 150° C and below.

Intermetallic compounds are mechanically strong, brittle and electrically conductive. Temperature and power cycling can cause failure as a result of differential thermal expansion between the intermetallic and the surrounding metal, and can reduce mechanical bond strength as a result of voiding of the surrounding metal which usually accompanies intermetallic formation. Gold-aluminum intermetallics are stronger than pure metals and their growth is enhanced by increased temperature and temperature cycling [Philosky and Ravi (1973)].

Newsome and Oswald (1976) conducted experiments on the effect of temperature on the intermetallic formation in two thick-film systems (Owens-Illinois and EMCA 212B), and in a conventional nichrome, nickel, and gold thin film system, by conditioning them at 25, 75, 125, 150, 160, 175, and 200° C for 168 hours in an air circulation oven. The data from energy dispersive x-ray spectrometer (EDS) analysis shown in Fig. 2.2 indicated that the purple phase of the intermetallic did not appear until 175° C, even though a significant thickness of Au_4Al was formed in the EMCA 212B at 75° C. Table 2.1 lists the observations in the various phases. The process of intermetallic growth was accompanied with volumetric expansion of the intermetallics which are formed at temperatures in the neighborhood of 160 to 175° C, creating severe mechanical stress cracks at the vertical interface with the gold. Kirkendall voiding and mechanical stress due to volumetric expansion (due to intermetallic formation) was found to increase in the presence of a larger volume of gold under the bonds. Philosky (1971) listed nine different rate constants for five gold-aluminum compounds. Gerling (1984) published an extensive compilation of reported activation energies for various types of gold-aluminum rich couples. Aluminum rich gold-aluminum compounds have a high melting point and are relatively stable.

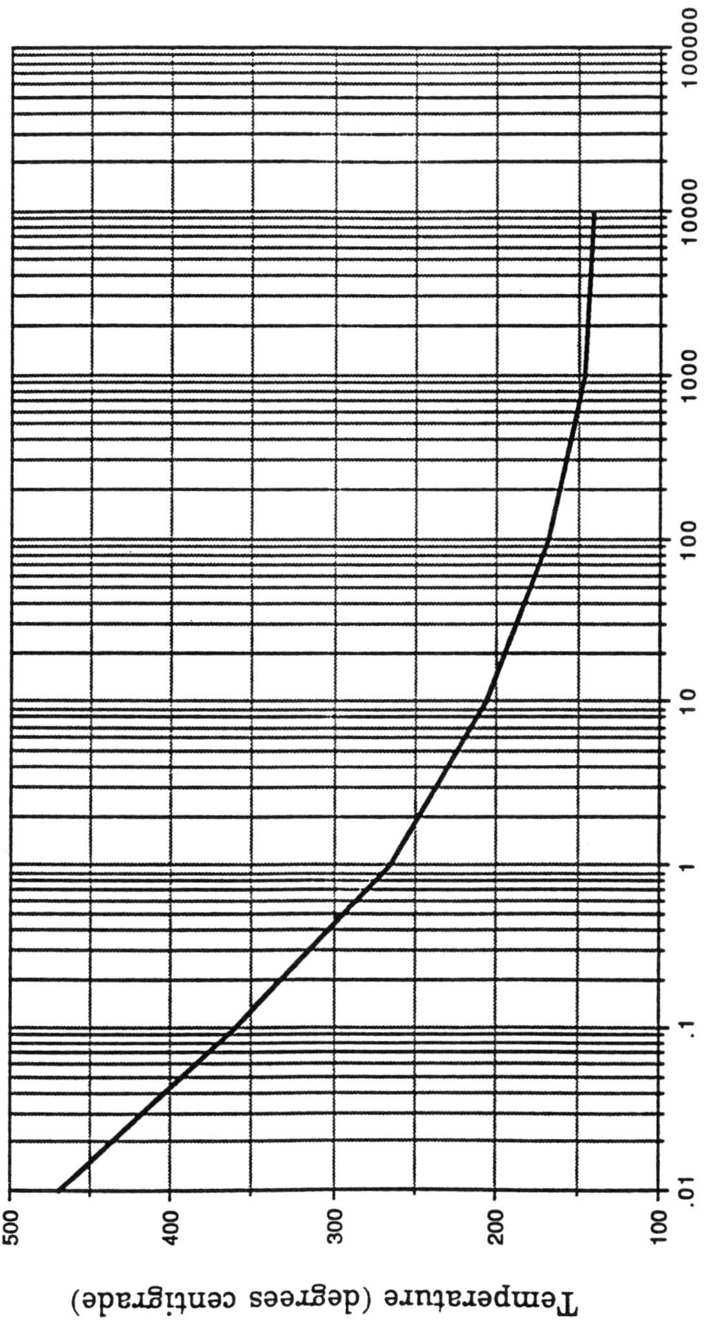

Figure 2.1 The formation of significant amounts of purple plague can take years in the equipment operating range of −55 to 125° C [DM data (1990)].

	EMCA 212B							Owens-Illinois 99⁺							Honeywell Thin Film						
	25	75	125	150	160	175	200	25	75	125	150	160	175	200	25	75	125	150	160	175	200
Purple ($AuAl_2$)					■	■	■					■	■	■					■	■	■
White ($AuAl$)						■	■						■	■						■	■
Tan (Au_2Al)																					
Tan (Au_5Al_2)			■	■	■	■	■			■	■	■	■	■			■	■	■	■	■
Tan (Au_4Al)		■	■	■	■	■	■														

■ = Appearance of the intermetallic phase
A blank box indicates that the intermetallic phase did not appear at that temperature.

Figure 2.2 Phase identification versus temperature for EMCA 212B, Owens-Illinois 99^+, and Honeywell thin film systems [Newsome (1976)].

Table 2.1 Summary of Optical and SEM Findings for the Three Gold Systems [Newsome and Oswald (1976)]

	EMCA 212B	Owens-Illinois 99	Honeywell Thin Film
Physical Characteristics	Rough, pitted gold surface. Irregular interface between gold and aluminum causes non-uniform intermetallic growth.	Rough gold surface but not as pitted as EMCA 212B. Intermetallic growth is more uniform than for EMCA 212B.	Uniform surface. Gold, 1/3 the thickness of thick film conductors. Interface between gold and aluminum is uniform as well as intermetallic growth.
Constituents and Energy Dispersive Spectrometer (EDS) Analysis	Thick film ink with frit for mechanical adhesion to substrate. The frit contains silicon, bismuth, and cadmium as major constituents. The concentration of bismuth was found to be higher at the intermetallic diffusion front than in either the intermetallic or the gold. Au_4Al, Au_5Al_2, AuAl, and $AuAl_2$ intermetallic phases identified with the AuAl and $AuAl_2$ not detectable for low temperatures.	Fritless thick film ink with copper additive for wetting the substrate. No other detectable impurities. The concentration of copper follows the same pattern as the bismuth in the EMCA 212B. Au_5Al_2, AuAl, and $AuAl_2$ intermetallic phases present but no Au_4Al. AuAl and $AuAl_2$ appear at the higher temperatures.	Gold over nickel and nichrome. No detectable impurities. Au_5Al_2, AuAl, and $AuAl_2$ intermetallic phases present in detectable quantities with AuAl and $AuAl_2$ appearing at the higher temperatures.

Table 2.1 Continued.

	EMCA 212B	Owens-Illinois 99	Honeywell Thin Film
Temperature: 25° C	No detectable intermetallics.	No detectable intermetallics.	No detectable intermetallics.
75° C	Significant thickness of gold rich intermetallic visible Au_4Al (tan).	No detectable intermetallics.	No detectable intermetallics.
125° C	Visible lateral voiding and annular cracks at gold interface. Visible elevation of bond above surface of gold due to volumetric expansion of intermetallic. Second gold rich phase appears Au_5Al_2 (tan).	First gold rich intermetallic phase appears. Identified as same phase which appears at this temperature for the EMCA 212B.	First gold rich intermetallic phase appears. Identified as same phase which appears at this temperature for the EMCA 212B.
150° C	Consumption of gold continues with associated increase in lateral voiding, annular cracks and bond elevation. Au_4Al narrowed to thin line by growth of Au_5Al_2 when supply of gold to intermetallic has been limited by voiding.	Intermetallic continues to consume gold.	Gold under the bond consumed if lateral voiding has not occurred. Some annular cracking and lateral voiding observed.

Table 2.1 Continued.

	EMCA 212B	Owens-Illinois 99	Honeywell Thin Film
Temperature: 160° C	The gold under the bond is consumed if lateral voiding has not occurred. The width of the Au_4Al_2 is approximately equal. A fine line of a third intermetallic phase appears at the aluminum interface, identified by coloration as $AuAl_2$ (purple).	Annular cracking becomes visible with elevation of the bond above the surface to the gold. $AuAl_2$ appears to be present as a fine line at the aluminum interface.	Visible elevation of the bond above the gold surface with an increase in the annular cracking. $AuAl_2$ appears to be present as a fine line at the aluminum interface.
175° C	AuAl (white) phase appears and width of $AuAl_2$ increases to a clearly defined band. Extensive annular cracking and lateral voiding cause slowdown in the growth of the Au_4Al and Au_5Al_2. AuAl reduced to a narrow band adjacent to the diffusion front on all samples.	AuAl phase appears and width of $AuAl_2$ increases. Some lateral voiding found at gold interface where initial intermetallic diffusion front was irregular. Annular cracking continues to increase.	AuAl phase appears and width of $AuAl_2$ increases. The increase in annular cracking with growth of the intermetallics continues.

Table 2.1 Continued.

	EMCA 212B	Owens-Illinois 99	Honeywell Thin Film
Temperature: 200° C	Width of aluminum rich phases increases, growing into both the bond and gold rich phases. Lateral voiding appears at aluminum interface. The width of $AuAl_2$ is approximately three times that of the AuAl.	Width of aluminum rich phases increases, growing into both the bond and gold rich phases. Lateral voiding appears at aluminum interface. Gold under bond consumed if lateral voiding has not occurred. The width of the $AuAl_2$ is approximately twice that of the AuAl and one-half the width of the $AuAl_2$ in the EMCA 212B samples.	Width of aluminum rich phases increases, growing into both the bond and gold rich phases. The AuAl and $AuAl_2$ phases are each about one-half the width of the $AuAl_2$ phase in the EMCA 212B.

Newsome and Oswald (1976) observed Kirkendall voiding in the neighborhood of 150 to 200° C (150° C for the EMCA 212B and Honeywell thin film systems, and in the neighborhood of 175° C to 200° C for the Owens-Illinois system). The appearance of lateral voiding at the aluminum interface was observed in the neighborhood of 200° C. Kirkendall voiding in the present day packages derives from impurity or corrosion driven reactions and are temperature independent below temperatures around 150° C [Newsome and Oswald (1976)]. However, in the presence of halogenated species this process is accelerated at a lower temperature [Khan and Fatini (1986)].

Gold wires bonded to bare copper lead frames and to copper thick films in hybrids react forming three intermetallic phases: Cu_3Au, AuCu, Au_3Cu. Temperatures above 300° C accelerate the rate of intermetallic compound formation. The "time to decrease bond strength" is found to decrease with an increase in temperature [Hall et al. (1975)]. Temperature time studies on thermocompression lead frame bonds indicate a decrease in strength as a result of void formation [Pinnel and Bennett (1972), Feinstein and Bindell (1979), and Feinstein and Pagano (1979)]. Figure 2.3 shows the temperature dependence of time to decrease the copper to gold bond strength 40% below the "as bonded strength" for temperatures above 300° C. Pitt and Needles (1982) studied the gold thermosonic bonds to thick film copper and found little strength degradation at 150° C for up to 3000 hours, and no failures at 250° C for over 3000 hours. This suggests that the copper-gold bond is not advisable for microelectronic devices operating at high temperatures (above 300° C). The difference in the hardness and the material properties, including the coefficient of thermal expansion between the intermetallic and the surrounding metal, make the interface a potential site for failure.

Generally, the bond strength is more a function of temperature cycling than steady state temperature in the range of temperature between −55° C and 125° C, although the bond strength decreases as a function of temperature above 150° C [Newsome and Oswald (1976)] for gold-aluminum bonds and above 300° C [Hall et al. (1975)] for gold-copper bonds.

2.3 Die Fracture

The die, the substrate or the lead frame and the case of a microelectronic package typically have different thermal expansion coefficients.

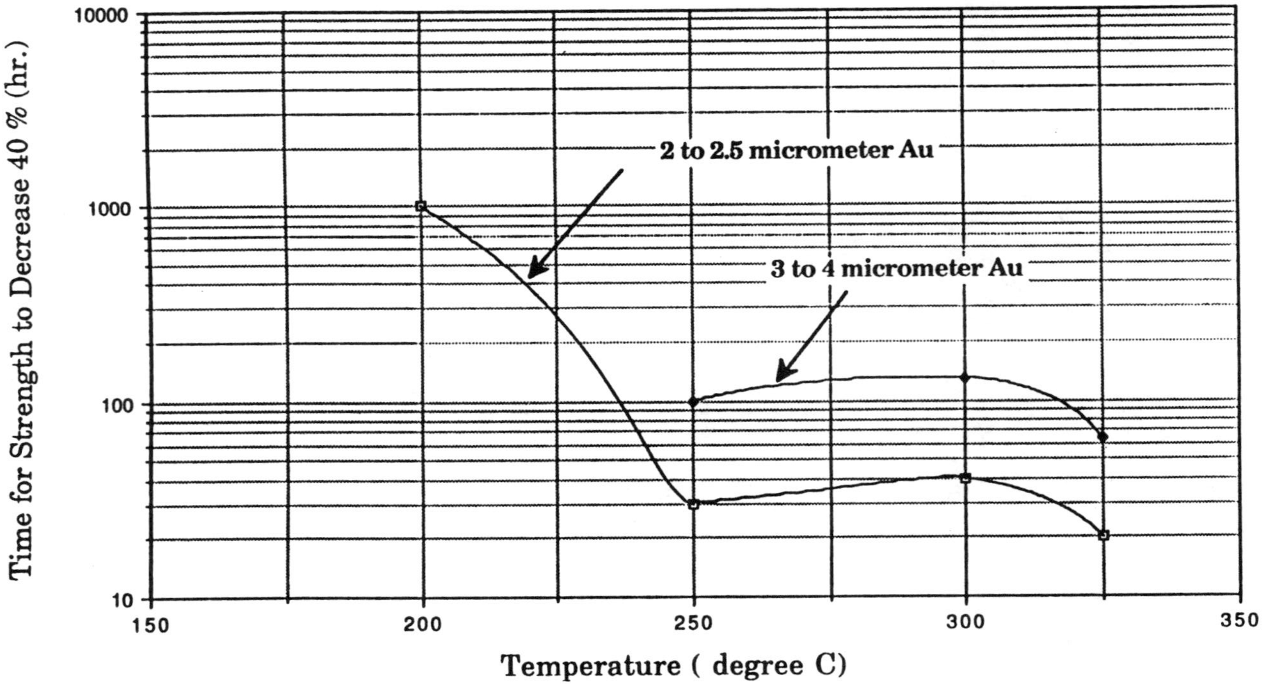

Figure 2.3 Temperature dependence of time to decrease copper gold bond strength 40% below the as bonded strength [Hall et al. (1975)].

For example, die are usually made of silicon, gallium arsenide or indium phosphide, while the substrate is typically alumina, berylia or copper having a coefficient of thermal expansion different from the die material. As the magnitude of the temperature cycle increases during temperature and power cycling, tensile stresses are developed in the central portion of the die and shear stresses are developed at the edges of the die. Ultimate fracture of the brittle die can occur suddenly without any plastic deformation when surface cracks at the center of the die or at the edge of the die reach their critical size and propagate during thermal cycling to the critical crack size.

Vertical cracking of the die is caused by tensile stresses and horizontal cracking is caused by high shear stresses at the edges. The brittle failure criterion can be represented by the size of a critical crack on the external die surface. Microcracks are developed in the die during manufacture. In some cases these microcracks may be large enough to cause brittle failure of the die. A pre-existing defect may develop into a crack under the influence of thermal cycling in the die. This crack may not be of critical size at the applied service stress but may grow to critical size gradually by stable fatigue propagation. Thus the rate of fatigue crack propagation per cycle is determined by the cyclic change in stress intensity factor, a measure of the stress at or around the crack tip.

The stress generated in the die during temperature cycling is dependent on the magnitude of temperature change. The number of cycles to failure in thermal fatigue is often given by Paris' Power Law. Paris developed a power law to predict the fatigue crack propagation based on stress intensity factor K at or near the crack tip in the plane of propagation. As the stress varies during thermomechanical cycling, K will proportionately vary as follows:

$$\Delta K = K_{\max} - K_{\min} \tag{2.13}$$

Then the rate of propagation, da/dN, the increase in crack length per cycle, N, will be given by Paris' Power Law [Paris et al. (1961)]:

$$\frac{da}{dN} = A(\Delta K)^n \tag{2.14}$$

where A is the material dependent coefficient, n is the material dependent exponent. The stress intensity factor for an infinite solid was given by Westergaard (1939) as

$$K_I = \sigma_{\mathrm{app}}(\pi a)^{1/2} \tag{2.15}$$

where K_I is the mode I stress intensity factor, σ_{app} is the mode I far-field applied stress, a is half of the penetration depth. The thermomechanical stress level in the die has been investigated by various authors including Bolger (1982) who represented the stress level by

$$\Delta\sigma_{\text{app}} = 10^{-6}k|\alpha_s - \alpha_d|\Delta T\sqrt{\frac{E_s E_a L}{X}} \tag{2.16}$$

where

k	=	geometric constant, dimensionless
α_s	=	coefficient of thermal expansion of substrate or case, 1/° C
α_d	=	coefficient of thermal expansion of die, 1/° C
E_a	=	tensile modulus of adhesive, N/m^2
E_s	=	tensile modulus of substrate, N/m^2
x	=	adhesive bond thickness, m
L	=	diagonal length of die, m
ΔT	=	temperature cycle magnitude, ° C

Substituting for the stress level and the stress intensity factor in the Paris' Power Law gives

$$\frac{da}{dN} = A(\Delta\sigma_{\text{app}}(\pi a)^{1/2})^n \tag{2.17}$$

where

A	=	die material constant
σ_{app}	=	mode I applied stress amplitude, N/m^2
a	=	half the penetration depth, m
n	=	die material exponent

Upon integrating this equation, the number of cycles to failure is found to be

$$N_f = \frac{2}{(n-2)A(\Delta\sigma_{\text{app}})^n \pi^{n/2}}\left(\frac{1}{a_i^{(n-2)/2}} - \frac{1}{a_f^{(n-2)/2}}\right) \tag{2.18}$$

where

A	=	die material constant
n	=	die material exponent
$\Delta\sigma_{\text{app}}$	=	mode I applied stress amplitude, N/m^2

a_i = initial crack size, m
a_f = final crack size, m
N_f = number of cycles to failure

Observe that a_f, the final crack length at failure, may be taken to be equal to the critical crack size.

Improper die to header bonds or unsuitable packaging conditions can increase the incidence of die cracks [Tan et al. (1987)]. Wafer backprocessing such as lapping or thinning can result in silicon flaws which are susceptible to fracture during temperature and power cycling [Hawkins et al. (1987)]. Die fracture and die adhesion fatigue are dependent primarily on the magnitude of the temperature cycle.

2.4 Die and Substrate Adhesion Fatigue

In a typical power or environmental temperature cycle, the die, the die attach, the die substrate and the package all experience temperature differences and temperature gradient differences. Because they also have different coefficients of thermal expansion, the bond between the die and the substrate can experience fatigue failure.

The most common die attach defects are voids. The presence of edge voids in the die attach induces high longitudinal stresses during power and environmental temperature cycling. These voids can act as microcracks which propagate during power cycling and result in debonding of the die from the substrate or the substrate from the case. Voids are responsible for weak adhesion, die lifting, increased thermal resistance, and poor power cycling performance. Solder and some epoxies have been shown to crack and detach during temperature cycling. Voids can form from melting anomalies associated with oxides or organic films on the bonding surfaces, outgassing of the die attach, trapped air in the bonds, and shrinkage of solder during solidification. Insufficient plating, improper storage, lack of cleaning, or even diffusion of oxidation prone elements from an underlying layer can generate voids during melting of die attaches. In other instances, dewetting of solder results in excessive voiding, especially when a solderable surface, a poor solderable underlying metal, or excess soldering time produces an intermetallic compound not readily wet by the solder. Even under ideal production conditions, voids are often present due to solvent evaporation or normal outgassing during cooling of organic adhesives. Although voids can form from a number of sources, they are normally limited to an acceptable level through process control. The actual package construction, the die attach unit materials, and the overall void concentration determine the actual effect of voiding on device reliability.

The formation of randomly distributed voids at the die substrate interface is generally unavoidable during the die attach process. Edge voids are most likely to produce die cracking due to high longitudinal stresses. The compressive nature of the stress near a center void greatly reduces the possibility for the crack to propagate. Die cracking statistics for two samples of equal size with edge and center voids subject to 10 cycles of thermal shock are summarized by Chiang and Shukla (1984). The devices with center voids show no cracks, while devices with edge voids show nearly 50% failure rate due to die cracks.

A large size void may reduce the thermal performance of the device by creating a large temperature gradient [Mahalingham et al. (1984)]. In general, small concentrations of random voids have little effect on the peak junction temperature. However, when a relatively large contiguous void is present, the heat must flow around the void creating a large temperature gradient in the silicon. Alternatively, if a large void is broken up into many smaller voids, the perturbation to heat flow is reduced.

The number of cycles to failure of the die attach as a result of plastic strain cycling is given by

$$N_f = 0.5 \left(\frac{\gamma_a}{\gamma_f} \right)^{1/c} \tag{2.19}$$

where N_f is the number of cycles to failure, γ_f is the fatigue ductility coefficient, defined as the shear strain required to cause failure in one load reversal, c is the Coffin-Manson coefficient, γ_a is the plastic strain amplitude given by [Smith et al. (1963), Manson and Hirschberg (1964), Coffin (1954), and Tavernelli and Coffin (1959)]

$$\gamma_a = \frac{L|\alpha_s - \alpha_d|\Delta T}{X} \tag{2.20}$$

where L is the diagonal die length (m), α_s is the substrate expansion coefficient (1/° C), α_d is the die expansion coefficient (1/° C), ΔT is the temperature excursion cycle (° C), and X is the height of the die attach (m). The fatigue failure of the substrate attach is similar to that of the die attach except that the dimensions of the materials are different. The number of cycles to failure of the substrate attach in fatigue is given by the Coffin-Manson relation [Bolger (1982); Bolger and Mooney (1984)] as

$$N_f = 0.5 \left[\frac{L_s(\alpha_c - \alpha_s)\Delta T}{h_{sa}\gamma_f} \right]^{1/c} \tag{2.21}$$

where L_s is the diagonal length of the substrate, (m), α_c is the thermal coefficient of expansion of the case (1/° C), α_s is the thermal coefficient of expansion of the substrate (1/° C), ΔT is the temperature excursion per cycle (° C), h_{sa} is the thickness of the substrate attach (m), γ_f is the fatigue ductility coefficient of the substrate attach defined by shear strain intercept at one load reversal, and c is the slope of the low cycle fatigue curve of log shear strain versus log cycles to failure [Bolger (1982); Bolger and Mooney (1984)].

2.5 Cracking in Plastic Packages

Thermal coefficient of expansion (TCE) mismatches drive the failure mechanisms of cracking of plastic packages. Silicon has a TCE of about 3 ppm/° C, while molding compound typically, have a TCE of about 20 ppm/° C below the glass transition temperature. Most lead frame materials are either alloy 42 (with a TCE of 4.7 ppm/° C) or copper (with a TCE of about 17 ppm/° C).

A typical process involves die attachment with adhesive cure at 270° C (polyimide adhesive) or 170° C (epoxy adhesive), followed by encapsulation and cure of the molding compound at 170° C. Negligible stress is established at these elevated temperatures, but the thermal coefficient of expansion differentials cause increasing stress as the surrounding temperature is lowered. The die attach produces a bending moment at room temperature, which puts the surface of the die into tension and the bottom of the die into compression. The magnitude of the tensile stress at the surface of the die is proportional to the thickness of the die. Encapsulation superimposes a compressive stress on the die and tends to place the molding compound under tension. Cracks therefore have a tendency to propagate in the molding compound. Cracks in packages are a reliability concern because they provide paths for the entry of contaminants. Shear stresses are applied on the surface of the chip due to differential thermal expansion between the plastic encapsulant and the passivation layer. This results in the fracture of the passivation layer which forfeits its action as an impurity getter and barrier. These stresses also cause lateral displacement of the interconnection lines on the surface of the chip and produce shorts between the overlying and underlying interconnection patterns in devices with two or more layers of interconnection [Schnable and Keen (1969)]. The effects of stresses applied by plastic are more pronounced in larger chips especially at the chip corners. Resilient conformal coating on the die and the use of plastic encapsulants with a lower coefficient of thermal expansion are some of the methods used to decrease thermal stress effects in the passivation layer due to thermal coefficient mismatches. Inayoski et al. (1979) reported damage of passivating film of plastic encapsulated LSI devices due to thermally induced stress at the encapsulant passivation layer interface. The location of aluminum corrosion coincided with defects in the passivating film typically caused by differential thermal stress from the encapsulant. Sims and Lawson (1979) found a phosphorus content of 1.8% optimum for plastic device reliability.

Cracks also initiate due to stress concentrations at the top edge of the die or at the bottom edge of the lead frame, or due to voids in the molding compound. The cracks propagate under the influence of temperature cycling. Nishimura et al. (1987) have predicted crack propagation using the relation

$$\frac{da}{dN} = C(\Delta K)^m \tag{2.22}$$

where a is the crack length (m), N is the number of cycles, ΔK is the stress intensity factor range, and C and m are constants. ΔK is function of the molding/curing temperature and the low temperature extreme. Package cracking can be controlled by downsetting the die below the plane of the leads so that there is roughly an equal thickness of molding compound both above and below the die. Modifications to the molding compound to control crack propagation characteristics include filler coatings to reduce delamination at the epoxy filler interface, and control of the filler size distribution.

Delamination is a failure mechanism which occurs between the die and molding compound and between the molding compound and the lead frame due to shrinkage after molding, especially in PSMCs (plastic surface mount components) which are typically subjected to temperatures in the range of 215–260° C and heating rates as high as 25° C/s during board assembly.

In addition to TCE driven stresses, there are two additional components to internal stresses inside the package. If the heating or cooling rates are much higher than 10° C/s, thermal stresses associated with internal temperature gradients can be appreciable. Moisture absorbed by the package molding compound evaporates under high temperatures creating an internal pressure which causes cracking (generally referred to as the "popcorn mechanism"). It is possible for PSMCs to absorb moisture during storage in normal ambients for vapor driven stresses to dominate the mechanics of the package. The main factors influencing the propensity to crack include: peak temperature reached during soldering, moisture content of the molding compound, dimensions of the die, thickness of the molding compound under the die, and the adhesion of the die to the lead frame. Fukuzawa et al. (1985) tied together all the factors affecting the failure mechanism of thermal cracking of plastic packages. The steam vapor pressure in the gap under the die pad causes the molding compound to delaminate from the pad, expand, and form a dome. A crack occurs where the maximum bending stress, σ_{max}, exceeds the fracture stress characteristic of the molding compound at elevated temperature, T_{elev}. The relation is represented mathematically as

$$\sigma_{\text{max}} > \sigma_{\text{crit}}(T_{\text{elev}}) \tag{2.23}$$

The maximum bending stress is first reached at the center of the long side of the die pad and is given by

$$\sigma_{\text{max}} = 6K\left(\frac{a}{t}\right)^2 P \tag{2.24}$$

where K is a dimensionless stress concentration factor which depends on the aspect ratio of the die pad ($K = 0.05$ for a square pad), where a is the length of the short side of the die pad (m), t is the thickness of the molding compound under the pad (m), and P is the vapor pressure in the cavity (N/m^2). For packages of different dimensions saturated at the same temperature and relative humidity followed by the same temperature shock, P equals the value of critical stress σ_{crit}, and the ratio (a/t) is the major determinant of whether the package will crack. Kitano et al. (1988) measured the water vapor absorption and water diffusivity of the molding compound and deduced the internal vapor pressure as a function of temperature from deflection measurements of the pressure dome. During the solder temperature shock there is not enough water vapor inside the gap under the die pad to account for this pressure. However, because the molding compound is hygroscopic, a larger amount of water vapor is available to generate steam pressure within a diffusion length of the cavity over the duration of temperature shock. The pressure P, generated inside the package due to thermal shock is represented by

$$P = H_1 P_{\mathrm{sat}}(T_{\mathrm{elev}}) \tag{2.25}$$

where H_1 is the relative humidity of the saturation ambient prior to the temperature shock, and T_{elev} is the peak temperature of the thermal shock. The equations for pressure and stress model the package cracking sensitivity as a function of temperature shock (in the neighborhood of temperatures 215–260° C).

2.6 Reversion or Depolymerization of Polymeric Bonds

Reversion or depolymerization is characterized by the breakage of polymeric bonds (say, in the encapsulant of plastic encapsulated devices) which turns the solid polymer into a gummy liquid comprising monomers, dimers, and lower molecular weight species. Reversion refers to the driving of the condensation reaction in a direction opposite to polymerization [Tummala and Rymaszewski (1989)]. Elevated temperature is an accelerator of failures; that is, temperature increases the reaction rate. However, most polymers used in the electronic packaging industry degrade at temperatures over 300° C.

Thermal degradation may also be encountered in power packages having brominated-epoxy flame retardants as encapsulants. Degradation involves scission of a highly linked polymer into constituent free radicals and monomers at elevated temperatures in the neighborhood of 320 to 350° C for most electronic materials. Figure 2.4 [Tummala and Rymaszewski (1989)] shows a differential scanning calorimetry plot of a typical epoxy molding resin for plastic packages. The degradation temperature is recognized as exotherm in the differential scanning calorimetry plot. Softening and melting temperatures for thermoplastics are a function of molecular weight, with the high temperature properties associated with the longer chain molecules. Thermoplastics with lower crosslink density generally have their glass transition temperature in a range of 150 to 200° C [Manzione (1990)]. Generally, the degradation temperature is much

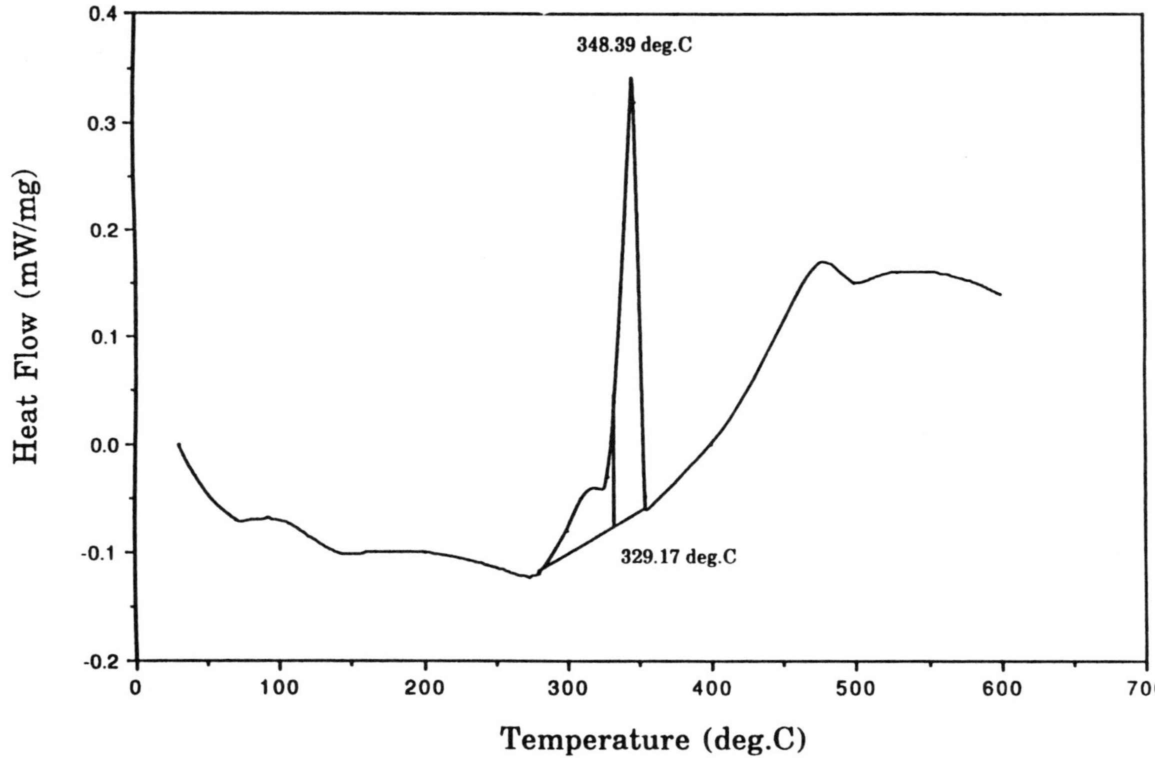

Figure 2.4 Differential scanning calorimetry plot of typical epoxy resin for plastic packages. The two degradation exotherms in the neighborhood of 320 to 350° C indicate degradation of bulk molding compound [Tummala (1989)].

higher than the semiconductor operational ranges and therefore has little impact on the packaging of these chips. The high temperature at which this failure phenomenon occurs makes it a recessive mechanism during normal device operation.

2.7 Constraint Cavitation of Conductor Metallization

Aluminum-silicon conductor lines in integrated circuits have a tendency to fail through nucleation and growth of voids. Voids assume two dominant morphologies: a wedge shaped form and a narrow crack shaped form. The conductor lines are typically passivated with a glass layer at approximately 427° C and cooled to room temperature. The thermal coefficient of expansion of glass is an order of magnitude smaller than that of the conductor so the cooling process imparts a tensile stress to the conductor line. This tensile field is not uniform and stress gradients cause diffusive mass transports. The passivated conductor lines are in a state of mechanical constraint which prevents large scale grain boundary sliding. With extensive mass transport and insufficient grain boundary sliding, void growth is possible. Typically, stress driven diffusive voiding of the conductor metallization has been characterized by Yost et al. (1988) as intergranular cracking without the extensive deformation usually associated with tensile creep. Relaxation due to dislocation glide and climb will occur throughout the conductor line network at a rate determined by the local stress, temperature and rate of resulting strain accommodation in the passivation layer and at the interface.

Yost et al. (1989) conceptually cooled 3 μm wide conductor lines rapidly from 400° C to 306, 213, 119, and 25° C. The effect of temperature on the rate of growth of voids is shown in Fig. 2.5 (where Y_D is the stress relaxation distance on both sides of the grain boundary, and $2b$ is the conductor width). As the growth temperature (i.e., the final temperature to which the metallization system along with the passivation is cooled from 400° C) increases, voids reach their final size, the size which is reached when far field stress is balanced by the stress relaxation due to voiding, more rapidly. The cracks are predominantly oriented perpendicular to the line length and usually nucleate in the center of the conductor and progress to open circuit failure. This type of cracking was found to be present without nitrogen embrittlement of the metallization and progressed in the absence of current and corrosive agents [Yost et al. (1988)].

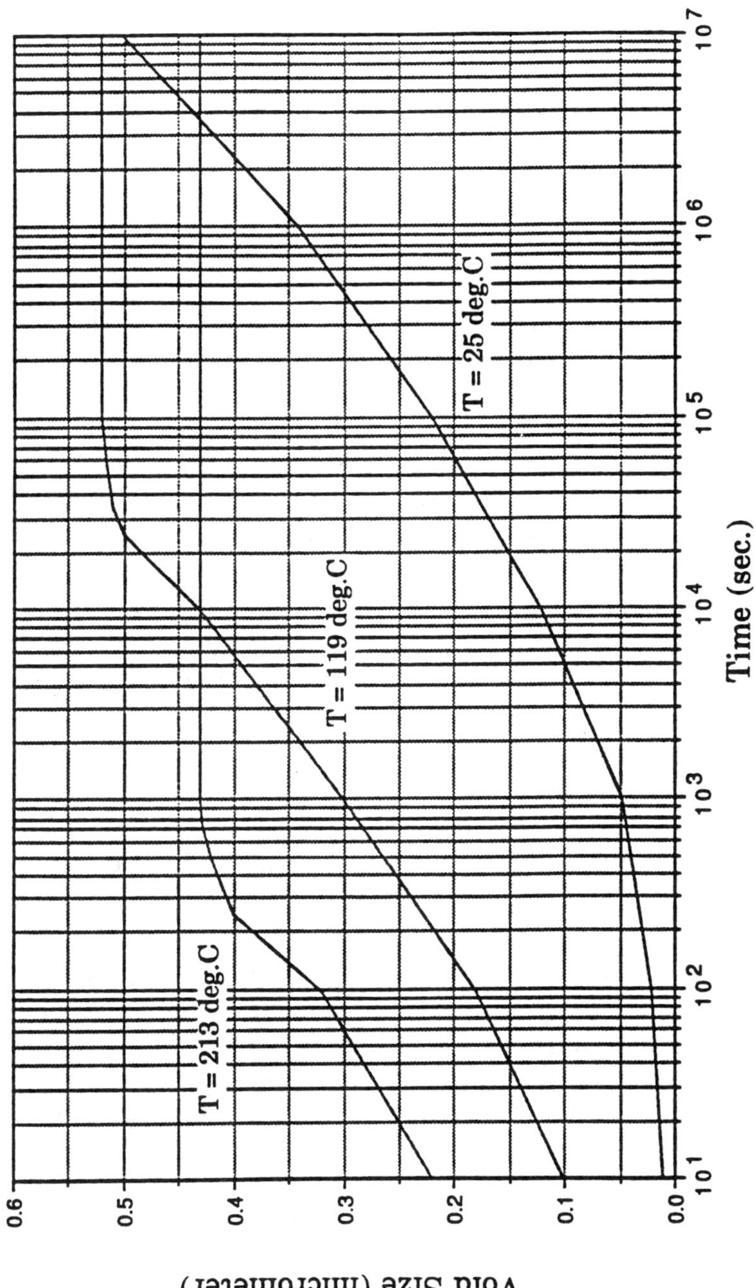

Figure 2.5 Growth kinetics of wedge voids at various aging temperatures [Yost et al. (1989)].

Yost et al. (1988) demonstrated analytically and experimentally that stress driven mass transport along the grain boundaries leads to the formation of either wedge shaped voids or slit (crack) shaped voids at the conductor center. The void shape was dependent upon the granular orientation of the metallization with respect to the applied stress. The temperature dependence of the surface diffusion coefficient has been modeled by

$$\log D_s = 5.07 - \frac{1.09 \times 10^4}{(273+T)} + \frac{2.89 \times 10^6}{(273+T)^2} - \frac{3.24 \times 10^8}{(273+T)^3} \quad (2.26)$$

where T is the temperature in degrees centigrade and D_s is the surface diffusion coefficient [Yost et al. (1988)]. The material stack-up used for the analysis was: silicon, BPSG, Al-Si, PSG. The time to failure decreases with the decrease in conductor line width. This behavior is a direct consequence of higher stress states which develop in narrower lines. The calculations also show an increasing growth rate as the temperature increases. The calculated increase is due to the diffusion coefficient which increases exponentially with temperature, and σ_∞ (the average far field stress) which decreases quadratically with temperature. In its present form, the model does not include the effect of the temperature dependent stress. The stress model includes the temperature dependent stress term derived from the differential thermal expansion between the aluminum and the glass passivation layer, but does not include any stress relaxation which would occur as a highly stressed conductor is aged.

There are at least two types of stress relaxation that can occur. A homogenous, or a mean field relaxation can take place throughout the entire conductor line and passivation layer. A local stress relaxation process can occur as a result of the void process itself. The extent of the latter relaxation process depends on the metal grain size and the geometry of the debond region. The inclusion of the stress relaxation effects would reduce the stress and significantly increase the time to failure at high temperatures. The rate of growth for crack like voids is slightly faster then wedge shaped voids. The predicted time to failure for crack growth in 3 μm conductor lines is 1.5 years at 27° C. While the time to failure for wedge like growth in 3 μm conductor lines is 2.3 years at 27° C. These results are consistent with the experience of the microelectronic community [Yost et al. (1988)]. The time to failure decreases with increase in temperature. The temperature

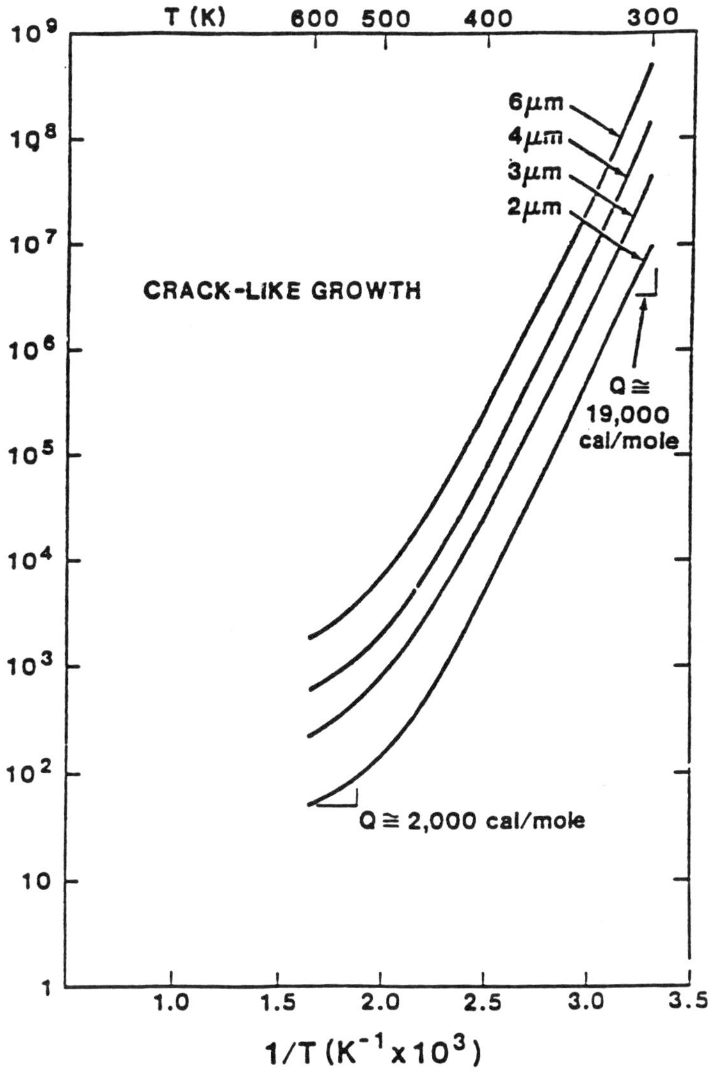

Figure 2.6 Time to failure for crack-like growth [Yost et al. (1989)]. The results presented here are very conservative, because the models do not take into account stress relaxation effects at high temperature.

dependence of time to failure for crack and wedge like growth is given by Figs. 2.6 and 2.7 respectively.

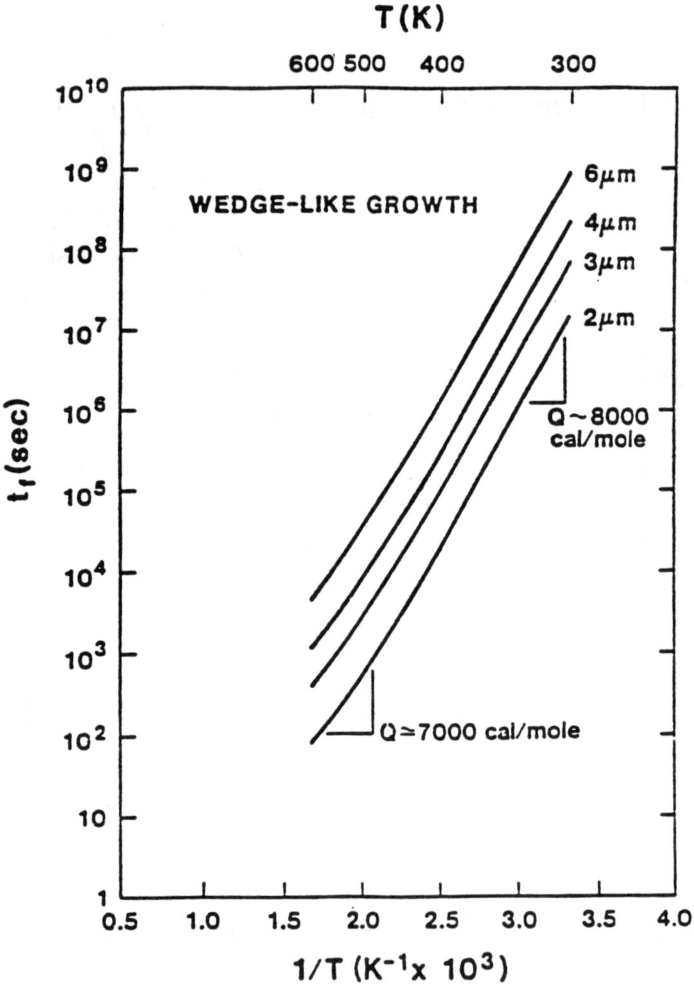

Figure 2.7 Time to failure for wedge-like growth [Yost et al. (1989)]. The results presented here are very conservative, because the models do not take into account stress relaxation effects at high temperature.

3 TEMPERATURE DEPENDENT CORROSION FAILURE MECHANISMS

3.1 Corrosion of Metallization and Bond Pads

Corrosion is typically defined as the chemical or electrochemical reaction of a metal with the surrounding environment. The mechanisms of corrosion can be divided into two types: dry, such as oxidation of aluminum in air, and wet, where the reaction occurs as in the presence of an electrolyte, a moist environment and an electromotive force. Dry corrosion is of minor importance in semiconductor devices, because the corrosion process is self passivating for the metal, forming a thin oxide film which prevents further oxidation. Wet corrosion, in the presence of an ionic contaminant and moisture, can provide a conductive path for electrical leakage between adjacent conductors, dendrite growth, or corrosion of device metallization or bond pads. The ions most commonly found on the die surface and that in the presence of water give rise to an electrolytic solution required to trigger the corrosion process include halogens (especially Cl^-) deriving their origin both from inadequate removal from the fabrication process residues and from the container if it is plastic, alkalines (especially Na^+) deriving from the diffusion ovens and more generally from the glass containers and the hands of operators and phosphorus, which unlike the above ions that are contaminants, is specially incorporated in the surface passivation glass to improve its mechanical characteristics and restrict the effects of the Na^+ present. The importance of passivation in microcircuit metallization corrosion as a function of temperature has been evaluated by Commizolli et al. (1980) who found that the dependence of corrosion current on temperature appeared linear between 60° C and 100° C. The corrosion current at 90% relative humidity using passivated chips was found to decrease with decrease in temperature between 60° C and 100° C.

Pecht and Ko (1990) modeled the time to failure due to corrosion as a sum of moisture ingress time and the time for corrosion attack and failure. The moisture ingress time for a hermetic package was determined based on the internal volume and the leak rate. The method to calculate the worst case operation-independent moisture ingress time was based on standard testing procedures.

After sufficient moisture ingress, a critical moisture content will be reached inside the package, and corrosion can initiate once the non-operating sealed package is exposed to temperatures below the dew point. At this time, the moisture inside the package condenses and combines with any ionic contaminant present to provide a conductive path between adjacent metallic conductors. The conductive path serves as a medium for the transfer of ions in the corrosion process. When the package is operating, the heat dissipated inside the package will typically elevate the temperature above the dew point. Consequently, the electrolyte will evaporate and no longer provide an electrolytic path between conductors [Pecht and Ko (1990)].

Corrosion typically begins when the temperature inside the package is below the dew point enabling condensation inside the package. During operation, heat dissipated by the die is often enough to raise the temperature to evaporate the electrolyte and alleviate humidity problems [RADC Report SOAR-3 (1985)]. Elevated temperature due to device power thus acts as a mechanism to slow the corrosion process.

When a packaged component is in the off mode, it can equilibrate with the ambient relative humidity. However when the packaged component reaches its operating temperature, the same moisture content results in a lower relative humidity. The low relative humidity at the operating temperature may prevent the formation of an electrolyte and may negate exposure to corrosion entirely.

The time to failure as a result of corrosion failure is given by

$$\tau_c = \left(\frac{K_1 K_2 K_3}{K_4}\right) \left[\frac{w^2 h n d F \rho}{4MVZ}\right] \tag{2.27}$$

where M (g) is the atomic weight of metal conductor of density d (g/cm^3), width w (cm), height h (cm), and chemical valence n. The quantity, ρ/Z (Ω cm), is the sheet resistance of the electrolyte, and V is the voltage applied. K_1 is the physical and chemical properties factor of the metallization materials, K_2 is the coating integrity index, K_3 is the mission profile correction factor, and K_4 is the environmental stress correction factor. The constants K_1, K_2, K_3, and K_4 and n are dimensionless.

The only temperature dependent factor is the environmental correction factor K_4, utilized to determine the time to failure for various temperature and humidity conditions. The K_4 term is modelled as

$$K_4 = \frac{(RH_R)^n \exp(E_a/KT_R)}{(RH)^n \exp(E_a/KT)} \quad (2.28)$$

where the subscript R designates a reference and

RH	=	relative humidity, percent
RH_R	=	reference relative humidity, percent
E_a	=	activation energy, eV
k	=	Boltzmann's constant, eV/K
n	=	a material constant
T_R	=	reference temperature, K

3.2 Stress Corrosion in Packages

Stress corrosion is a failure phenomenon which occurs around 300° C and above predominantly in power devices. Tummala and Rymaszewski (1989) found that electrochemical corrosion interacting with mechanical stress is a potential cause of failure. Failures are typically transgranular, and result from the acceleration of the fatigue process by corrosion of the advancing fatigue crack. For example, craze cracking of polymers occurs in halogenated solvent vapors. Adsorbed moisture films due to relative humidity exposure cause static fatigue of the lead zinc-borate sealing glass [Tummala and Rymaszewski (1989)]. The distinct mechanism is chemical attack on glass, and the time to failure is usually modelled by an Arrhenius equation of temperature which activates above 300° C [Tummala and Rymaszewski (1989)]. The rate of galvanic corrosion in the presence of a liquid electrolyte increases as temperature is increased due to a more rapid rate of electron transfer. Typically corrosion products in microelectronic devices such as $Al(OH)_3$ are derived from reaction processes which are monotonically increasing functions of temperature. Although the corrosion rate depends in part on the steady state temperature, corrosion rate also depends on the magnitude and polarity of the corrosion galvanic potential which are functions of the electrolyte concentration, pH, local flow conditions and aeration effects. Finally, as noted previously, elevated temperature acts as a means to slow the corrosion process.

4 TEMPERATURE DEPENDENT ELECTRICAL FAILURE MECHANISMS

4.1 Electrical Overstress

Alexander (1978), Canali et al. (1981), and Smith (1978) all point out that electrical overstress occurs when a higher than rated voltage or current induces a hot spot temperature beyond specifications for short periods of time. Hot spot development typically occurs at a semiconductor junction due to an increase in the current flow to accommodate the additional stress in the device. As the junction heats up, the increased temperature encourages increased current flow due to lowering of silicon resistance at higher temperature. This in turn, further heats up the junction. The fundamental cause of failure is joule heating due to power dissipation along current conducting paths. If the material reaches its melting temperature or its eutectic temperature, permanent damage may result. In a typical device, the greatest power dissipated per unit volume is either in the depletion region of the junction or in the lightest doped material. The temperature rises due to power dissipation, resulting in the decrease in resistivity of silicon.

Runayan (1965) plotted the resistivity of silicon as a function of temperature for a variety of doping concentrations. Silicon initially exhibits a positive temperature coefficient of resistivity but reaches a peak value (in the neighborhood of 160° C) and thereafter exhibits a large negative temperature coefficient of resistivity. As the temperature of a region rises, the initial increase in the resistance of the region tends to spread the current to the cooler region. If the situation continues, the hot spot temperature may exceed the intrinsic temperature of silicon beyond which the resistivity of silicon decreases greatly. This allows more current to pass through the hot spot, further increasing the temperature, and resulting in thermal runaway. That is, the temperature of the hot spot rises suddenly while the resistance of the device drops. The process continues until the silicon in the hot spot melts, destroying the silicon crystal structure. If the electrical transient continues, the interconnects are expected to melt as well. If the transient is accompanied by a high voltage, an electrical arc may occur when the metallization opens. The temperature in the vicinity of the arc can cause the metallization to vaporize. The energy required to raise the junction to an instable temperature can be calculated using a power model [Pancholy and Jhnoki (1978)].

4.2 Gate Oxide Breakdown

The two forms of oxide breakdown that can occur in a device result from electrostatic discharge (ESD) and electrical overstress, or from time dependent dielectric breakdown (TDDB). Both can occur during operation within rated conditions of voltage, temperature and power dissipation.

Electrostatic Discharge. An integrated device can be considered equivalent to a circuit having multiple paths to the ground. When one pin is grounded, potentials sufficient to cause dielectric breakdown, junction shorts or cracks between isolated regions may exist throughout the device.

Electrostatic pulses can arise from contact with air, skin, glass or charge-carrying particles. Electrostatic pulses can damage the gate oxides causing cracking of the device oxide or melting of small amounts of the device creating minute explosions on the device surface and resulting in voids, cratering and subsequent short or open circuits. Increased temperature causes a significant reduction in the electrostatic discharge resistance of the component [Kuo and Kuo (1983) and Hart et al. (1980)]. In extreme cases the die may vaporize due to electrostatic pulses.

Moss (1982) and Hart et al. (1980) point out that the electrostatic discharge failure voltage decreases with increase in temperature in the range of 25 to 125° C [Moss (1982), Hart et al. (1980)]. Amerasekera and Campbell (1987) have shown that simple low voltage electrostatic pulses can damage gate oxides to varying degrees in NMOS without actually causing complete gate oxide breakdown. Damage shows up in the form of reduction in saturation drain current, and gate voltage. Scherier (1978) has shown that MOSFETs are most susceptible to damage due to electrostatic discharge. Total breakdown of n-channel MOSFETs occurs around 100 V, while TTL NAND gate devices show degradation around 1.0 kV. Latent damage can also occur with microcracks remaining dormant until aggravated.

Devices are typically protected against ESD by protection structures shorting the ESD pulse voltage to ground or to the supply voltage to limit the current entering the critical junction. The damage voltages for protected devices are in the range of 3000 to 9000 V, depending on protection, as compared to as low as 100 V for unprotected devices. However, on-chip protection can reduce device performance.

Time Dependent Dielectric Breakdown. TDDB is the formation of low resistance dielectric paths through localized defects in dielectrics such as thermally grown oxides or other oxides in MOS devices. Thin FET dielectric materials exhibit this breakdown failure mechanism, which depends on electric field and the distances between the conductors in electronic packages, which are defined by technology limits or electrical requirements. Failures typically occur at weaknesses in the oxide layer due to poor processing or uneven oxide growth. Anolick and Nelson (1979) modeled the time to failure by

$$t_f(F) = Ae^{\Delta H/kT} e^{\gamma(V_B(F)-V_A)} \tag{2.29}$$

where

$t_f(F)$	=	time to failure for F percent of the population
γ	=	voltage form factor (determined by life testing), V^{-1}
ΔH	=	activation energy, eV
k	=	Boltzmann constant, 8.617×10^{-5} eV/K
T	=	steady state temperature, K
$V_B(F)$	=	breakdown voltage for F percent of population, V
V_A	=	applied voltage, V
A	=	constant

Time dependent dielectric breakdown has been a significant failure mechanism in metal-gate MOS integrated circuits, silicon-gate transistors and integrated circuits with two or more levels of polysilicon and circuits with trench capacitors. Crook (1978, 1979) and Schnable (1988) have reported that, typically, TDDB has been found to have a very low thermal activation energy (typically 0.3 eV) but a very large voltage acceleration (10^7 megavolt/cm). The projected failure rate due to time dependent dielectric breakdown under temperature stress was found to be too low to be of any importance for 10 V and 55° C. At 10 V and 200° C operation, the projected lifetime was found to be 30×10^6 years, which, according to McPherson and Baglee (1985), Boyko and Gelarch (1989), and Swartz (1986), is a failure rate of 10^{-2} failures in 10^9 years For this reason, device screening procedures involve stressing the devices at higher voltages and room temperature, rather than normal biases and higher temperatures.

4.3 Second Breakdown

The mechanism of second breakdown has been considered by Beatty et al. (1976), Chen et al. (1983), Hower and Reddi (1970), and Hu and Chi (1982). It is typical of power transistors and is characterized by two different failure mechanisms: forward second breakdown, and the reverse second breakdown. Forward second breakdown is temperature dependent, while reverse second breakdown is not. The failure mechanism of second breakdown is characterized by an abrupt reduction in voltage, and increased localized current density. The consequence is high temperature and permanent device damage or degradation.

Forward second breakdown. Forward second breakdown is typically a result of thermal runaway at a point in the transistor. Current density and temperature increase jointly until the thermal runaway occurs. A power transistor can be considered to be an aggregate of many elementary transistors that operate under different conditions. When a power pulse is applied to the transistor, the junction temperature rises. Because the resistivity of the silicon increases with temperature, the temperature rise causes the collector current to increase, which, in turn causes the dissipated power to increase:

$$dP_1 = V_{CE} dI_C \tag{2.30}$$

$$dP_2 = \frac{dT}{R_{th}} \tag{2.31}$$

If $dP_1 < dP_2$, a thermal equilibrium is reached. If, on the contrary, $dP_1 > dP_2$, then temperature and current increases enhance each other and thermal runaway occurs, causing transistor destruction.

Usually temperatures above 160° C are required to cause device failure due to second breakdown. The major cause of device destruction is localized heating. Second breakdown has been found to occur in vertical power MOSFETs.

Reverse second breakdown. Reverse second breakdown is a phenomenon which occurs in power transistors during switching operation on an inductive load and is characterized by a precipitous drop in voltage and a rapid increase in current. This is an internal phenomenon. The second breakdown voltage increases mildly with increase in temperatures in the range of 25 to 150° C. The breakdown voltage increases from 650 V to 680 V when the temperature increases from 25 to 150° C [Beatty et al. (1976)].

4.4 Ionic Contamination

Ionic contamination causes reversible degradation phenomena such as threshold voltage shift and gain reduction due to the presence of mobile charge ions within the oxide or at the device-oxide interface (typically above 100° C [Schnable (1988)]). A survey of published data shows that about 30% of the microelectronic MOS device failures are a result of ionic contamination [Brambilla et al. (1981), Johnson (1976), and Johnson (1983)]. Ionic contamination is caused by mobile ions in semiconductor devices. Contamination can arise during packaging and interconnect processing, assembly, testing, screening, and during operation. Sodium, chloride and potassium ions are the most common contaminants. The most dangerous contaminant is sodium, which, in extremely small quantities (10^{11} per cm^2), can cause failures. The amount and distribution of the alkali ions (Na^+, Li^+, K^+) in or near the FET gate dielectric region will influence the device threshold voltage by superposition of these ionic charges on externally applied FET device voltage. Extra positive charge at the Si/SiO_2 interface induces extra negative voltage in the n-channel, resulting in a decrease in the threshold voltage of the device. In the case of a p-channel device (PMOS), the direction of the field at the gate oxide repels the positively charged ions away from the gate area; p-channel devices are therefore less sensitive to ionic decontamination than n-channel devices.

The mobility of the ions is temperature dependent [Hemmert (1981a, b)]. For this reason a high temperature storage bake and exposure to high temperature during burn-in are found to screen out the ionic contamination failures [Hemmert (1981a, b), Bell (1980)]. In particular, the position of ionic charges relative to the Si-SiO_2 interface greatly influences the effectiveness in altering the threshold voltage V_T. A uniformly distributed contaminant (Na^+) within the oxide is redistributed to the Si channel surface due to the influence of an electric field E and the thermal activation energy. The change in threshold voltage as a function of steady state temperature T, has been modeled by Wager and Cook (1984):

$$\Delta V_T \propto E^{1/2} t^{1/2} e^{-\Delta H/kT} \tag{2.32}$$

where E is the electric field strength, t is the time under bias, ΔH is the activation energy and k is the Boltzmann constant. Wager and Cook (1984) also point out that higher chip temperature would result in more rapid ΔV_T change and shorter times to saturation.

MOSFETs (n-channel) are covered with phosphosilicate glass (PSG) films to stabilize the threshold voltage against changes resulting from ionic contaminants (sodium). Hemmert (1981a) investigated the temperature bias kinetics for sodium ion drift in phosphosilicate glass. Microelectronic devices may be subject to a variety of defects such as exposed gate oxide, extraneous etched holes contiguous to the gate which subject the phosphosilicate glass film to either a high local ionic concentration or physical damage. The temperature at which ionic drift occurs in defective devices is not as large as compared to non-defective devices. Hemmert (1981b) conducted experiments to measure ionic stability on a 1.5 mm MOS consisting of gate oxide, phosphosilicate glass film and Al-Cu metallurgy. He found that below 250° C, the plot of voltage versus temperature shows a polarization plateau which is a saturation PSG polarization voltage. The activation energy ΔH was observed to be 2.0 eV. This value is uncharacteristically high as compared to the value usually reported for sodium related failure mechanisms ($\Delta H = 1.0$ eV) assuming a temperature dependent defect level. The cumulative percent defect predictions were found to be more accurate assuming temperature dependent defect level. In the temperature range of 250–350° C flatband voltage shifts were observed. This temperature called the break temperature, occurs when the contribution of flatband voltage shift to the ionic drift and polarization are equal. The maximum variation in the flat band voltage is related to the mobile charge by the relation

$$Q_m = C_{ox} \Delta V_{FB} \tag{2.33}$$

where C_{ox} and Q_m represent the oxide capacitance and the mobile charge per unit area respectively. The threshold voltage V_T and the flatband voltage V_{FB} are related by the equation

$$V_T = V_{FB} + 2\phi_F + \gamma\sqrt{2\phi_F} \tag{2.34}$$

where ϕ_F is the difference between the Fermi level and the intrinsic Fermi level in the bulk of the semiconductor.

It is evident that the threshold voltage depends on the flatband voltage. Thus, the usual assumption that the defect level is temperature independent was found to be invalid. A common characteristic of contamination failures is the reversibility under high temperatures in the neighborhood of 150 to 200° C, which partially or fully restores device characteristics [Hemmert (1981a, b) and Bell (1980)]. However, this just reflects a disordering of charge accumulations resulting from ion mobility and applied bias. The problem of ionic contamination has been solved to a large extent by use of high purity materials and chemicals for processing, use of HCl during oxidation [Robinson and Heiman (1971)], and use of PSG (phosphosilicate glass) or BPSG (borophosphosilicate glass) over the polysilicon to getter the ions [Schnable (1988)]. Passivation layers can provide additional protection against ingress of alkali ion contamination in completed devices [Schnable (1988)].

4.5 Surface Charge Spreading

Surface charge spreading is largely observed in MOS and memory devices, and, as pointed out by Edwards (1982), Blanks (1980), and Stojadinovic (1983), involves the lateral spreading of ionic charge from the biased metal conductors along the oxide layer or through moisture on the device surface. An inversion layer outside the active region of the transistor is developed due to the charge forming a conduction path between the two diffused regions or by an extension of the p-n junction through a high leakage region resulting in leakage currents between neighboring conductors. The rate of charge spread increases with temperature. Surface charge spreading failure is usually observed at temperatures around 150° C to 250° C and as indicated by Lycoudes and Childers (1980) is a type of wear-out mechanism. There are no existing physics of failure models for prediction of surface charge spreading failures.

4.6 Slow Trapping

Slow trapping is a failure mechanism largely observed in standard MOS transistors and certain types of memory devices, programmed by the transport of charge from the source or the drain through the gate oxide to the gate interface. High temperatures combined with a high electric field provide electrons with enough energy to cross the Si-SiO_2 interface [Nicollian (1974), Woods and Rosenberg (1980)]. Interstitial states at the Si-SiO_2 interface trap electrons and hold them in the oxide, thereby permanently shifting the threshold voltage of the device. The presence of electrons permanently trapped at the oxide interface decreases the speed at which the device can be programmed by creating a field which opposes further electron flow through the oxide interface. The failure mechanism of slow trapping decreases the circuit speed and causes functional failures. At this writing there are no existing physics of failure models for the prediction of slow trapping failures. Gottesfeld and Gibbons (1984) report that typically temperatures in the neighborhood of 175° C or greater result in failures due to slow trapping.

4.7 Hot Electrons

Hot electrons, or "holes," are charge carriers that acquire energies from very high electric fields in excess of those that would be indicated by lattice or ambient temperature. Hot electrons is a phenomenon largely prevalent in high density, small geometry, MOS memory devices because of small conduction channels and high voltages. It is accelerated at temperatures below 0° C [Stojadinovic (1983), Woods and Rosenberg (1980), Woods (1981)]. Electrons may gain high potential energy due to high electric fields present in the conduction channel and cross the potential barrier at the substrate-oxide interface. Impact ionization creates electron-hole pairs which have sufficient energy to cross the Si-SiO_2 interface as well. Threshold voltage shifts or transconductance degradation can occur as renegade charge concentrations build up over time [Ning et al. (1979)]. These charges may be channel electrons or holes themselves that have gained energy from the channel field, carriers that may have been generated by impact ionization or carriers that may have been thermally generated in the substrate. Penetration of the oxide by hot electrons can lead to excess gate and substrate currents [Takeda et al. (1983)].

Figure 2.8 shows the effect of ambient temperature on the hot carrier mechanism [Matsumoto et al. (1981)]. The temperature dependence of substrate current at fixed bias can have three origins: temperature dependent channel current, temperature dependent local electric field and temperature dependent ionization rate. The low field mobility follows $-3/2$ power dependence on the temperature

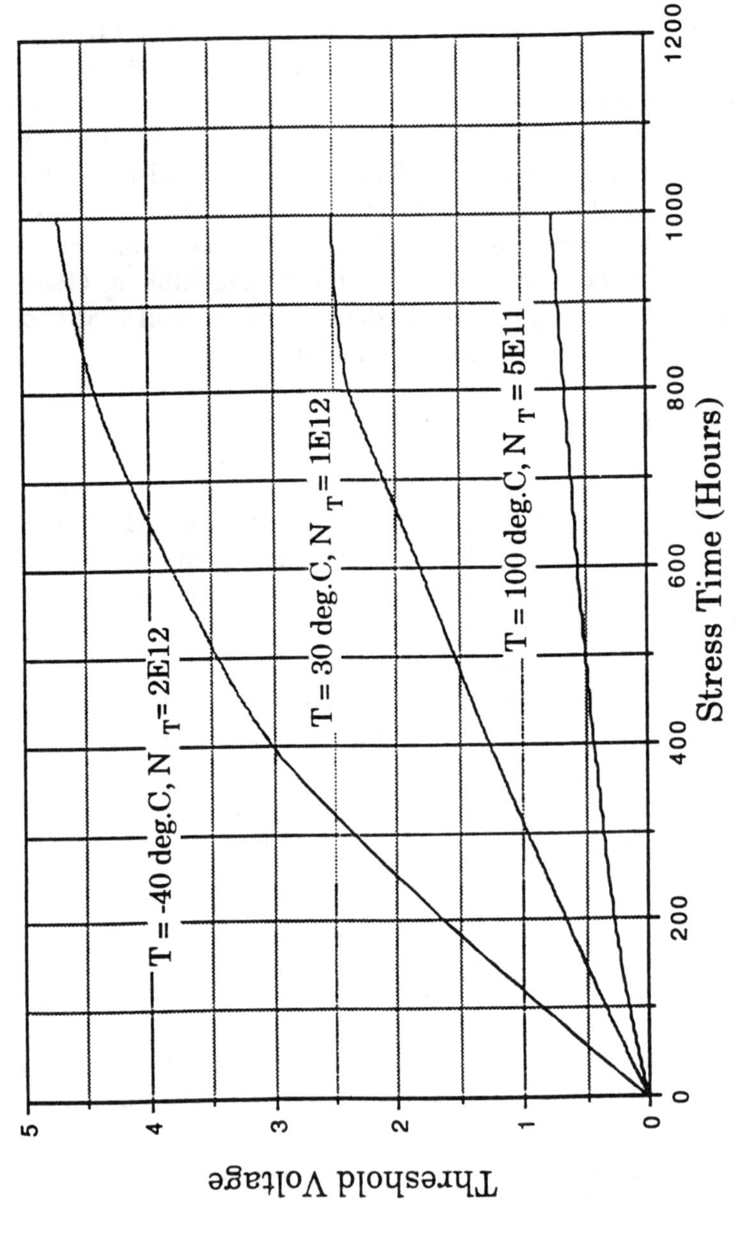

Figure 2.8 Change in threshold voltage due to hot electrons is much greater at lower temperature than at higher temperatures [Matsumoto et al. (1981)].

for typical n-channel MOSFETs [Sze (1981)]. The reduced mobility increases the saturation voltage which lowers the maximum. channel electric field at fixed drain bias and results in lower substrate current [Ko (1982) and Hu and Chi (1982, 1983)]. Tam and Hsu (1983)] proposed that the device degradation rate can be related to the substrate current by

$$\Delta V_t = C_1[C_2[I_s]^{\alpha}]^{\beta} \tag{2.35}$$

where C_1 is a structure related factor, C_2 is a material related factor, α is the energy factor (typically 2.9), and β is the factor from general kinetics (typically 0.6). As indicated by Hsu and Chiu (1984) the device degradation due to hot electrons very mildly decreases with the increase in temperature for temperatures between 20 and 100° C [Hsu and Chiu (1984)]. The power law dependence at higher temperatures follows the temperature dependence of substrate current as shown in Fig. 2.9 and Fig. 2.10. A plot of device lifetime as a function of temperature and substrate current is shown in Fig. 2.11. A calculation of energy and proportionality factors in Tam and Hsu's (1983) equation indicates that they are constant over the temperature range suggesting that the physical mechanism for device degradation is the same over the temperature range from 20 to 100° C. The variation of lifetime vs. substrate current for temperatures 20° C and 100° C shown in Fig. 2.11 indicates that hot electron degradation is temperature independent in this range (i.e., 20 to 100° C).

The increase in the ambient temperature decreases the observed ΔV_T shift from the channel hot electron effect. For a given device design, an equation describing this behavior is provided by Matsumoto et al. (1981):

$$\Delta V_T = AN_T(1 - e^{-BN_{inj}}) \tag{2.36}$$

where N_T is the effective oxide trap density, N_{inj} is the density of injected electrons and A and B are constants. From this equation, it is evident that the hot electron threshold voltage shift increases with both the electrons injected into the oxide and the availability of traps to capture the electrons. Both these quantities decrease with increasing temperature, giving rise to reduced threshold shift due to hot electron trapping at elevated temperatures [Matsumoto et al. (1981)].

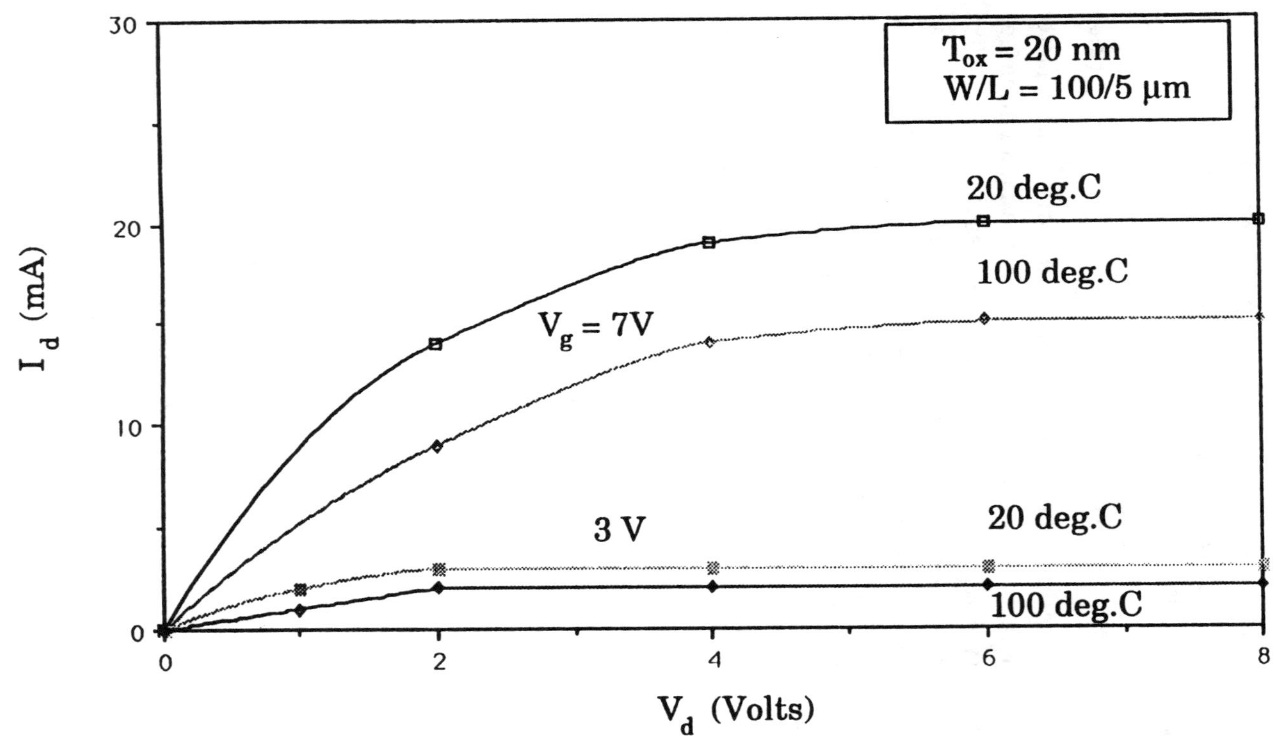

Figure 2.9 Drain characteristics for a $5\mu m$ device at 20 and 100° C: $V_g = 3V$ and $7V$, $V_t = 0.9V$ [Hsu and Chiu (1984)]. The variation in device characteristics between 20 and 100° C indicates that degradation due to hot electrons is almost temperature independent in this range.

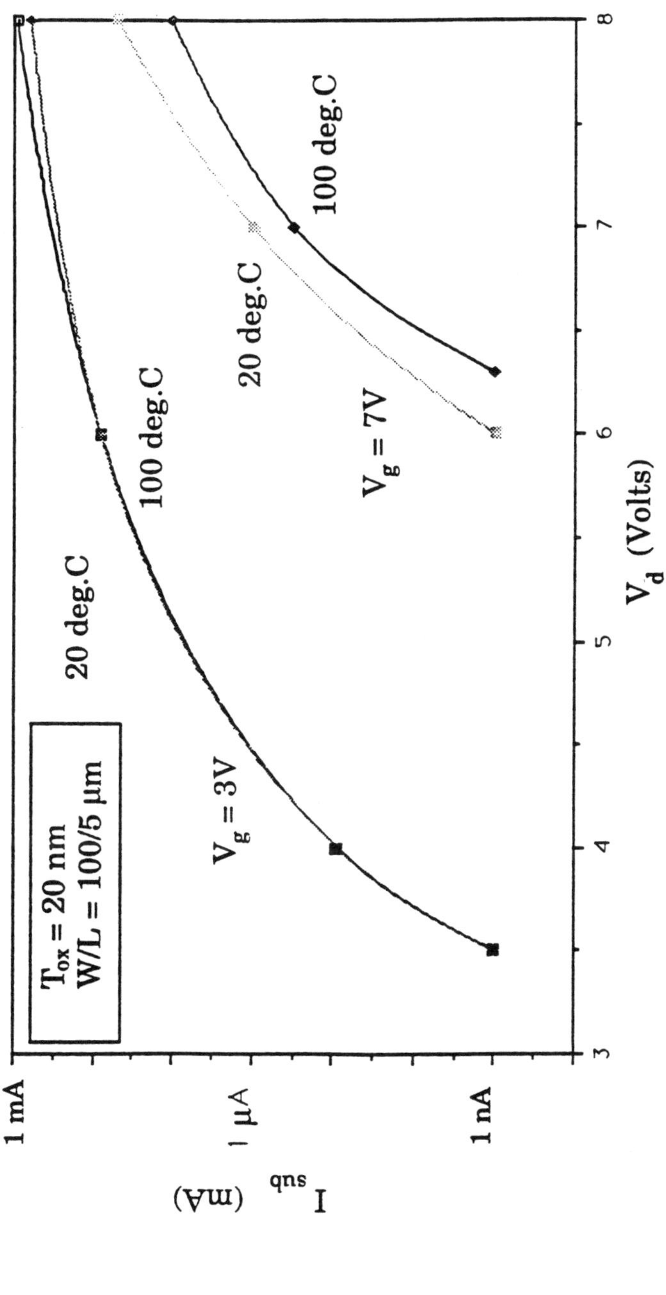

Figure 2.10 Substrate characteristics for a $5\mu m$ device at 20 and 100° C: $V_g = 3V$ and $7V$, $V_t = 0.9V$ [Hsu and Chiu (1984)]. The variation in device characteristics for temperatures between 20 and 100° C indicates that degradation due to hot electrons is temperature independent in this range.

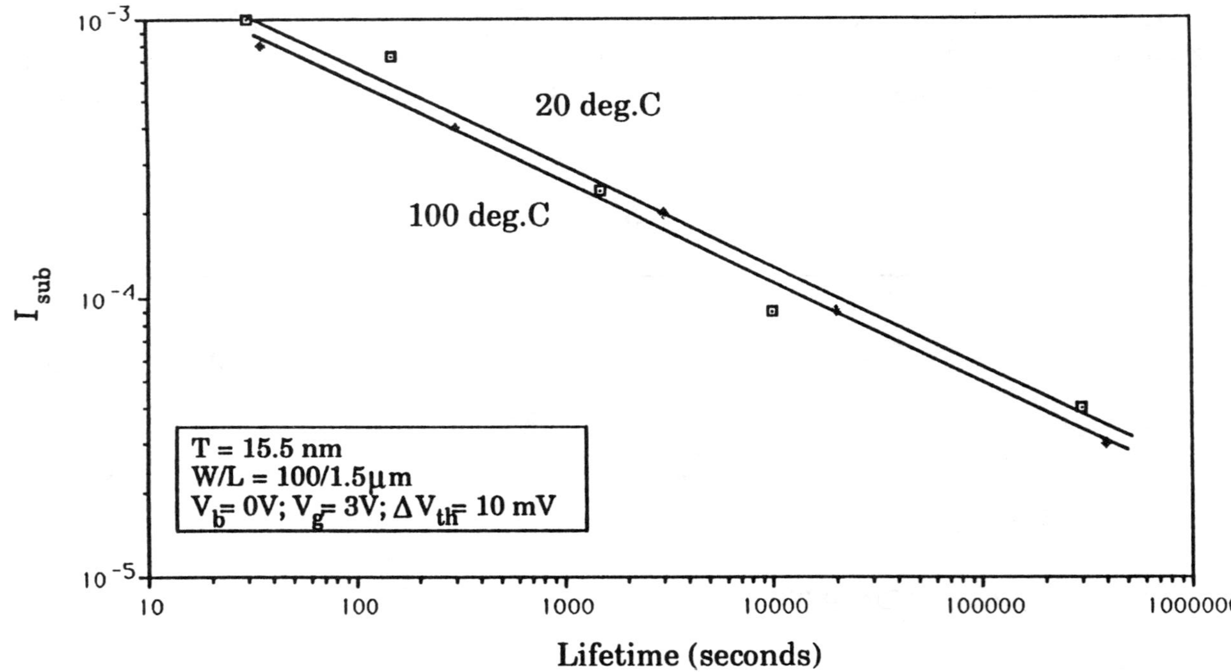

Figure 2.11 Substrate current dependence of device lifetime at 20 and 100° C [Hsu and Chiu (1984)]. The small variation in device lifetime for temperature variation between 20 and 100° C clearly indicates that degradation due to hot electrons is temperature independent in this range.

4.8 Hillock Formation

Hillocks in die metallization can form as a result of electromigration or to extended periods under temperature cycling conditions (thermal aging) [LaCombe et al. (1982) and Thomas and Calabrese (1983)]. Hillock formation as a result of electromigration often occurs upstream (in the sense of the electron flow) from the area of voiding. Hillocks also grow fastest at the down stream edge closest to the source of migrating metallization. Both voiding and hillock formation occur sometimes on top of each other. These phenomena of simultaneous voiding and hillock formation occur at temperatures in the neighborhood of 140 to 200° C [Thomas and Calabrese (1983)].

Hillock formation due to extended periods under temperature cycling conditions (thermal aging) is believed to be due to a self diffusion process which occurs in the presence of strains within the metallization [LaCombe et al. (1982)]. These strains may be due to a mismatch in the thermal expansion coefficients of the metallizations (Au), underlying titanium-tungsten (TiW) refractory layer, Si, and SiO_2. Hillock growth is more extensive in films deposited on room temperature substrates than on heated substrates indicating that the effect may vary with grain size. Coating of the metal with silicon nitride prevents failures from occurring and voids and hillocks from forming for at least 500 hours at 360° C. Hillocks form at random in aluminum films when heated to temperatures around 400° C during fabrication. Hillocks can cause electrical shorts between the adjacent lines and fracture of the overlying dielectric film. In double level metallized devices, hillocks can result in shorts between the underlying and the overlying metal layers. Hillocks can also be caused by electromigration and can result in thin dielectric sites which are susceptible to subsequent breakdown of the intermetal dielectric. Hillock formation is largely remedied by use of an alloyed metal such as Al-Cu and improvements in the technologies of passivation and packaging. Hillock formation is more a function of ΔT and temperature gradient and is mildly dependent on temperature (in the neighborhood of 400° C).

4.9 Contact Spiking

Contact spiking is the penetration of metal into the semiconductor in contact regions at temperatures typically above 400° C. This failure mechanism is accompanied by solid state dissolution of the semiconductor material into the metal or the alloy of the metallization system with the metal semiconductor interface moving vertically or laterally into the semiconductor. Contact spiking in chips is observed at high chip temperatures or localized high contact temperatures. Localized high temperatures may cause failure of the chip to substrate bond, thus increasing the thermal resistance or producing a thermal runaway of the device. Localized high temperatures may also result in a large magnitude of electrical overstress.

Contact spiking can occur in device fabrication when the device is exposed to high temperatures. Chang et al. (1988), Farahani et al. (1987), and Texas Instruments (1987) all point out that this failure mechanism can be minimized by using silicon containing aluminum alloys such as Al–1% Si or using barrier metals such as Ti-W. Metal spikes penetrate into the electrically active region of the device, causing increased leakage current or shorting. It is not possible to characterize such interdiffusion failure mechanisms by an activation energy because of their irregular behavior.

Migration of Al along Si defects has been observed in NMOS LSI devices (logic gates), due to contact migration, also known as electrothermomigration. Failures are accelerated by elevated ambient temperature in the neighborhood of 400° C. As indicated by DeChairo (1981), Christou (1982), Bellamy and Kemmerling (1978), and Christou et al. (1980), contact migration is a major cause of failure in GaAs devices. This failure mechanism may be dominant during VLSI manufacture and packaging when the temperatures exceed 400° C.

4.10 Electromigration

Electromigration is the result of high current density (typically of the order of 10^6 amperes/cm^2 in aluminum) [Schnable (1988)] in metallization tracks, which produces a continuous impact on the metal grains in the metallization causing the metal to pile up in the direction of the current flow and produce voids in an upstream direction with respect to the current flow. In thin film conductors, electromigration induced damage usually appears in the form of voids and hillocks. Voids can grow and link together to cause electrical discontinuity in conductor lines which leads to open circuit failure. Hillocks can also grow and extrude out materials to cause short circuit failure between adjacent conductor lines in the same level or in adjacent levels in multi-level interconnecting structure. Alternatively, they can break through the passivation or the protective coating layers and lead to subsequent corrosion induced failures. Void induced open circuit failure usually occurs at an earlier time than extrusion induced short-circuit failure in thin-film conductors using Al-based metallurgies.

The phenomenon of electromigration induced mass transport is attributed to atomic flux during the passage of current through a polycrystalline thin-film conductor as a result of electromigration in the lattice and at the grain boundaries. The contribution due to electromigration in the lattice is [Huntington and Grone (1961)]

$$J_1 = \frac{1}{kT} N_1 D_1 j \rho e Z_1^* \tag{2.37}$$

The contribution of electromigration in the grain structure is given by [Ho and Howard (1974)]:

$$J_b = \frac{1}{kT} \frac{N_b \delta}{d} D_b j \rho e Z_b^* \tag{2.38}$$

where

N	=	atomic density, atoms/m^3
D	=	diffusivity, m^2/s
j	=	current density, amperes/m^2
ρ	=	electrical resistivity, Ω cm
eZ^*	=	effective charge, esu
k	=	Boltzmann's constant, 8.617×10^{-5} eV/K
T	=	steady state temperature, K
δ	=	effective boundary width to mass transport (10Å), angstroms
d	=	average grain size, angstroms

The subscripts 1 and b denote the electromigration parameters of the lattice and grain boundary, respectively. The relative contributions of the two components can be estimated from the ratio of J_1 and J_b, given by [Ho and Kwok (1989b)]

$$\frac{J_1}{J_b} = \frac{N_1}{N_b} \frac{d}{\delta} \frac{D_1}{D_b} \frac{Z_1^*}{Z_b^*} \tag{2.39}$$

The measured values of Z_1^* and Z_b^* usually do not differ by more than an order of magnitude. For thin films with a 1 μm grain size at $0.5T_m$, where T_m is the melting temperature of the materials [Ho and Kwok (1989b)],

$$\frac{N_1}{N_b} \simeq 1 \qquad \frac{d}{\delta} \simeq 10^3 \qquad \frac{D_1}{D_b} \simeq 10^{-7} \qquad \frac{J_1}{J_b} \simeq 10^{-4} \tag{2.40}$$

Thus at moderate temperatures (in the neighborhood of $0.5T_m$), the atomic flux from electromigration in the lattice is vanishingly small as compared with that from the grain boundaries, so that the grain boundary electromigration becomes a dominant mode of mass in the neighborhood of $0.5T_m$ [Ho and Kwok (1989a)]. The mass transport in the grain boundaries in most metals is by a vacancy diffusion mechanism [Kwok et al. (1981, 1984)]. The vacancy flux in the grain boundary with contributions from electromigration and diffusion via grain boundary are given by [Ho and Kwok (1989a)]:

$$J_v = -D_v \nabla C_v + J_b \tag{2.41}$$

where

J_v = vacancy flux, $1/\text{cm}^2$-s
D_v = vacancy diffusivity in the grain boundary, cm^2/s
C_v = local variation of the vacancy concentration, cm^{-3}

and where J_b is obtained from Eq. (2.38). The local variation of the vacancy concentration is given by the equation [Ho and Kwok (1989b)]

$$\frac{dC_v}{dt} = -\nabla J_v + \frac{C_v - C_v^0}{\tau} \tag{2.42}$$

where

C_v^0 = vacancy concentration at thermal equilibrium, cm^{-3}
τ = average lifetime of the vacancy, s

Here τ is determined by the efficiency of the source and the sinks in creating the annihilating vacancies.

Under a steady state condition where $dC_v/dt = 0$, Ho and Kwok (1989b) point out that

$$C_v - C_v^0 = \tau \nabla J_v \tag{2.43}$$

Hence, the extent of deviation of the vacancy concentration from equilibrium is proportional to the vacancy flux divergence. Whenever there is a spatial variation in any of the parameters affecting the grain boundary electromigration, which may include structural defects in the form of non-uniform thickness of the conductor, or temperature gradient, a divergence in the atomic flux occurs, giving rise to a local vacancy supersaturation or depletion (void formation). Thus, at temperatures in the neighborhood of $0.5T_m$ of the metallization material, electromigration is more temperature gradient dependent, and has a mild dependence on steady state temperature.

The temperature gradient dependence of electromigration failures has been confirmed by Lloyd et al. (1988) who, while conducting electromigration tests on the lifetime of Cr/Al-Cu conductors covered with polyimide passivation, noticed that the electromigration failure location was typically nearer to the location of maximum temperature gradient, whereas the location of non-temperature gradient induced failure was randomly distributed. Lloyd et al. (1988) modeled the temperature of the stripe carrying current to produce significant joule heating as the balance between the heat generated in the stripe and the heat conducted away from the stripe to the surrounding thermal sinks such as the substrate, the passivation and the portion of the metal stripe which is not conducting heat. The heat balance relation is given by [Lloyd et al. (1988)]:

$$\rho j^2 = K\left(\frac{d^2\Delta T}{dx^2}\right) - h\Delta T \tag{2.44}$$

per unit volume, where ρ is the metal resistivity (Ω cm), K is the metal's thermal conductivity (W/m$-$K) and h is the parameter which characterizes the thermal efficiency of the heat sinks (W/m K). The term on the left of Eq. (2.44) is the heat generated due to current passing through the metal element. The first term on the right is the heat conduction along the metal stripe away from the heating element and the last term is the heat conduction to the environment. If the heat conduction through the oxide (of thickness l_{ox}) is the only consideration, the values of h is of the order [Lloyd et al. (1988)]

$$h = \left(\frac{K_{ox}K}{l_{ox}l}\right) \tag{2.45}$$

where K_{ox} is the thermal conductivity and l is the stripe thickness. The heat sink is assumed to be at the ambient temperature. The solution to the heat conduction equation for a stripe with the origin in the center was given by Lloyd et al. (1988) as

$$T = A\left[1 - \left(\frac{\cosh(Bx)}{\cosh\left(\frac{BL}{2}\right)}\right)\right] \tag{2.46}$$

where

$$B = \sqrt{\left(h - \frac{\rho_o \alpha j^2}{K}\right)} \tag{2.47}$$

where $\rho = \rho_o(1 + \alpha\Delta T)$ and

$$A = \frac{\rho_o j^2}{B^2 K} \tag{2.48}$$

The location of failure was argued to be near the position of maximum flux divergence. The atomic flux being given by the equation [Lloyd et al. (1988)]:

$$J = \frac{DF}{KT} \tag{2.49}$$

where D is the diffusivity, F is the driving force for diffusion, the point of maximum flux divergence was calculated to be

$$\frac{dJ}{dx} = \frac{dC}{dt} = \left(\frac{dJ}{dT}\right)\left(\frac{dT}{dx}\right) \tag{2.50}$$

and the condition of failure was given as

$$\frac{d^2J}{dx^2} = \frac{d}{dx}\left(\frac{dJ}{dT}\right)\left(\frac{dT}{dx}\right) = 0 \tag{2.51}$$

The location of electromigration failure calculated from the foregoing equation was found to be near the location of maximum temperature gradient [Lloyd et al. (1988)].

Multilayered interconnection metallizations with sensitive electrical circuits have caused the definition of electromigration failure to change [Onduresk et al. (1988)]. Many studies on single layer metallizations have used opening of the conductor as a criterion for failure, ignoring the functional failures (including resistance change of the metallization) which occur at much smaller resistance changes. The open criterion for layered metal systems may not even be achievable in a reasonable test environment if one of the layers is not susceptible to electromigration. Typically, the use of percentage resistance criterion allows the definition of failure at any point during electromigration tests.

Predicted interconnection failure rates at use conditions of current density and temperature typically vary several orders of magnitude, as they strongly depend on accelerated test data, model parameter selection and high temperatures coupled with high current density pulses. More specifically, the failure time depends inversely on temperature gradient and current density.

Onduresk et al. (1988) derived a void formation model for multilayered metallizations that allows calculation of void length from measured resistance and temperature coefficients. The temperature coefficient of resistance was investigated to verify that the refractory layer remains undamaged throughout the voiding process. The temperature coefficient C for metal is defined by

$$R = R_i[1 + C(T - T_i)] \tag{2.52}$$

where R is the resistance at temperature T and R_i is the resistance at some fixed initial temperature T_i. In a simplified case, for a void extending through the aluminum layer, all of the current must pass through the refractory metal layer. The total resistance of the metal stripe is the sum of the initial component due to the refractory material metal/aluminum sandwich plus the additional series refractory metal resistance which results from the void. This is represented by

$$R(\nu) = R_1 + R_2 \tag{2.53}$$

where

$$R_1 = R_{im}\frac{L - \nu(t)}{L}[1 + C_m(T - T_i)] \tag{2.54}$$

and

$$R_2 = R_{ir}\frac{\nu(t)}{L}[1 + C_r(T - T_i)] \tag{2.55}$$

Taking the derivative of Eq. (2.55) with respect to temperature gives

$$\frac{\partial R(\nu)}{\partial T} = R_{im}C_m\frac{L - \nu(t)}{L} + R_{ir}C_r\frac{\nu(t)}{L} \tag{2.56}$$

where R_{im} is the initial resistance of the composite, R_{ir} is the initial resistance of the refractory metal layer, C_m is the initial composite temperature coefficient, C_r is the refractory metal temperature coefficient, L is total metal line length, and ν is the void length. Deviation from this model can occur due to the intermetallic compound formation.

Variation of flux between the different regions of the conductor may lead to failure due to depletion in some regions; the time to failure is inversely proportional to the flux gradient within that region. The flux gradient in terms of the temperature gradient is given by Oliver and Bower (1970)

$$\frac{dU}{dX} = \frac{N\epsilon I D_*}{k} e^{-E/kT} f(\rho, Z, T) \frac{dT}{dX} \tag{2.57}$$

where

N	=	density of ions, atoms/m^3
ϵ	=	electronic charge, esu
ρ	=	electrical resistivity, Ω cm
D_*	=	diffusion coefficient for ions in the conductor, m^2/s
k	=	Boltzmann's constant, 8.617×10^{-5} eV/K
T	=	average absolute conductor temperature, K
E	=	activation energy, eV
Z	=	effective ionic valency representing the net effect of momentum exchange and electric field forces, dimensionless

Schwarzenberger et al. (1988) demonstrated the importance of the temperature gradient as a source of electromigration flux divergence in metallization tracks and the necessity of controlling the temperature profile in the integrated circuit to optimize its lifetime.

A general, but not universal, expression for the mean time to failure (MTF) (or t_{50}, which is the time to reach failure of 50% of a group of identical conductor lines) is

$$MTF = Aj^{-n}e^{E/kT} \tag{2.58}$$

where

A	=	parameter depending on sample geometry
j	=	current density, amperes/m^2
n	=	experimentally determined exponent
E	=	activation energy, eV
k	=	Boltzmann's constant, 8.617×10^{-5} eV/K
T	=	steady state temperature, K
MTF	=	mean time to failure, hours

Note this is not a failure rate expression, as the failure times are typically observed to follow a lognormal distribution. The electromigration lifetime test is carried out under a set of accelerated test conditions at elevated temperatures and with high current density stressing. The data are then extrapolated to device operating conditions, with current density stressing below 5×10^5 A/cm^2, by the Arrhenius-like empirical equation of Eq. (2.58) (originally formulated by Black (1969a, b)). Values of "n" that have been reported are

$n = 1$ to 3	Chabra and Ainslie (1967)
$n = 1.5$	Attardo (1972)
$n = 1.7$	Danso and Tullos (1981)
$n = 2$	Black (1983)
$n = 6$ to 7	Blair et al. (1970)

Table 2.2 shows the estimated MTF of Ti: W/Al and Ti: W/Al + Cu film conductors at 85° C for 5×10^5 and 2×10^5 A/cm^2 current densities at 100% duty cycle. Even at the worst case, $n = 1$, it has been anticipated [Ghate (1981)] that the actual time to failure will be greater than those predicted in Table 2.2. For current densities of 2×10^5 A/cm^2 even temperatures of 125° C will not lead to failure in less than 10 years with a typical exponent of $n = 1.7$.

From the foregoing discussion of electromigration, it is implied that the basic requirement for reducing the electromigration damage is to reduce the local divergence of atomic flux. This can be accomplished, in principle, by reducing the magnitude of the atomic flux and/or the inhomogeneity of the parameters controlling the mass transport. The magnitude of atomic flux is determined by the electromigration driving force and the grain boundary diffusivity. Thus, to reduce the atomic flux, the choice is to reduce the driving force and/or the diffusivity. Reduction of the driving force has some basic difficulties since it requires either a change in the scattering process responsible for the effective charge or a reduction in the current density. Because the current density is dictated by the device functional requirements, the only choice is to reduce the grain boundary diffusivity. The most common approach is by solute addition, which also results in improvement in conductor properties in the form of grain structure modification. The common examples are the addition of Cu to Al stripes or other solute elements such as Mn, Mg, and, Ti [Ho and Kwok (1989)].

4.11 Metallization Migration

Metal migration occurs between biased lands under conditions conducive to electrocrystallization. Dendritic growth is a common cause of failure. Conditions for metal migration include a level of current density at the tip of the dendrite (around 10 A/cm^2) through spheroidal and parabolic diffusion, sufficient liquid medium such as condensed water, applied voltage which exceeds the sum of anodic and cathodic potentials in equilibrium with the electrolyte and materials with defects which allow water condensation to satisfy the current density requirement.

For a given material and pore frequency distribution, the ionic current density or the rate of mass transport per unit area is proportional to the fraction of pore area containing condensed water. DiGiacomo (1982) proposed a model to predict failure as a function of environmental conditions and physical properties of the package and verified the predictions on the basis of migration failure data from accelerated tests.

Table 2.2 Estimated MTF Values for Electromigration at 80° C.

Current Density	Exponent	Ti: W/Al (years)	Ti: W/Al +Cu (years)
5×10^5 A/cm^2	$n = 1.0$	4	12
	$n = 1.5$	5	17
	$n = 2.0$	8	24
2×10^5 A/cm^2	$n = 1.0$	10	30
	$n = 1.5$	23	68
	$n = 2.0$	50	152

The fractional area in which condensation occurs can be represented by

$$A_{\text{fractional}} = \frac{1}{2} erfc\left(\frac{1}{2\sigma}\right) \ln\left(\frac{r_{\text{average}}}{r}\right)^2 \tag{2.59}$$

where

$$r = \frac{2\gamma v}{kT \ln\left(\frac{p}{p_o}\right)} \tag{2.60}$$

$$erfc\left(\frac{1}{2\sigma}\right) = 1 - \frac{2}{\sqrt{\pi}} \int_0^{1/2\sigma} e^{-u^2} du \tag{2.61}$$

and

γ	=	surface tension, N/m
k	=	Boltzmann's constant, 8.617×10^{-5} eV/K
$A_{\text{fractional}}$	=	fractional area of condensation, m^2
v	=	molar volume, m^3
r	=	pore radius, m
r_{average}	=	average radius, m
p	=	saturated vapor pressure above meniscus, N/m^2
p_o	=	saturated vapor pressure above flat surface, N/m^2
p/p_o	=	H, relative humidity at which condensation occurs, percent
σ	=	$\ln(r_{50}/r_{16})$, dimensionless
u	=	a dummy valuable for integration

The average current density is expressed as a function of ionic concentration, electric field, diffusivity, temperature, fractional condensed area, and overpotential using the Butler-Volmer equation of electrode kinetics, and assuming diffusion control [DiGiacomo (1982), Barton and Bochris (1962), Price et al. (1958), and Adamson (1990)]:

$$j = \frac{(Z_f)^2 CDE}{2kT} erfc\left(\frac{1}{2\sigma}\right) \ln\left(\frac{kT(\ln H) r_{\text{average}}}{2\gamma v}\right)^2$$

where

j	=	average current density, amperes/m^2
C	=	ionic concentration in bulk, m^{-3}
r	=	pore radius, m
r_{average}	=	average pore radius, m
Z	=	valence of the metal ions, dimensionless
F	=	Faraday's constant
D	=	diffusivity of metal ions, m^2/s
E	=	the overpotential; proportional to V_{pot}/r_d; where v is the voltage applied and r_d is the radius of curvature of the dendrite, V
γ	=	surface tension, N/m
v	=	molar volume, m^3
p	=	saturated vapor pressure above meniscus, N/m^2
p_o	=	saturated vapor pressure above flat surface, N/m^2
H	=	p/p_o, relative humidity at which condensation occurs, percent
σ	=	ln (r_{50}/r_{16}), dimensionless
T	=	steady state temperature, K
k	=	Boltzmann's constant, 8.617×10^{-5} eV/K

To produce a metal dendrite, the current density at the whisker's tip should be orders of magnitude higher than the average current density [DiGiacomo (1982)]. The growth is possible only through parabolic and spheroidal diffusion focussing of the ionic current on the dendrite tip, which can be expressed in terms of the radius of curvature:

$$j_{\text{tip}} = \frac{(ZF)^2 CD}{2kT} \frac{V_{\text{pot}}}{r_d} erfc\left(\frac{1}{2\sigma}\right) \ln\left(\frac{kT(\ln H) r_{\text{average}}}{2\gamma v}\right)^2$$

for dendritic growth $j_{\text{tip}} > j_{\text{critical}}$, the critical value of current density or a mass transport rate of $\partial Q_{\text{tip}}/\partial t > \partial Q_{\text{critical}}/\partial t$. The mass transport rate is represented as $Q_{\text{tip}} = j_{\text{tip}}/ZF$. Integrating the equation with respect to time and substituting for $D = D_o \exp(-\Delta H/kT)$ gives the critical value of mass transport rate, that is, the number of ions that must be transported to achieve dendritic growth across the gap, and can be found to be $\partial Q_c/\partial t = \ell\rho/Mt$ (where ℓ = distance between electrodes; ρ = density of dendrite; M = atomic weight).

The time to failure can be given as [DiGiacomo (1982)]

$$t_f = \frac{\left(\frac{\rho}{MC}\right)\left(\frac{2r_d}{ZFD_o}\right)\ell}{\beta(V_{\text{pot}} - V_T)\frac{1}{kT}e^{-\Delta H/kT} erfc\frac{1}{2\sigma}\ln\left(\frac{kT\ln H r_{\text{average}}}{2\gamma v}\right)^2}$$

where

ρ	=	density of the dendrite, g/m^3
M	=	atomic weight, g
C	=	ionic concentration, $1/m^3$
Z	=	ionic charge or the valence of the metal ions, dimensionless
D_o	=	diffusivity coefficient, m^2/s
ℓ	=	distance between electrodes, m
r_d	=	dendrite radius, m
r_{average}	=	average pore radius, m
γ	=	surface tension, N/m
v	=	molar volume, m^3
p	=	saturated vapor pressure above the meniscus, N/m^2
p_o	=	saturated vapor pressure above the flat surface (H approaching p/p_o is the relative humidity at which condensation occurs, N/m^2)
σ	=	$\ln[r_{50}/r_{16}]$, dimensionless
β	=	fraction of metal surface at the anode which is susceptible to metal oxidation and therefore promoting metal migration
V_{pot}	=	potential difference applied, V
V_T	=	critical value of potential difference for dendritic growth, V

The average current j is expressed as a function of ionic concentration, electric field diffusivity, temperature and condensed area. β is the fraction of metal surface at the anode which is susceptible to metal oxidation and therefore promoting metal migration in regions satisfying current density requirements. The value of β varies from metal to metal because of differences in device passivation and solubility and also with testing time and temperature. DiGiacomo (1982) performed tests on Ag migration between tinned Ag-Pd leads in encapsulated packages consisting of a polyimide-polyimide surface coating, Al cap and silicone rubber or epoxy backseal. The test was carried out with temperature-relative humidity conditions under bias. A failure was defined as a resistance of less than $5 \times 10^6\,\Omega$.

Metallization migration involves the formation of aluminum growths beneath the silicon dioxide layer. Metallization migration occurs during deposition of conductors on silicon dice due to the combined effects of elevated temperature and electrical stress. It has been reported by Bart (1969) and Lane (1970) that triangular shaped aluminum growths formed beneath the silicon dioxide layer when conductor metallization is deposited at high temperature on silicon substrate. Lane (1970) found that triangles were formed within minutes after deposition of an 8000 angstrom thick aluminum layer on silicon at temperatures from 500 to 577° C. The time for growth formation was found to decrease with increasing temperature. Growth caused local electrical short circuits. The high temperatures (500 to 577° C) at which this failure phenomenon occurs makes it a recessive mechanism in the normal operation of microelectronic devices.

5 EFFECT OF HYDROGEN (H_2), HELIUM (He) AMBIENTS ON DEVICE METALLIZATION VERSUS TEMPERATURE

Reductions in the solid-state transport have been found to be possible through interaction of the active gases with thin film metal conductors and bi-metallic diffusion couples, (see Table 2.3). Studies indicate that hydrogen ambient results in an improvement in the electromigration rates, and stabilizes the intermetallic formation due to interdiffusion.

Pasco and Schwarz (1983) considered several pathways for the gaseous ambients to affect mass transport mechanisms in metallization including external surface diffusion, grain boundary diffusion of metal atoms as well as diffusion of vacancies into the bulk from the defects present on the external surfaces, internal surfaces and from dislocations. They studied the effects of heating rates in pure hydrogen and He-8 1/2% H_2 relative to an inert helium environment on the electromigration of Al-2% Cu conductors. For a heating rate of 1° C/min there was a decrease in the electromigration rate by a factor of 10 for the hydrogen versus helium ambient. A decrease in the electromigration rate by a factor of five is achieved by only an

Table 2.3 Summary of Kinetic Parameters for Electromigration and Extent of Damage Sustained by Al-2% Cu Thin Stripes (current stressed at 3×10^6 A/cm^2). (An almost constant activation energy at different heating rates suggests that the phenomenon is temperature independent in hydrogen and helium ambients)

Heating Rate	He (Helium)	He-8 1/2% H_2	H_2
1 K/min	$Q = 0.67$ eV $A = 7.0 \times 10^3$ s^{-1} $\Delta R/R_o$ failure $= 0.67$	$Q = 0.67$ eV $A = 1.4 \times 10^3$ s^{-1} $\Delta R/R_o$ failure $= 0.96$	$Q = 0.67$ eV $A = 6.7 \times 10^2$ s^{-1} $\Delta R/R_o$ failure $= 1.2$
5 K/min	$Q = 0.67$ eV $A = 1.4 \times 10^3$ s^{-1} $\Delta R/R_o$ failure $= 0.28$	$Q = 0.67$ eV $A = 4.2 \times 10^3$ s^{-1} $\Delta R/R_o$ failure $= 0.56$	$Q = 0.67$ eV $A = 1.1 \times 10^4$ s^{-1} $\Delta R/R_o$ failure $= 0.62$

8 1/2% H_2 in the ambient. At higher heating rates (5° C/min) an increase in the degree of damage before the onset of failure increases with increasing hydrogen constant. Table 2.4 shows the effect of heating rate on the effects of hydrogen and helium ambients [Pasco and Schwarz (1983)]. The activation energy is found to be constant, while the pre-exponential "A" varies, indicating that the ambient effects are independent of heating rate (*because the activation energy is constant*).

The process of electromigration is a combination of two processes: nucleation of the voids, and the growth of voids. Davis (1967) has shown that the nucleation of dislocation loops with the quenching in the vacancy supersaturation is a low temperature process. At low temperatures in the neighborhood of 25° C, the process of nucleation of voids is transport limited, and at moderately high temperatures (in the neighborhood of 100° C) the process is

Table 2.4 Effect of H_2, He, Ar, N_2 Ambients versus Temperature

Atmosphere	Stripe Composition	Failure Mechanism/ Failure Definition	Observation	Reference
Argon (Ar), Hydrogen (H_2)	Aluminum	Electromigration /Resistivity	Electromigration rate ($dR/R_o dt$) decreased by a factor of 10 for hydrogen (H_2) atmosphere compared to argon (Ar) at $j = 0.7 \times 10^6$ A/cm^2 and 200° C.	Shih and Ficalora (1981)
Hydrogen (H_2), Helium (He)	Aluminum (Al), Aluminum-Copper (Al-Cu)	Electromigration /Resistivity	Al stripes: H_2 improves (decreases) electromigration rate by a factor of 13 to 16 as compared to Helium (He). Al-Cu Stripes: H_2 improves electromigration rate by factor of 4 to 7 compared to Helium (He) environment.	Sardo (1981)
Nitrogen (N_2), Hydrogen (H_2), and H_2O	Aluminum (Al)	Hillock formation /Reflectivity	Hillock formation at elevated temperatures during film deposition was greatly decreased for high partial pressures of H_2 and H_2O relative to that of N_2.	McLeod and Hartsough (1977)

Table 2.4 Continued

Atmosphere	Stripe Composition	Failure Mechanism/ Failure Definition	Observation	Reference
Hydrogen incorporated during E-beam deposition	Aluminum (Al)	Hillock formation	TTF of 40 hours at 7×10^6 A/cm^2 for hydrogen incorporated samples. An anneal at 550° C produced fewer hillocks in the E-beam film.	Meyer (1969)
Hydrogen (H_2), Argon (Ar), Oxygen (O_2), Nitrogen (N_2), Helium (He), and Air	Au-Al thin film diffusion couple	Intermetallic compound formation	Au-Al intermetallic compound formed at 350 to 400° C for all environments except H_2.	Shih and Ficalora (1978)
Hydrogen (H_2), Air	Cu-Sn, Ag-Sn, Ni-Sn thin film diffusion couple	Intermetallic compound formation	Rates of intermetallic formation reduced dramatically in H_2 environment compared to air. Reduction of rate by a factor of 3 to 10 observed.	Shih and Ficalora (1979)

Table 2.4 Continued

Atmosphere	Stripe Composition	Failure Mechanism/ Failure Definition	Observation	Reference
Oxygen (O_2), Argon (Ar)	Au-Al thin film diffusion couple	Intermetallic compound formation	After 250° C anneal in 0.5 atm or either O_2 or Ar, SEM photography was used to evaluate the character of the intermetallic formation. Grain boundary diffusion appears to be the dominant failure mechanism in O_2 environment. Surface diffusion dominated in Ar environment.	Shih and Ficalora (1978)

limited by a low driving force due to the effectively lower supersaturation of vacancies. Dislocation loops can be precursors to formation of microscopic voids in electromigration. Under current stressing, there is a transport of ions along the grain boundary generally in the direction of the electron flow for Al alloys. In addition, there is a flow of vacancies in the opposite direction producing local vacancy supersaturation. These vacancies can cluster to form dislocation loops and microscopic voids given sufficient time at lower temperatures. The effect of the active ambient is to promote the uniform nucleation and increase in the nucleation rates. The ambient increases the rate of electromigration damage to a point where the growth of voids becomes the rate limiting factor. This increase in the nucleation rate is actually beneficial since the presence of many small voids is less likely to be harmful than a few large voids. Void growth is decreased by the hydrogen ambient through a number of mechanisms [Pasco and Schwarz (1983)]. Adsorbed hydrogen can become bound to vacancies decreasing their migration rates or pinning them or effectively reducing their concentration. In addition, the segregation of hydrogen to grain boundaries will decrease the grain boundary energy and the grain boundary diffusion rates. These effects decrease the overall transport rates and electromigration rates due to the hydrogen ambient [Pasco and Schwarz (1983)]. Higher heating rates decrease the time for nucleation and shift the electromigration damage to higher temperatures, which is reflected in the lowered pre-exponential "A" for higher heating rates (Table 2.5) [Pasco and Schwarz (1983)]. In addition, hydrogen promotes uniform nucleation, and, therefore, in all cases increases the amount of the damage that can be sustained before the rapid onset of failure.

Black (1978) observed the cracking of the passivating glass overlayer as a result of hillock formation during electromigration damage could lead to sudden increase in the electromigration rates due to the release of the compressive stresses in the conductor. These sudden increases in the electromigration rate could be avoided if the device is packaged in hydrogen ambient.

Table 2.5 Steady State Temperature Dependence of Device Failure Sites (Normal Operation Assumed −55° C to 125° C)

Failure Site	Failure Mechanism	Dominant Temperature Dependence	Nature of Steady State Temperature Dependence	Reference
Wire	Flexure fatigue	ΔT	Independent of steady state temperature function under normal operation.	Gaffeny (1968), Villela and Nowakowski (1970), Ravi and Philosky (1972), Phillips (1974), Pecht et al. (1989), Harman (1974)
Wire bond	Shear Fatigue	ΔT	Independent of steady state temperature.	Philosky (1970), Pecht et al. (1989), Newsome and Oswald (1976), Philosky (1971), Gerling (1984), Khan and Fatini (1986), Pinnel and Bennett (1972), Feinstein and Bindell (1979), Feinstein and Pagano (1979), Pitt and Needles (1982)

Table 2.5 Continued

Failure Site	Failure Mechanism	Dominant Temperature Dependence	Nature of Steady State Temperature Dependence	Reference
Wire bond	Kirkendall voiding	T	Independent of steady state temperature below 150° C; independent of steady temperature above lower temperatures ($T < 150°$ C) in presence of halogenated compounds.	Newsome and Oswald (1976), Philosky (1971), Gerling (1984), Khan and Fatini (1986), Pinnel and Bennett (1972), Feinstein and Bindell (1979), Pitt and Needles (1982), Villela and Nowakowski (1971)
Die	Fracture	ΔT, ∇T	Primarily dependent on temperature cycle.	Tan et al. (1987), Hawkins et al. (1987)
Die adhesive	Fatigue	ΔT	Independent of steady state temperature under normal operation.	Chiang and Shukla (1984), Mahalingham et al. (1984)
Encapsulant	Reversion	T	Independent of steady state temperature below 300° C (glass transition for typical epoxy molding resin for plastic packages.	Tummala and Rymaszewski (1989)

Table 2.5 Continued

Failure Site	Failure Mechanism	Dominant Temperature Dependence	Nature of Steady State Temperature Dependence	Reference
Encapsulant	Cracking	ΔT	Independent of steady state temperature (T) below the glass transition temperature of the encapsulant. It is a ΔT, ∇T driven mechanism.	Nishimura et al. (1987), Fukuzawa et al. (1985), Kitano et al. (1988)
Package	Stress corrosion	dT/dt	Mildy steady state temperature dependent under normal operation.	Tummala and Rymaszewski (1989)
Die Metallization	Corrosion	dT/dt	Only occurs above dew point temperature. Mildly steady state temperature dependent under normal operation.	Pecht et al. (1990), Commizolli et al. (1980), Inayoski et al. (1979), Sims (1979), White (1969), Schnable (1969)

Table 2.5 Continued

Failure Site	Failure Mechanism	Dominant Temperature Dependence	Nature of Steady State Temperature Dependence	Reference
Die Metallization	Electromigration	∇T	Steady state temperature dependent above 150° C.	Schnable (1988), Onduresk et al. (1988), Oliver and Bower (1970), Schwarzenberger et al. (1988), Chabra and Ainslie (1967), Attardo (1972), Danso and Tullos (1981), Black (1983), Blair et al. (1970), Ghate (1981), Patridge et al. (1982), LaCombe and Parks (1986), Canali et al. (1984), Kinsborn et al. (1978)
	Hillock formation	T	Hillocks in die metallization can form as a result of: electromigration or due to extended periods under temperature cycling conditions (thermal aging). Extended periods at or in the neighborhood of 400° C produce hillocks.	Lacombe et al. (1982), Thomas and Calabrese (1983), Amerasekera and Campbell (1987)

Table 2.5 Continued

Failure Site	Failure Mechanism	Dominant Temperature Dependence	Nature of Steady State Temperature Dependence	Reference
Die Metalization	Metallization migration	T	Independent of steady state temperature below 500° C.	DiGiacomo (1982), Bart (1969), Lane (1970)
	Contact spiking	T	Independent of steady state temperature below 400° C.	Chang et al. (1988), Farahani et al. (1987), Texas Instruments (1987), DeChairo (1981), Christou et al. (1980), Christou (1982), Bellamy and Kimmerling (1978)
	Constraint cavitation	T	Temperature steady state dependent above 25° C.	Yost et al. (1988), Yost et al. (1989)
Die	Electrical overstress	T	Independent of steady state temperature below 160° C (i.e., the temperature at which the thermal coefficient of resistance changes sign).	Alexander (1978), Canali et al. (1981), Smith (1978), Runayan (1965), Pancholy and Jhnoki (1978)

Table 2.5 Continued

Failure Site	Failure Mechanism	Dominant Temperature Dependence	Nature of Steady State Temperature Dependence	Reference
Device Oxide	Slow trapping	T	Steady state temperature dependent above 175° C.	Nicollian (1974), Woods and Rosenberg (1980), Gottesfeld and Gibbons (1984)
	Electrostatic discharge	T	ESD voltage (i.e., resistance to ESD) reduces with temperature increase (from 25 to 125° C). Note a dominant mechanism in properly protected devices.	Kuo and Kuo (1983), Hart et al. (1980), Moss (1982), Amerasekera and Campbell (1986, 1987), Scherier (1978)
	Time dependent dielectric breakdown	T^{-1}	Steady state temperature dependence is very weak; TDDB is a dominant function of voltage.	Anolick and Nelson (1979), Crook (1979), Crook (1978), Schnable (1988), McPherson and Baglee (1985), Boyko and Gelarch (1989), Swartz (1986), Lee et al. (1988)

Table 2.5 Continued

Failure Site	Failure Mechanism	Dominant Temperature Dependence	Nature of Steady State Temperature Dependence	Reference
	Ionic contamination	T^{-1}	Steady state temperature dependent above 200° C.	Brambilla et al. (1981), Johnson (1976), Hemmert (1981a, b), Bell (1980), Wager and Cook (1984)
Device	Forward second breakdown	T	Independent of steady state temperature below 160° C.	Beatty et al. (1976), Chen et al. (1983), Hower and Reddi (1970), Hu and Chi (1982)
	Reverse second breakdown	T^{-1}	Insignificant dependence of steady state temperature; the breakdown voltage increases from 650 V to 680 V when the temperature increases from 25 to 150° C.	Beatty et al. (1976), Chen et al. (1983), Hower and Reddi (1970)

Table 2.5 Continued

Failure Site	Failure Mechanism	Dominant Temperature Dependence	Nature of Steady State Temperature Dependence	Reference
Device	Surface charge spreading	T	Steady state temperature dependent above 150° C.	Edwards (1982), Blanks (1980), Stojadinovic (1983), Lycoudes and Childers (1980)
Device substrate/ oxide	Hot electrons	T^{-1}	Steady state temperature dependence decreases above −55° C. Temperature independent in range of 20 to 100° C.	Stojadinovic (1983), Woods (1980), Ning et al. (1979), Takeda et al. (1983), Matsumoto et al. (1981), Ko (1982), Hu (1983), Sze (1982), Tam and Hsu (1983), Hsu and Chiu (1984)

6 TEMPERATURE DEPENDENCE ON OPERATIONAL DEVICE FREQUENCY

Experimental studies conducted by Lewis and Awkward (1990), statistically evaluated the dependence of propagation delay on temperature. The device type investigated was a 54S140, dual quad-input nand gate with a valid TTL high level output voltage (V_{OH}) of 2.5 volts and a specified typical low-to-high level output propagation delay of t_{PLH} = 4ns with a maximum time for propagation from "low" to "high" t_{PLH} = 6.5ns at an ambient temperature of 25° C with a resistor load of 93 Ω and a capacitive load of 50pF. From the experimental studies, the low-to-high level output propagation delay times of the devices was determined at 25° C and 125° C. Although t_{PLH} of the devices increased with temperature, the values of t_{PLH} = 125° C remained within the specified value of 10ns delineated in MIL-M38510/81C which addresses the 54S140. Empirical evidence also indicated a direct relationship between an increase in t_{PLH} at elevated temperature and a decrease in peak output voltage at high frequency but within specifications.

Test data show that there is a variation in t_{PLH} between 25° C and 125° C and at 125° C, even for devices with the same date code. The mean t_{PLH} at 25° C is 4.59ns with a standard deviation of 0.21ns. The mean t_{PLH} at 125° C is 7.13ns with a standard deviation of 1.29ns. Assuming a linear dependence of temperature (T) on the mean (M) and standard deviation (σ),

$$M = 0.0254T + 3.935 \tag{2.62}$$

and

$$\sigma = 0.0108T - 0.06 \tag{2.63}$$

The value of t_{PLH} for five of the 54S140 devices was tested for peak output voltage at 25° C and 125° C to determine the frequency at which the output falls below a valid high. The high-level output voltage (V_{OH}) is specified at 2.5 V. The data indicate that an increase in temperature results in a decreased maximum operating frequency, but within manufacturing guidelines.

7 SUMMARY

An overview of the effects of temperature on microcircuits can be categorized as:

- Effects due to changes in material characteristics with temperature that affect electronic functions or compatibility with the circuit parameters;

- Effects due to temperature change and temperature rate of change that produce mechanical stresses, causing dimensional changes, fatigue and fracture failures; and

- Effects in combination with other failure accelerators that affect corrosion, electronic wearout, electrostatic discharge, and similar events.

Table 2.5 presents an overview of the various failure mechanisms and the related temperature dependencies. The failure mechanisms are observed to be dependent on temperature, the magnitude of temperature change, and temperature gradient (∇T) in the range of -55 to 125° C all of which are given in Table 2.5. A few of the failure mechanisms are dependent on temperature in this range of operating temperatures. However, the use of the Arrhenius temperature equation as a generic relationship is not appropriate. In fact, temperature acts as a strong decelerator for time dependent dielectric breakdown and hot electrons. If those were the dominant failure mechanisms for a device, lowering the temperature would decrease reliability. Thus, these failure mechanisms cannot be modeled by the Arrhenius relationship.

As discussed in this chapter, the temperature influence on reliability can, in some cases, be inarguable, especially in cases where temperature can be directly related to parameter drift, premature aging, and even catastrophic failure. These cases can be examined in this light even though examinations of case histories indicate that the relationship of temperature to reliability may be more a case of exposing the incompatibilities between operational requirements and the design or manufacturing processes. In other words, the product, as designed, is not suitable for operation in the desired environment without changes. Failure mechanisms such as reversion or depolymerization, contact spiking, and metallization migration occur at high temperatures encountered during fabrication or assembly well above normal operating temperatures. In such cases certain questions must be addressed before taking action:

- Will lower temperature by itself avoid or de-accelerate the experienced failures? If so, by how much should the temperature be lowered and how is the conclusion reached?
- Will lowering the magnitude of the temperature change avoid the experienced failures? If so, by how much should it be lowered and how is the conclusion reached?

Absolute temperature is often judged to be the cause of device failures. However, blaming steady state temperature may shift emphasis away from actual design or manufacturing deficiencies. In some cases, where temperature may be identified as an important ingredient in the explanation for unreliability, the product as designed is not suitable for operation in the desired environment without its being changed. Each action has its own costs, and the designer or user must fully understand the implications of each to produce cost effective reliable equipment.

NOMENCLATURE

A	a constant
A	cross-sectional area, m^2
A	die material constant
a	penetration depth, m
a	length of short side of die pad, m
b	thickness, m
C	a temperature coefficient, 1/° C
C	structural and material related factors
C	a constant
C	capacitance, farads
C'	a shear fatigue property
C	Coffin-Manson coefficient
D	diffusion constant, m^2/s
D	diffusivity, m^2/s
d	density, g/m^3
d	average grain size, angstroms
E	modulus of elasticity, N/m^2
E	electric field strength, N/m
E	activation energy, eV
e	charge on electron, 4.80325×10^{-10} evu

F	Faraday's constant
G	shear modulus of elasticity, N/m^2
H	relative humidity, percent
H	activation energy, eV
h	height or thickness, m
h	parameter characterizing thermal efficiency of heat, $W/m^2\ {}^\circ C$
j	current density, amperes/m^2
K	rate constant
K	stress intensity factor, $N/m^2 - m^{1/2}$
K	thermal conductivity, $W/m^2\ K$
k	Boltzmann constant, 8.617×10^{-5} eV/K
k	geometric constant
L	length, m
l	stripe thickness, m
M	atomic weight, g
MTF	mean time to failure, hours
m	shear fatigue property (with primed subscript)
N	atomic density, atoms/m^3
N	number of cycles, dimensionless
n	material constant
n	chemical valence, dimensionless
p	vapor pressure, N/m^2
Q	parameter defined by Eq. (2.4), $N/m^2\ {}^\circ C$
Q	mobile charge per unit area, esu
R	electrical resistance, ohm
RH	relative humidity, percent
r	radius, m
r	pore radius, m
r	dendrite radius, m
T	steady state temperature, K
t	time, seconds
t	thickness of molding compound, m
V	voltage, volts
v	molar volume, m^3
W	width, m
w	width, m
X	intermetallic layer thickness, m

x	height of the die attach, m
x	horizontal dimension parameter for wire-wirebond assembly, m
Z	a parameter defined by Eq. (2.10), 1/m
Z	effective ion valency, dimensionless
eZ^*	effective charge, esu

Greek

α	coefficient of thermal expansion, 1/° C
α	an energy factor, dimensionless
β	kinetic factor, dimensionless
β	fraction susceptible to oxidation
γ	strain amplitude, dimensionless
γ	surface tension, N/m
γ	voltage form factor, V^{-1}
Δ	changes in a variable
δ	effective boundary width, angstroms
ϵ	parameter defined by Eq. (2.2), dimensionless
ϵ	charge in electron, 4.80325×10^{-5} esu
λ	angle, radians
ν	Poisson's ratio, dimensionless
ρ	radius of curvature, m
ρ	resistivity, Ω cm
σ	stress, N/m^2
τ	stress, N/m^2
τ	average vacancy lifetime, seconds

Subscripts

a	amplitude
a	activation energy
app	applied
b	grain boundary
crit	critical value
d	die
d	dendrite
elev	elevated value

FB	flatband
f	failure
f	fatigue
f	final condition
I	mode
i	initial condition
inj	injected electons
m	melting
max	maximum condition
min	minimum condition
OH	high level output value
o	initial value
ox	oxide
p	bond pad material
p	shear fatigue property of bond pad material
pot	potential
R	a reference
r	refractory
S	shear
s	substrate
s	surface
sa	substrate attach
sat	saturation pressure
T	threshold
T	trap density
tip	tip

REFERENCES

Adamson, A W (1990). *Physical Chemistry of Surfaces*, Interscience Pubs, New York, NY.

Alexander, D R (1978). An electrical overstress failure model for bipolar semiconductor components, *IEEE Trans Comp Hybrids and Manf Tech*, **1**, 345–353.

Amerasekera, E A and Campbell, D S (1986). Electrostatic pulse breakdown in NMOS devices, *Qual and Rel Eng Intl*, **2**, 107–116.

Amerasekera, E A and Campbell, D S (1987). *Failure Mechanisms in Semiconductor Devices*, John Wiley and Sons, New York, NY, 12–96.

Anolick, E S and Nelson, G R (1979). Low field time dependent dielectric integrity, *Proc 17th Ann IEEE Rel Phys Symp*, 1.

Attardo, M H, cited in J R Venables and R G Lye (1972). A statistical model for electromigration-induced failures in thin film conductors, *Proc 10th Ann IEEE Rel Phys Symp*, New York, NY, 159.

Bart, J J (1969). Analysis of chemical and metallurgical change in microcircuits, *IEEE Trans Elec Dev*, **ED-16**, 351.

Barton, J L and Bochris, J O (1962). *The Electrolyte Growth of Dendrites from Ionic Solutions*, Royal Society of London, Series A, **268**, 485–505.

Beatty, B A, Krishna, S, and Adley, M S (1976). Second breakdown in power transistors due to avalanche injection, *IEEE Trans Elec Dev*, **ED-23** (8), Aug, 851–857.

Bell, J J (1980). Recovery characteristics of ionic drift induced failures under time/temperature stress, *18th Ann Proc Int Rel Phys Symp*, 217–219.

Bellamy, W C and Kimmerling, L C (1978). Premature failure in Pt-GaAs IMPATTs-recombination assisted diffusion as a failure mechanism, *IEEE Trans Elec Dev*, **25**, 746–752.

Black, J R (1969a). Electromigration — A brief survey and some recent results, *IEEE Trans Electron Dev*, **ED-16**, 338.

Black, J R (1969b). Electromigration failure modes in aluminum metallization for superconductor devices, *Proc IEEE*, **57**, 1587.

Black, J R (1978). Electromigration of Al-Si alloy films, *Proc IEEE Intl Rel Phys Symp*, 233–240.

Black, J R (1983). Physics of Electromigration, *Proc IEEE Intl Rel Phys Symp*, 142–149.

Blair, J C, Ghate, P B, and Haywood, C T (1970). Electromigration induced failures in aluminum film conductors, *Appl Phys Lett*, **17**, 281.

Blanks, H S (1980). Temperature dependence of component failure rate, *Microelec and Rel*, **20**, 219–246.

Bloomer, R L F, Johnson, M J, Kent, S, Mepham, B, Smith, S, Sonnicksen, R M, Walker, L E, and Motorola Inc (1989). Failure mechanisms in through hole packages, *Elec Mat Handbook, ASTM Int*, 976.

Bolger, J C (1982). Polyimide adhesive to reduce thermal stress in LSI ceramic packages, *14th Natl SAMPE Tech Conf*, Oct 12–14.

Bolger, J C and Mooney, C T (1984). Die attach in Hi-rel DIPS: Polyimides or low chloride epoxies, *IEEE Trans Comp Hyb Manf Tech*, **CHMT-7** (4), Dec.

Boyko, K C and Gelarch, D L (1989). Time dependent dielectric breakdown of 210A oxides, *27th IEEE Ann Rel Phys Symp*, 1–8.

Brambilla, P, Fantini, F, Malberti, P, and Mattana, G (1981). CMOS reliability: A useful case history to revise extrapolation effectiveness, length, and slope of the learning curve, *Micro and Rel*, **21**, 191–201.

Canali, C, Fatini, F, Gaviraghi, S, and Senin, A (1981). Reliability in TTL-LS devices, *Microelec and Rel*, **21**, 637–651.

Chabra, D S and Ainslie, N G (1967). Open circuit failures in thin film conductors, IBM Component Div, E Fishkill Facility, New York, NY, Tech Report 22.419, July.

Chang, P H, Hawkins, R, Bonifield, T D, and Melton, L A (1988). Aluminum spiking at contact windows in Al/Ti-W/Si, *Appl Phys Lett*, **52**, 25 Jan, 272–274.

Chen, D Y, Lee, F C, Blackburn, D L, and Berning, D W (1983). Reverse-bias second breakdown of high power Darlington transistors, *IEEE Trans Aerosp Syst*, **AES19** (6), Nov, 840–846.

Chiang, S S and Shukla, R K (1984). Failure mechanism of die cracking due to imperfect die attachment, *34th Elec Comp Conf*, 195–202.

Christou, A, Cohen, E, and McPherson, A C (1980). Failure modes in GaAs power FETs: Ohmic contact electromigration and the function of refractory oxides, *19th Ann Proc Intl Rel Phys Symp*, 134–139.

Christou, A (1982). Report on the 1982 GaAs device workshop, *20th Ann Proc Intl Rel Phys Symp*, 276–277.

Coffin, L F, Jr (1954). *Trans ASME*, **76**, 931.

Commizolli, R B, White, L K, Kern, W, and Schnable, G L (1980). Corrosion of aluminum IC metallization with defective surface passivation layer, *19th Ann Proc Rel Phys Symp*, 282.

Crook, D A (1978). Techniques for evaluating long term oxide reliability at wafer level, *IEDM Tech Digest*, Dec, 444–448.

Crook, D A (1979). Method of determining reliability screens for time dependent dielectric breakdown, *17th Ann Rel Phys Symp*, 1–7.

Danso, K A and Tullos, T (1981). Thin film metallization studies and device lifetime prediction using Al-Si, Al-Cu-Si conductor test bars, *Microelec Reliab*, **21**, 513.

Dantowitz, A, Hirshberger, G, and Pravidlo, D (1971). Analysis of aeronautical equipment environmental failures, Report No. AFFDL-TR-71-32, May.

Davis, T L (1967). Nucleation rate of vacancy cluster in aluminum, *J. Applied Physics*, **38**, 3756–3760.

DeChairo, L F (1981). Electrothermomigration in NMOS LSI devices, *19th Ann Proc Intl Rel Phys Symp*, 212–277.

DELCO Data (as of 11-2-88). ETDL LABCOM-from 636 nos ICs analyzed 513 nos bipolar and 123 nos MOS.

Devanay, J R (1989). Failure mechanisms in active devices, *Elec Mat Handbook, ASTM Intl*, 1007.

DiGiacomo, G (1982). Metal migration (Ag, Cu, Pb) in encapsulated modules and time to fail model as a function of the environment and package properties, *IEEE Proc of IRPS*, 27–33.

DM Data Inc (1990). System design guidelines for reliability, 96.

Edwards, D G (1982). Testing for MOS IC failure modes, *IEEE Trans Rel*, 9–17.

Farahani, M M, Turner, T E, and Barnes, J J (1987). Evaluation of titanium as a diffusion barrier between aluminum and silicon for 1.2 mm CMOS integrated circuits, *J Electrochem Soc*, **134** (11), Nov, 2835–2845.

Feinstein, L G and Bindell, J B (1979). The failure of aged Cu-Au thin films by Kirkendall porosity, *Thin Solid Films*, **62**, 37–47.

Feinstein, L G and Pagano, R J (1979). Degradation of thermocompression bonds to Ti-Cu-Au and Ti-Cu by thermal aging, *29th Proc Elec Comp Conf*, Cherry Hill, NJ, 346–354.

Fukuzawa, I, Ishiguro, S, and Nanbu, S (1985). Moisture resistance degradation of plastic LSIs by reflow soldering, *Proc 23rd Ann Intl Rel Phys Symp*, 192–197.

Gaffeny, J (1968). Internal lead fatigue through thermal expansion in semiconductor devices, *IEEE Trans Elec Dev*, **ED-15**, 617.

Gerling, W (1984). Electrical and physical characterization of gold-ball bonds on aluminum layers, *34th Proc IEEE Elec Comp Conf*, New Orleans, May 14–16, 13–20.

Ghate, P B (1981). Aluminum alloy metallization for integrated circuits, *Thin Solid Films*, **83**, 195–205.

Gottesfeld, S and Gibbons, L (1984). Reliability characterization of high speed logic IC's, *RCA Review*, **45** (2), Jun, 179–193.

Hall, P M, Panousis, N T, and Menzel, P R (1975). Strength of gold plated copper leads on thin film circuits under accelerated aging, *IEEE Trans on Parts, Hybrids, and Packaging*, **PHP-11** (3), Sep, 202–205.

Harman, G G (1974). Metallurgical failure modes of wire bonds, *12th Intl Rel Phys Symp*, 131–141.

Hart, A, Teng, T T, and McKenna, A (1980). Reliability influences from electrical overstress on LSI devices, *18th Ann Proc Intl Rel Phys Symp*, 190–196.

Hawkins, G, Berg, H, Mahalingham, M, Lewis, G, and Lofgran, J (1987). Measurement of silicon strength as affected by wafer back processing, *25th Ann Proc Rel Phys*, 216–223.

Hemmert, R S (1981a). Temperature dependent defect level for an ionic failure mechanism, *19th Ann Proc Intl Rel Phys Symp*, 172–174.

Hemmert, R S (1981b). Recoverable ionic contaminant induced failures on N-channel memory products, *Microelec and Rel*, **21**, 63–77.

Ho, P S and Howard, J K (1974). *J Appl Phys*, **45**, 3229.

Ho, P S and Kwok, T (1989a), Electromigration on metals, *Rep Prog Phys*, **52**, 301–348.

Ho, P S and Kwok, T (1989b). Electromigration on metals, *Rep Prog Phys*, **52**, 301–348.

Hower, P L and Reddi, V G K (1970). Avalanche injection and second breakdown in transistors, *IEEE Trans Elec Dev*, **ED-17** (4), Apr, 320–335.

Hsu, F C and Chiu, K Y (1984). Temperature dependence of hot electron induced degradation in MOSFETs, *IEEE Electron Dev Lett*, **EDL-5** (5), May.

Hu, C (1983). Hot electron effects in MOSFETs, *IEDM Tech Digest*, 176.

Hu, C and Chi, M (1982). Second breakdown of vertical MOSFETs, *IEEE Trans Elec Dev*, **ED-29** (8), Aug, 1287–1293.

Hu, J, Pecht, M, and Dasgupta, A (1991). A probabilistic approach for predicting thermal fatigue life of wirebonding in microelectronics, *J Elec Packaging*, American Society of Mechanical Engineers, **113**, 275.

Huntington, H B and Grone, A R (1961). *J Phys Chem Solids*, **20**, 76; quoted from Ho, P S and Kwok, T (1989), Electromigration on metals, *Rep Prog Phys*, **52**, 301–348.

Inayoski, H, Nishi, K, Okikawa, S, and Wakashima, Y (1979). Moisture induced aluminum corrosion and stress on the chip in plastic encapsulated LSIs, *Proc 17th Rel Phys Symp*, 113–117.

Johnson, G M (1976). Accelerated testing highlights CMOS failure modes, EAS-CON-76 Record, 142-A142-I, Sep.

Johnson, J E (1983). Die bond failure modes, *IEEE Proc Intl Rel Phys Symp*, 150–153.

Khan, M M and Fatini, H (1986). Gold-aluminum bond failure induced by halogenated additives in epoxy molding compounds, *Proc of ISHM*, 420–428.

Kinsborn, E, Melliar-Smith, C M, English, A T, and Chynoweth, T (1978). Failure of small thin film conductors due to high current density pulses, *16th Ann Proc Intl Rel Phys Symp*, 248–254.

Kitano, M, Nishimura, A, Kawai, S, and Nishi, K (1988). Analysis of package cracking during reflow solder process, *Proc 26th Ann Intl Rel Phys Symp*, 90–95.

Ko, P K (1982). Hot electron effects in MOSFETs, PhD Dissertation, EECS Dept, UC Berkeley, Jun.

Kuo, W and Kuo, Y (1983). Facing the headache of early failures: State-of-the-art review of burn-in decisions, *Proc IEEE*, **71**, Nov, 1257–1266.

Kwok, T, Ho, P S, and Yip, S (1984). *Surf Sci*, **144**, 44.

Kwok, T, Ho, P S, Yip, S, Ballufi, R W, Bristow, P D, and Brokman, A (1981). Evidence for vacancy mechanism in grain boundary diffusion in bcc iron: Molecular-dynamics stydy, *Phys Rev Lett*, **47**, 1148.

LaCombe, D J and Parks, E L (1986). The distribution of electromigration failures, *24th IEEE Ann Rel Phys Symp*, 1–6.

LaCombe, D J, Denning, D C and Christou, A (1982). A new failure mechanism in thin gold films at elevated temperatures, *20th Proc Intl Rel Phys Symp*, 81–87.

Lane, C H (1970). Aluminum metallization and contacts for integrated circuits, *Met Trans*, 713.

Lee, J, Chin, I C, and Hu, C (1988). Statistical modeling of silicon dioxide reliability, *26th Ann Intl Rel Phys*, 131–138.

Lewis, R and Awkward, K (1990). Methods for determining a maximum operating frequency for TTL gates, *Proc 1990 Rel and Maint Symp*, 372–377.

Lloyd, J R, Shatzkes, M and Challaner, D C (1988). Kinetic study of electromigration failure in Cr/Al-Cu thin film conductors covered with polymide and the problem of the stress dependent activation energy, *IEEE Intl Rel Phys Symp*, 216–225.

Lycoudes, N and Childers, C G (1980). Semiconductor instability failure mechanism review, *IEEE Trans Rel*, **29**, 237–247.

Mahalingham, M, Nagarkar, M, and Lofgran, L (1984). Thermal effects of die bond voids in metals, ceramic and plastic packages, *IEEE*, New York, NY.

Manno, P T (1983). RADC failure rate prediction methodology—Today and tomorrow, *Image Sequence Processing and Dynamic Scene Analysis*, NATO ASI Series, **F3**.

Manson, S S and Hirschberg, M H (1964). *Fatigue, An Interdisciplinary Approach*, Syracuse Univ Press, Syracuse, NY, 133.

Manzione, L T (1990). *Plastic Packaging of Microelectronic Devices*, Van Nostrand Reinhold, New York, NY.

Matsumoto, H, Sawada, K, Ansai, S, Hirayama, M, and Nagasawa, K (1981). Effect of long term stress on IGFET degradations due to hot electron trapping, *IEEE Trans on Elec Dev*, **ED-28** (8), 923.

McLeod, P S and Hartsough, L D (1977). High rate sputtering of aluminum for metallization of integrated circuits, *J Vac Sci Tech*, **14**, 263–265.

McPherson, J W and Baglee, D A (1985). Acceleration factors for thin gate oxide stressing, *23rd IEEE Ann Rel Phys Symp*, 1–5.

Meyer, D E (1969). *Trans TMS-AIME*, **245**, 593–599; quoted from Pasco, R W and Schwarz, J A (1983), Application of a dynamic technique to the study of electromigration kinetics, *21st Proc of Rel Phys Symp*, 10–23.

Moss, R Y (1982). Caution-electrostatic discharge at work, *IEEE Trans Comp Hyb and Man Tech*, **5**, 512–515.

Newsome, J L and Oswald, R G (1976). Metallurgical aspects of aluminum wire bonds to gold metallization, *14th IRPS*, 63–74.

Nicollian, E H (1974). Interface instabilities, *12th Ann Proc Intl Rel Phys Symp*, 267–272.

Ning, T H, Cook, P W, Dennard, R H, Osburn, C M, Schuster, S E, and Yu H W (1979). 1mm MOSFET VLSI technology Part IV: Hot electrons design constraints, *IEEE Trans on Electron Dev*, **ED-26** (4), 346–353.

Nishimura, A, Tatemichi, A, Muira, H, and Sakamoto, T (1987). Life estimation for IC plastic packages under temperature cycling based on fracture mechanics, *IEEE Trans on Comp, Hybrids, Manf Tech*, **CHMT-12**, 637–642.

Oliver, C B and Bower, D E (1970). Theory of the failure of semiconductor contacts by electromigration, *8th Ann Proc Rel Phys Symp*, 116–120.

Onduresk, J C, Nishimura, A, Hoang, H H, Sugiura, T, Blumenthal, R, Kitagawa, H, and McPherson, J W (1988). Effective kinetic variations with duration for multilayered metallizations, *26th Ann IRPS.*

Pancholy, R K and Jhnoki, T (1978). CMOS/SOS gate protection networks, *IEEE Trans on Elec Dev*, **ED-25** (8), Aug, 917–925.

Paris, P C, Gomez, M P, and Anderson, W E (1961). A rational analytical theory of fatigue, *The Trend in Engineering*, **13**, Univ of Washington, Seattle, WA, 9–14.

Pasco, R W and Schwarz, J A (1983). Application of a dynamic technique to the study of electromigration kinetics, *21st Proc Rel Phys Symp*, 10–23.

Patridge, J, Marques, A and Camp, R (1982). Electromigration, thermal analysis and die attach — A case history, *20th IEEE Ann Proc Rel Phys Symp*, 34–44.

Pecht, M, Barker, D, Dasgupta, A, and Hu, J (1990). A proabilistic approach for predicting thermal fatigue life of wirebonding in microelectronics, *J Elec Mat.*

Pecht, M, Lall, P and Dasgupta, A (1989). A failure prediction model for wire bonds, *Proc 1989 Intl Symp on Hyb Micro*, 607–613.

Pecht, M and Ko, W (1990). A corrosion rate equation for microelectronic die metallization, *Intl J for Hybrid Microelec*, **13** (2), Jun.

Phillips, W E (1974). Microelectronic ultrasonic bonding, G G Harman, ed, Nat Bur Stands (US), Spec Publ 400-2, 80–86.

Philosky, E (1970). Intermetallic formation in gold-aluminum systems, *Solid State Elec*, **3**, 1391–1399.

Philosky, E (1971). Design limits when using gold-aluminum bonds, *Proc 9th Ann IEEE Rel Phys Symp*, 11–16.

Philosky, E M and Ravi, K V (1973). On measuring the mechanical properties of aluminum metallization and their relationship to reliability problems, *11th Ann Proc Rel Phys Symp*, 33–40.

Pinnel, M R and Bennett, J E (1972). Mass diffusion in polycrystalline copper/electrodeposited gold planar couples, *Metallurgical Trans*, **3**, Jul, 1989–1997.

Pitt, V A and Needles, C R S (1982). Thermosonic gold wire bonding to copper conductors, *IEEE Trans Comp, Hybrids, and Manf Tech*, **CHMT-5** (4), Dec, 435–440.

Price, P B, Vermilyea, D A, and Webb, M B (1958). On the growth and properties of electric whiskers, *Acta Metallurgica*, **6**, 524–531.

Ravi, K V and Philosky, E M (1972). Reliability improvement of wire bonds subjected to fatigue stresses, *10th IEEE Ann Proc Rel Phys Symp*, Las Vegas, NV, 143–149.

Robinson, P H and Heiman, F P (1971). Use of HCl Gettering in silicon device processing, *J Electrochem Soc*, **118** (1), 141–143.

Rome Air Development Center, Reliability Analysis Center (1985). IC quality grades: Impact on system reliability and life cycle cost, **SOAR-3**.

Runayan, W R (1965). *Silicon Semiconductor Technology*, McGraw-Hill, New York, NY.

Sardo, G M (1981). Masters Thesis, Solid State Science and Technology, Dept Chem Eng & Matl Sci, Syracuse Univ, Syracuse, NY; quoted from Pasco, R W and Schwarz, J A (1983), Application of a dynamic technique to the study of electromigration kinetics, *21st Proc Rel Phys Symp*, 10–23.

Scherier, L A (1978). Electrostatic damage susceptibility of semiconductor devices, *16th Ann Proc Rel Phys Symp*, 151–153.

Schnable, G L and Keen, R S (1969). Aluminum metallization-advantages and limitations for integrated circuit applications, *Proc IEEE*, **57**, 1570.

Schnable, G L (1988). Failure mechanisms in microelectronic devices, *Microelec and Rel*, **1**, Artech House, Inc, 25–87.

Schwarzenberger, A P, Ross, C A, Evetts, J E and Greet, A L (1988). Electromigration in the presence of a temperature gradient: Experimental study and modelling, *J Elec Matl*, **17**, 473–478.

Semiconductor Reliability News (1990). What is the latest ratio of failure causes, Nov, 4.

Shih, D Y and Ficalora, P J (1978). The reduction of Au-Al intermetallics formation and electromigration in hydrogen environments, *Proc IEEE Intl Rel Phys Symp*, 268–272.

Shih, D Y and Ficalora, P J (1979). The effect of hydrogen on the rate of formation of intermetallics in the Cu-Sn, Ag-Sn, and Ni-Sn systems, *Proc IEEE Intl Rel Phys Symp*, 87–90.

Shih, D Y and Ficalora, P J (1981). Effect of oxygen and argon on the interdiffusion of Au-Al thin film couples, *Proc IEEE Intl Rel Phys Symp*, 253–256.

Sims, S P and Lawson, R W (1979). Influence of plastic encapsulants and passivation layers on corrosion of thin Al films subjected to humidity stress, *Proc 17th Ann Rel Phys Symp*, 102–112.

Sims, S P (1979). Procurement specifications requirements for protection against electromigration failures in Al metallizations, *Micro and Rel*, **19**, 207–218.

Smith, J S (1978). El overstress failure analysis in microcircuits, *16th Ann Proc Intl Rel Phys Symp*, 41–46.

Smith, R W, Hirschberg, M H and Manson, S S (1963). NASA TN D-1574, Apr.

Stojadinovic, N D (1983). Failure physics of integrated circuits (a review), *Microelec and Rel*, **23**, 609–707.

Swartz, G A (1986). Gate oxide integrity of NMOS transistor arrays, *IEEE Trans on Electron Dev*, **ED-33** (11), 1826–1829.

Sze, S M (1981). *Physics of Semiconductor Devices*, 2nd Edition, Wiley, New York, NY, 30–47.

Takeda, E, Nakagome, Y, Kume, H, Suzuki, N, and Asai, S (1983). Comparison of characteristic of N-channel and P-channel MOSFETs for VLSI, *IEEE Trans Electron Dev*, **ED-30** (6), 675–680.

Tam, S and Hsu, F C (1983). A physical model for hot electron induced degradation in MOSFETs, *IEEE Semiconductor Interface Specialist Conf*, Ft Lauderdale, FL, Dec.

Tan, S L, Chang, K W, Hu, S J, and Fu, K K S (1987). Failure analysis of die attachment on static random access memory (SRAM) semiconductor devices, *J Elec Mat*, **16** (1), Jan, 7–11.

Tavernelli, J F and Coffin, L F (1959). *J Trans ASM 51*, 438.

Texas Instruments (1987). Advanced CMOS logic applications enginnering and logic marketing dept, Texas Instruments General Purpose Logic Reliability Bulletin, SDLT053.

Thomas, R W and Calabrese, D W (1983). Phenomenological observations on electromigration, *21st Proc Intl Rel Phys Symp*, 1–9.

Tummala, R R and Rymaszewski, E J (1989). *Microelectronics Packaging Handbook*, Van Nostrand Reinhold, New York, NY.

Tummala, R R (1975). Stress corrosion of low temperature solder glass, *Noncrystalline Solids*, **19** (1), 263–272.

Villela, F and Nowakowski, M F (1970). Investigation of fatigue problems in 1-mil diameter thermocompression and ultrasonic bonding of aluminum wire, *NASA Technical Memorandum*, NASA TM-X-64566.

Villela, F and Nowakowski, M F (1971). Thermal excursion can cause bond problems, *9th IEEE Ann Proc Rel Phys Symp*, Las Vegas, NV, 172–177.

Wager, A J and Cook, H C (1984). Modeling the temperature dependence of integrated circuit failures, *Thermal Management Concepts in Microelec Packaging*, Intl Soc of Hybrid Microelec, 1–45.

Westergaard, H M (1939). Bearing pressure and cracks, *J Appl Math*, A-49–A-53.

Westinghouse Electric Corp (1987). Summary chart of 1984/1987 failure analysis memos.

White, M L (1969). Encapsulation of integrated circuits, *Proc IEEE*, **57**, 1610.

Woods, J (1981). Reliability and degradation of silicon devices and integrated circuits, Chap 4, *Reliability and Degration: Semiconductor Devices and Circuits*, M J Howes and D V Morgan, eds, John Wiley and Sons (Wiley Interscience).

Woods, M H and Rosenburg, S (1980). EPROM reliability; Part I and II, Electronics, *Micro and Rel*, **53**, 133–141.

Yost, F G, Amos, D E, and Romig, A D (1989). Stress-driven diffusive voiding of the aluminum conductor lines, *Intl Rel Phys Symp*, 193–201.

Yost, F G, Romig, A D, Jr, and Bourcier, R J (1988). Stress driven diffusive voiding of aluminum conductor lines: A model for time dependent failure, *Sandia National Laboratories*, Report SAND 88-09 46, NTIS DE 89-0010507.

SOURCES OF ADDITIONAL INFORMATION

Ajiki, T, Sugumoto, M, Higuchi, H, and Kumada, S (1979). A new cyclic biased THB test for power dissipating ICs, *Intl Rel Physics Symp*, 118–126.

Baker, E (1972). Some effects of temperature on material properties and device reliability, *IEEE Trans Parts, Hybrids, and Pack*, **PHP-8** (4), 4–14.

Billinton, R and Allen, R N. *Reliability Evaluation of Engineering Systems: Concepts and Techniques*, Longman Scientific and Technical Pub Co, Burnt Mill, Harlow, Essex, UK.

Carslaw, D S and Jaeger, J C (1959). *Conduction of Heat in Solids*, Oxford Press, London, 255–262.

Chu, R C and Simons, R E (1988). Evolution of cooling technology in medium and large scale computers — An IBM perspective, *Proc Intl Symp on Heat Transfer in Elec and Microelec Equip*, Intl Center for Heat Transfer, Dubrovnik, Yugoslavia, Hemisphere Pub Co, Aug 29–Sep 2.

Farrow, R H and Parker, G W (1972). The failure physics approach to IC reliability, *Micro and Rel*, **11**, 151–157.

Hannemann, R (1988). Thermal control for mini- and microcomputers: The limits of air cooling, *Proc Intl Symp on Heat Transfer in Elec and Microelec Equip*, Intl Center for Heat Transfer, Dubrovnik, Yugoslavia, Hemisphere Pub Co, Aug 29–Sep 2.

Hoang, H (1988). Effects of annealing temperature on electromigration performance of multilayered metallizations systems, *26th Ann Symp IRPS*.

Hocheng, H and Jia, C L (1991). Reliability monitoring in drilling electronic circuit boards, *J Elec Packaging*, **113**, 263.

James, K (1977). Reliability study of wire bonds to silver plated surfaces, *IEEE Trans on Parts, Hybrids, and Packaging*, **PHP-13**, 419–425.

Jellison, J L (1975). Effect of surface contamination on thermocompression bondability of gold, *IEEE Trans on Parts, Hybrids, and Packaging*, **PHP-11** (3), Sep, 206–211.

Khajezadah, H and Rose, A S (1977). Reliability evaluation of trimetal integrated circuits in plastic packages, *15th Ann Rel Phys Symp*, 244-249.

Kraus, A D and Bar-Cohen, A (1983). *Thermal Analysis and Control of Electronic Equipment*, Hemisphere Pub Co, New York, NY.

Lerner, I and Elpride, J M. Effects of several parameters on corrosion rates of Al conductors in integrated circuits, *J Elecrochem Soc, Solid State Sci and Tech*, **129** (10), 2270.

Machiels, F, Lijbers, G, Allaire, R, and Poiblaud, G (1991). Derating of results according to various models and prediction of lifetimes of plastic SMDs under humidity stress conditions, *5th Intl Conf Quality in Elec Comp Failure Prevention, Detection and Analysis*, Bordeaux, France, Oct 7–10, 869–874.

Majni, G and Ottavini, G (1979). Au-Al compound formation by thin film interactions, *J Crystal Growth*, **47**, 583–588.

Manchester, K E and Bird, D W (1980). Thermal resistance: A reliability consideration, *IEEE Trans Comp, Hybrids, and Manf Tech*, **CHMT-3** (4), Dec, 580–587.

MOTOROLA (1988). TMOS power MOSFET reliability report, Motorola Literature Distribution, Phoenix, AZ.

Mirman, I B (1991). Effects of peeling stresses in bimaterial assembly, *J Elec Packaging*, **113**, 431.

Morgan, H S (1991). Thermal stresses in layered electrical assemblies bonded with solder, *J Elec Packaging*, **113**, 350.

Murthi, A K and Shewchen, J (1982). The effect of hydrogen ambients on failure mechanisms in CMOS metallization, *IEEE Proc IRPS*, 55–65.

Nakayama, W and Bergles, A E (1988). Cooling electronic equipment: Past, present, and future, *Proc Intl Symp on Heat Transfer in Elec and Microelec Equip*, Intl Center for Heat Transfer, Dubrovnik, Yugoslavia, Hemisphere Pub Co, Aug 29–Sep 2.

Shirley, G C and Hong, C E (1991). Optimal acceleration of cyclic THB tests for plastic packages devices, *29th Intl Rel Physics Symp*, 12–21.

Steppan, J J, Roth, J A, Hall, L C, Jeanotte, D A, and Carbone, S P (1987). A review of corrosion failure mechanisms during accelerated tests, *J Electrochem Soc*, **134** (1), 175–190.

Suhir, E (1991). Stress relief in solder joints due to the application of a flex circuit, *J Elec Packaging*, **113**, 240.

Suhir, E (1988). Thermal stress failures in microelectronic components – review and extension in *Advances in Thermal Modeling of Electronic Components and Systems*, A Bar-Cohen and A D Kraus, Eds, Hemisphere Pub Co, New York, NY.

Workman, W (1968). Failure modes of integrated circuits and their relationship to reliability, *Micro and Rel*, **7**, 257–264.

Wunch, D C (1978). Applications of electronic overstress models to gate protective networks, *IEEE Proc Rel Phys*, 47–55.

Chapter 3

METHODOLOGIES FOR THERMAL ANALYSIS OF ELECTRONIC COMPONENTS AND SYSTEMS

Gordon N. Ellison
Tektronix, Inc.
Beaverton, Oregon

1 INTRODUCTION — THE ROLE OF THERMAL ANALYSIS IN THE DESIGN PROCESS

The principal purpose of this chapter is to address the utilization of various methodologies useful in the thermal design aspects of electronics packaging. The role of thermal analysis as an integral part of the actual design process must be considered. Design engineers may choose to totally circumvent the utilization of analysis tools or go so far as to perform a detailed design of nearly every component of the mechanical system. The actual choice should be based on a rational judgement of the resources available for the project, that is, required design completion date, access to engineering design support, available computational tools and platforms, prototype machine shop support, and testing resources.

The extent to which resources are utilized is usually determined by economic constraints. One cannot justify allocation of expensive resources on a low volume, low cost product. On the other hand, a product intended to generate large revenues is generally compatible with detailed design studies whereby efficient manufacturing and maintenance is attainable. In all cases, however, the recipient of the product, the customer, deserves a safe instrument, and certainly the best possible value. The role of the design engineer, therefore, becomes one of judiciously selecting resources to meet these goals.

It is possible to consider various analysis methods as belonging to one of three categories: (1) preliminary, (2) intermediate, and (3) detailed. A preliminary analysis is readily accomplished. This may require only a simple judgement that relies upon the success of an earlier, but very similar design, which, upon construction and testing, was found to perform satisfactorily. More generally, however, actual thermal analysis of several system design alternatives is required. In this preliminary design phase, detailed dimensions and component heat dissipation rates are not of concern. Usually the intent is to develop estimates of the overall cooling requirements of the various components. From the results of the analysis, the designer will determine the feasibility of closed system cooling, ventilation cooling, or even the necessity of forced air or liquid cooling. The actual selection of the cooling system will be based on the "goodness of fit" of the required cooling system and the requirements of product specification such as system enclosure dimensions, level of acceptable fan noise, availability of liquid coolant if required, and other issues, depending on the type of product and intended customer. Although highly detailed thermal models are not usually expected in this phase, the results should be sufficiently accurate to permit selection of the cooling methodology and the required conductive and convective heat transfer paths.

An intermediate level of design activity would be expected as a follow-up to the preliminary effort. The overall dimensions and layout of the product would be settled here. The thermal analyst would have sufficient feedback from other engineering practitioners (industrial design, circuit design, circuit board design, and manufacturing) to refine and incorporate refined vent and/or fan placement, estimates of power distribution, actual placement of some critical components such as power transistors on heat sinks, and perhaps the first design layout of one or more critical, specially designed, devices such as ceramic-based hybrid designs. Actual materials are probably not selected. Details such as these will be settled upon completion of this level of design activity.

A detailed design activity may, on occasion, be bypassed with the judgement that sufficient information has been determined to insure the likelihood of the design attaining the required goals. More likely, however, the thermal analyst will have some concerns as to the accuracy of some of the earlier modeling judgements and analysis assumptions. Modeling judgements are usually accommodated by merely resorting to developing a more detailed model. This is particularly the case when conductive heat transfer is the dominant transport mechanism in the problem of concern. The equations are well known, but models can be complex and rather demanding of the toolset capability. Many problems are not amendable to a very accurate solution. These are usually those that involve complex fluid flow and convective heat transfer mechanisms. While the literature reflects more and more well-done simulations of these problems, the realistically complete problem, with many components of varying geometry and style on an air-cooled printed circuit board, still remains to be accurately solved in any adequate generality. However, the analyst may still make good use of analysis tools by performing "what if?" simulations, where semi-quantitative results are useful in making design judgements that may otherwise be only estimated and then, possibly, in the form of educated guesses. Furthermore, if the minimum and maximum values of the more inaccurate analysis parameters are known, the thermal design characteristics may sometimes be successfully bracketed.

A not-to-be forgotten issue in the design activity is the role of experimentation. While many engineers are quick to shun the role of experimentation in favor of popular and complex analysis methodologies, they forget that much of the basic data such as the all-important heat transfer coefficient is based on experiment. It is often very effective to make a quick, but accurate, measurement on some small piece of equipment design, and use this experimental result as a "component value" in the analytical model, thereby producing more meaningful analysis results. Certainly the last step in any thermal design activity must be a carefully conducted thermal test, whereby the product may be guaranteed to operate in the reliable and safe manner that the customer deserves.

Novice practitioners of electronics thermal analysis are easily overwhelmed by the many worthwhile tools at their disposal. Success is assured only by a judicious selection and application of the appropriate analysis procedure. The acquisition of the knowledge to successfully apply a procedure to any given problem usually requires the expenditure of financial resources to purchase the tool and even more costly, the time to adequately learn to use it. It is the purpose of this article to provide some guidance as to the selection of the appropriate methodology required to solve a variety of thermal design problems.

2 AN OVERVIEW — THERMAL TRANSPORT IN ELECTRONIC SYSTEMS

Realistic expectations of the thermal analysis of electronic components are possible only with some appreciation of the variation and complexity of geometry of the various components that comprise any electronic system. Figure 3.1 illustrates some of the many package styles used to house heat dissipating circuit chips. A careful analysis of any one of these packages is in itself a large task. However, a qualitative consideration of the heat flow mechanisms can be accomplished with the aid of simple electro-thermal analog circuits.

Heat flow paths for a dual-in-line package mounted on a printed circuit board are shown in Fig. 3.2. Several thermal resistances are indicated in the drawing. Each resistance quantifies the temperature rise per watt of heat transferred between two temperatures. The actual heat dissipator consists of a multitude of transistor junctions just below the surface of the chip. The heat is conducted from

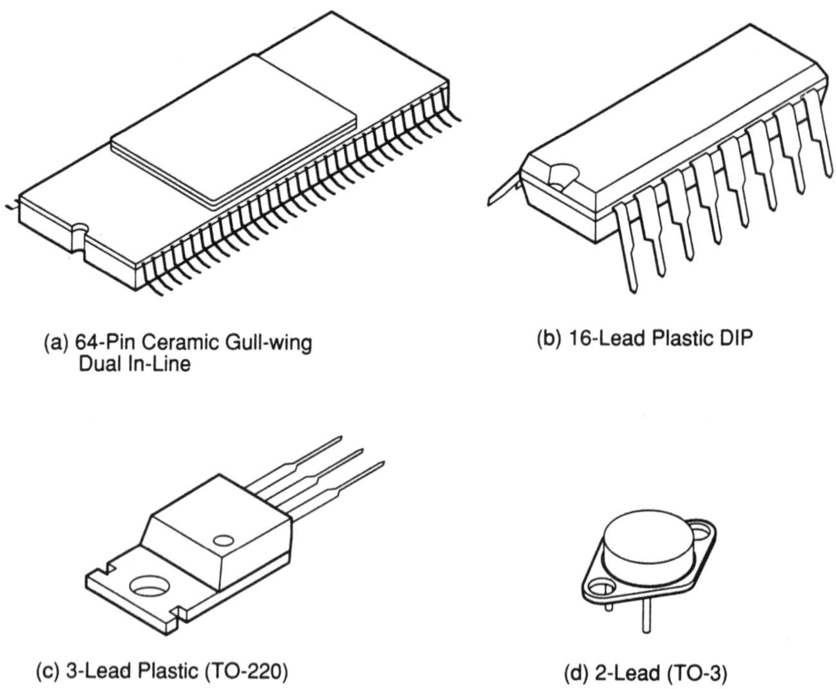

Figure 3.1 Variety of IC packages.

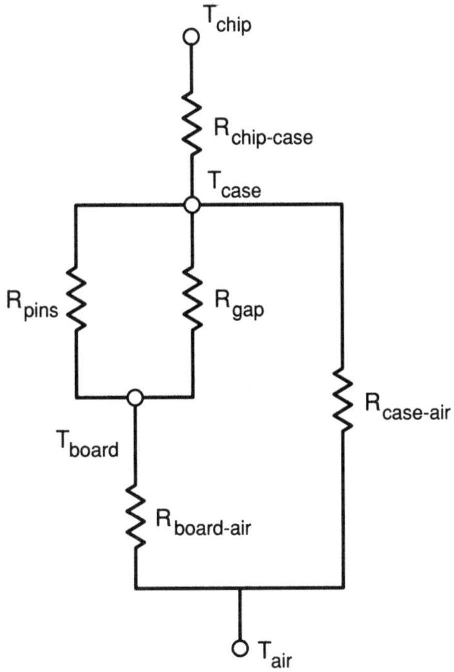

T_{chip} = chip junction temperature
T_{case} = package case temperature
T_{board} = circuit board temperature in vicinity of chip package
T_{air} = temperature of ambient air local to package

Figure 3.2 Heat-flow paths for dual in-line chip packages [Courtesy of Tektronix, Inc.].

the chip junction(s) to the chip package mounting surface through $R_{\text{chip-case}}$, the overall thermal resistance for heat conduction through the semi-conductor chip and chip bonding material. Heat transfer from the package is distributed between a direct path, $R_{\text{case-air}}$, to the surrounding ambient from the package surface and a less direct path to the ambient that involves the circuit board. The path from package to board can be via conduction through an air gap element, R_{gap}, and conduction through the element, R_{pins}, for the package pins which are often soldered into the board. The board then transmits heat to the surrounding ambient via $R_{\text{board-air}}$. Both $R_{\text{case-air}}$ and $R_{\text{board-air}}$ may contain combined convection and radiation elements.

Power transistor packages (TO-220 and TO-3 in Fig. 3.1) present a different heat transfer situation in that the major heat transfer surface, the package base, is often mounted directly to a finned heat sink. The simplest electro-thermal analog that adequately represents this problem is shown in Fig. 3.3. A resistance, $R_{\mathrm{chip-case}}$, represents both the die and the die-package interface resistance as in most other packages. The resistance, $R_{\mathrm{case-sink}}$, for this application simulates the package-to-heat sink mounting conditions. This may be a dry, but not perfectly smooth and flat surface, or it may be a thermal grease/mica washer/thermal grease sandwich. The heat sink thermal resistance, $R_{\mathrm{sink-air}}$, may represent a simple flat piece of metal, a more complex stamping, an enclosure panel, or a finned extrusion.

Heat transfer from heat dissipating components is enhanced by the use of a variety of heat sinks, of which a few are shown in Fig. 3.4. The devices shown in (a) through (d) are intended to be attached to a single component. The extrusion (e) may be used for one or multiple dissipators. While all of these devices may be used in natural convection applications, forced convection greatly improves the situation. It is common for commercial heat sinks to be anodized to maximize radiative heat transfer. This can be expected to significantly

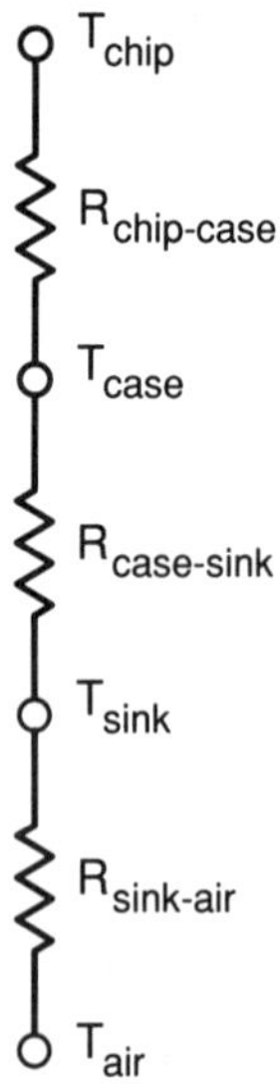

Figure 3.3 Heat-flow paths for power transistor chip package mounted on heat sink [Courtesy of Tektronix, Inc.].

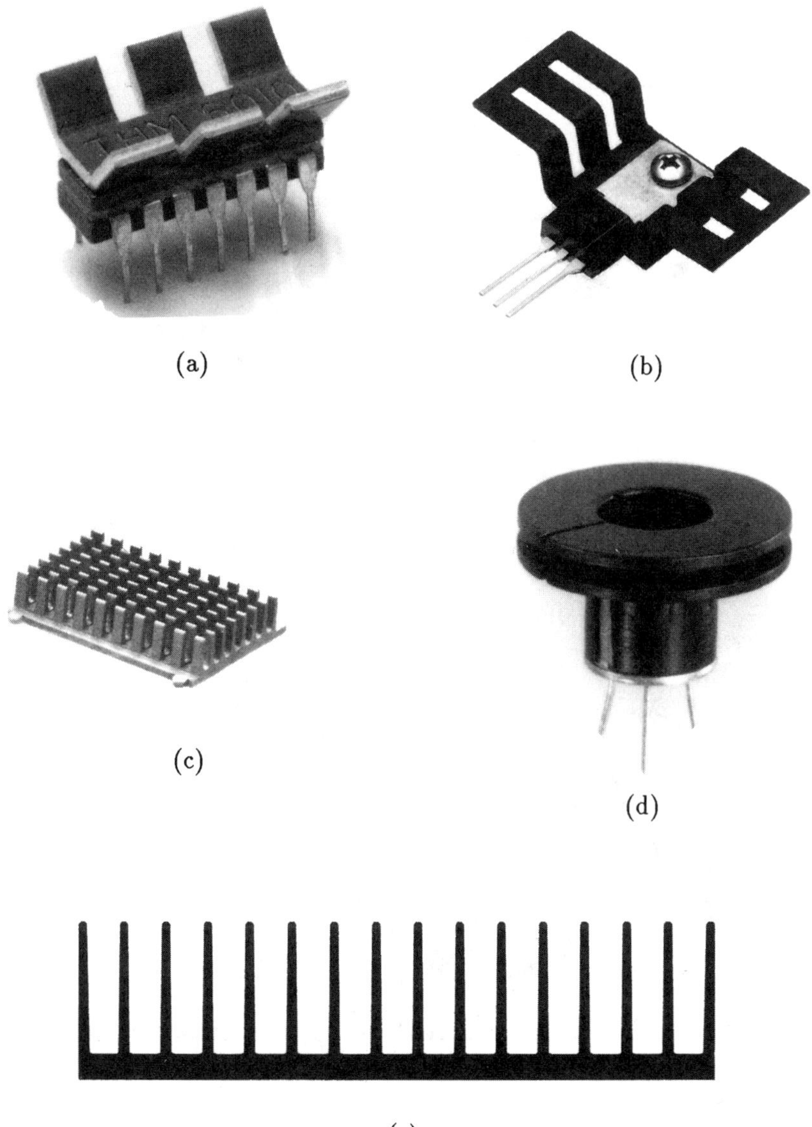

Figure 3.4 Component heat sinks (a) bond-on DIP heat sink for 14- or 16-pin packages, (b) power transistor heat sink for TO-202 packages, (c) pin grid array heat sink, (d) press-on sink for TO-5, 8 packages, (e) extrusion, end view [Courtesy of Thermalloy, Inc.].

(a)

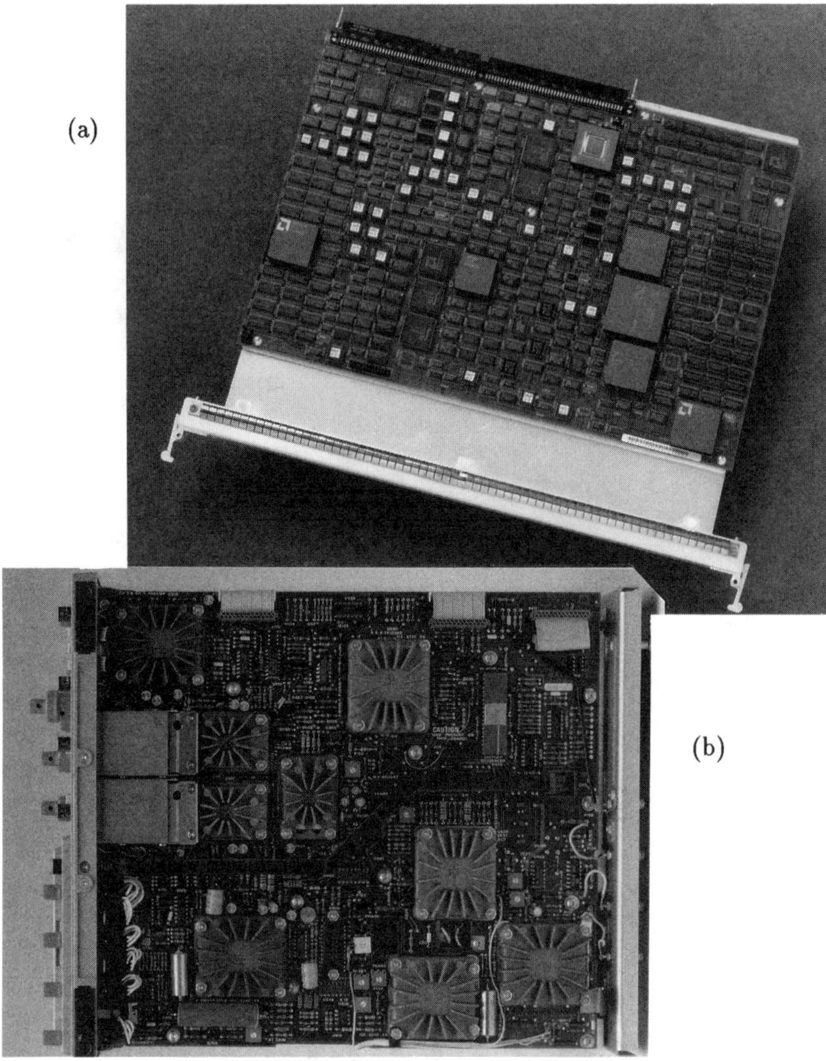

(b)

Figure 3.5 Circuit boards: (a) computer graphics workstation board, (b) portable oscilloscope main board [Courtesy of Tektronix, Inc.].

improve the performance of natural convection/radiation cooled heat sinks, particularly the simple stamped variety. Geometries with more complex fin patterns produce quite significant inter-fin radiation interference that somewhat negates the total performance.

Printed circuit boards from two different electronics systems are shown in Fig. 3.5. The large board (a) is from a computer graphics workstation and reflects the uniformity of component style and

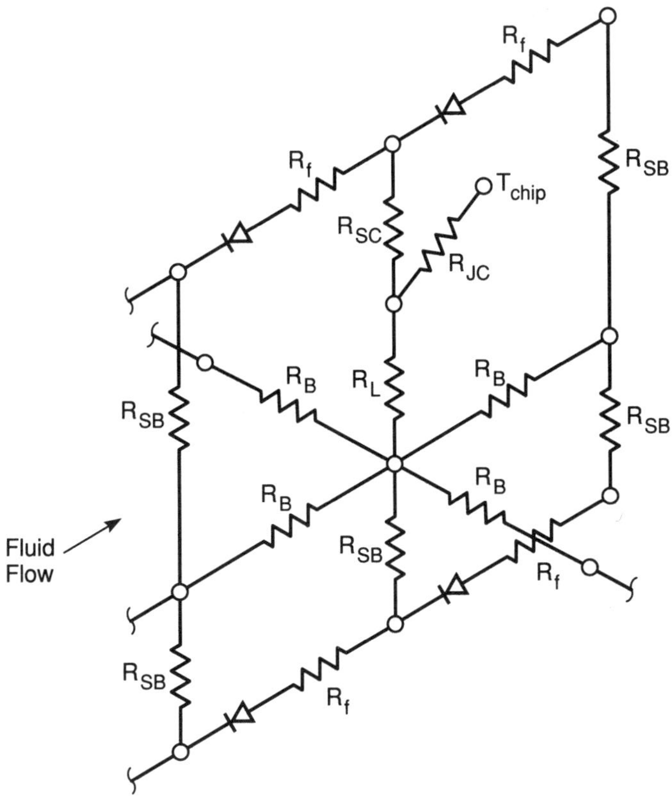

R_B = board conduction resistance
R_l = component - board interface resistance
R_{SC} = component surface - fluid resistance
R_{SB} = board surface - fluid resistance
R_f = resistance for fluid temperature rise
R_{CC} = chip to case resistance

Figure 3.6 Thermal analog representation of printed circuit board heat transfer problem [Courtesy of Tektronix, Inc.].

placement common to computer hardware. The small board (b) is from a portable oscilloscope and is a good representation of the irregular geometry and complex nature of laboratory and diagnostic electronic systems.

Thermal modeling and temperature prediction issues associated with printed circuit boards are extremely complex at best. An electro-thermal analog representation of the heat transfer problem for a section of a printed circuit board (hereafter abbreviated PCB) is illustrated in Fig. 3.6. Elements R_{JC} and R_L are chip junction to case and chip case to board resistances, respectively. The PCB heat conduction paths are represented as elements R_B. Elements, R_{SC}, and R_{SB}, account for surface losses via convection, from the component and board surfaces to the cooling medium. Fluid temperature differences are established by the R_f elements. The latter have diode symbols to serve as a reminder that any given fluid node is predominantly influenced by upstream portions of the field.

Electronics enclosure designs are available in nearly as many varieties as there are designers. The basic external form factor is generally a reflection of the intended use of the product. Personal computers ideally occupy a small portion of desk space, portable oscilloscopes come in relatively small packages (depending on performance features), and workstations may fit beneath a desk. Figure 3.7 illustrates a workstation-type system, and a high performance portable oscilloscope.

A thermal analyst's view of the system packaging problem is illustrated by the hypothetical computer graphics workstation in Fig. 3.8. The overall cooling of the majority of the electronics is accomplished by drawing air through bottom side/front vents, over the electronics card cage, and out through an exhaust fan. Thermal paths for the PCB-mounted components in the card cage would, of course, follow the paths illustrated in the preceding paragraphs. The sealed power supply would also have one or more PCBs. In addition, a large extruded heat sink with power transistors is mounted on the rear power supply panel with fins extending outside the overall system enclosure.

The complete design solution of the packaged system requires evaluation of the coolant flow rate and resulting fluid and surface temperatures. A schematic diagram of a simplified pressure/air flow analog is shown in Fig. 3.9. The solution of such a problem, by matching the circuit pressure loss characteristics to the fan specification, results in an estimate of the volumetric flow rate in each indicated leg of the model. The flow rates are used in solution of the thermal problem.

The thermal analog representation of the generic workstation is shown in Fig. 3.10. The thermal resistance elements, R_f, for the various air channels are calculated from the previously computed air flow rates. The determination of the various convective path elements, R_S, for flat panel heat transfer to the air streams are straightforward, though not necessarily precise calculations. Although the disk drive and card cage dissipations, Q_{DD} and Q_{CC}, respectively, originate within solid volumes and dissipate to the air streams via convection, a practical consideration is to bypass these complex details and treat the sources as originating in the air streams.

(a)

(b)

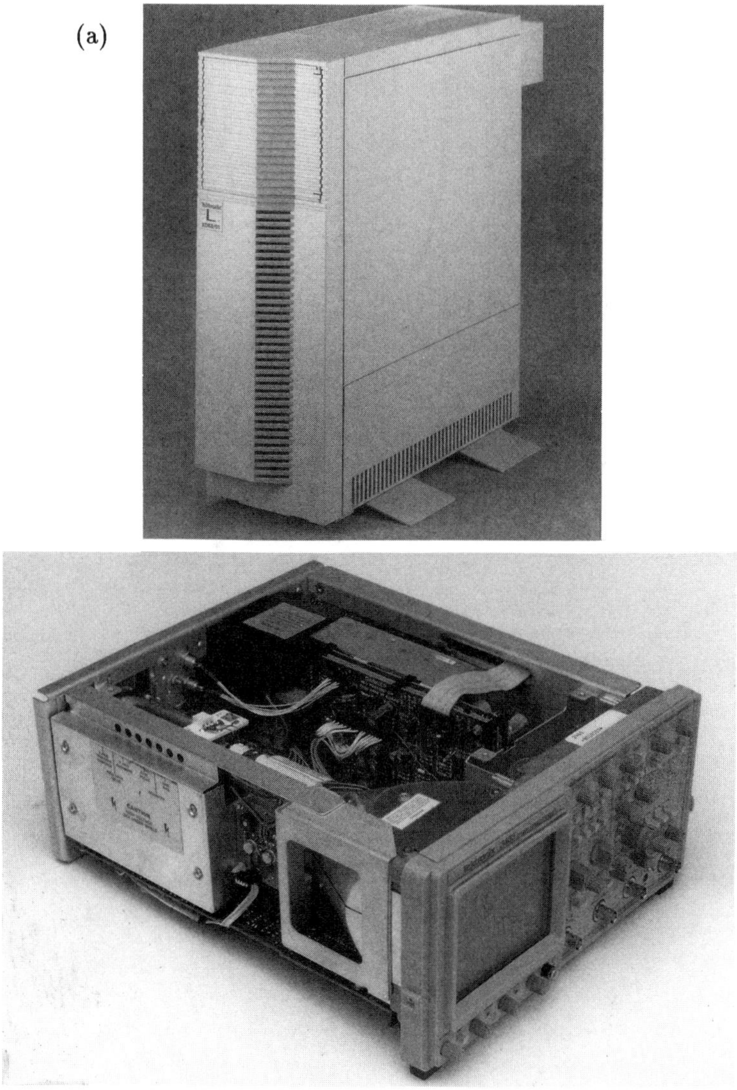

Figure 3.7 Packaged electronic systems: (a) computer workstation, (b) portable oscilloscope, 2400 Series [Courtesy of Tekronix, Inc.].

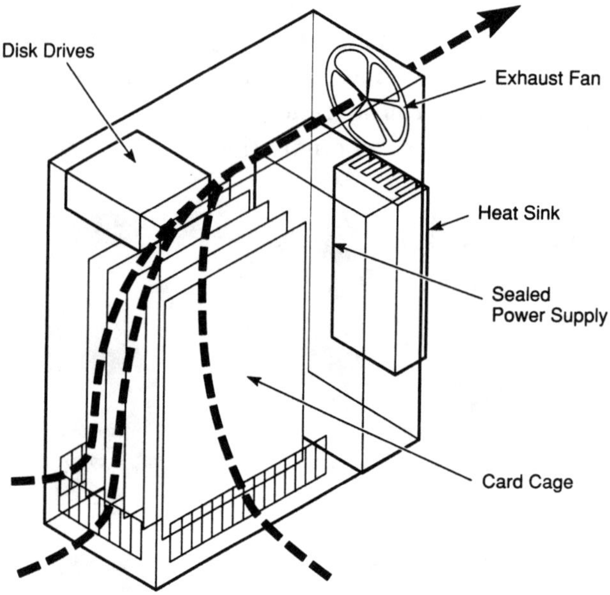

Figure 3.8 Generic computer workstation. Air flow paths are illustrated [Courtesy of Tektronix, Inc.].

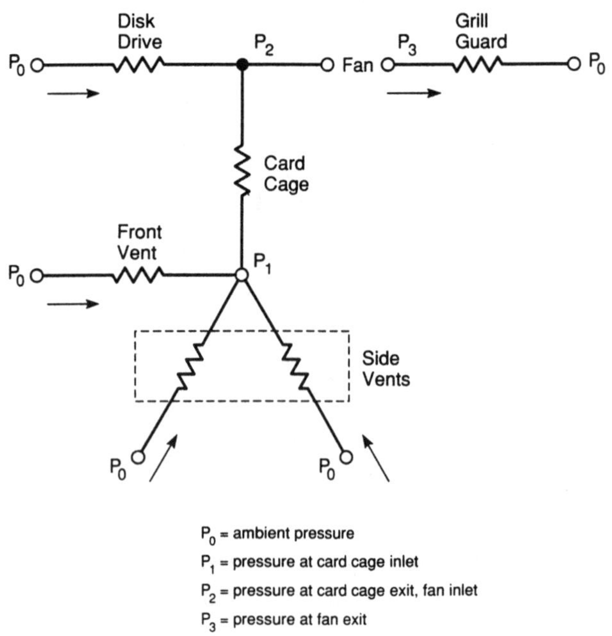

Figure 3.9 Pressure/air flow analog for generic workstation in Fig. 3.8 [Courtesy of Tektronix, Inc.].

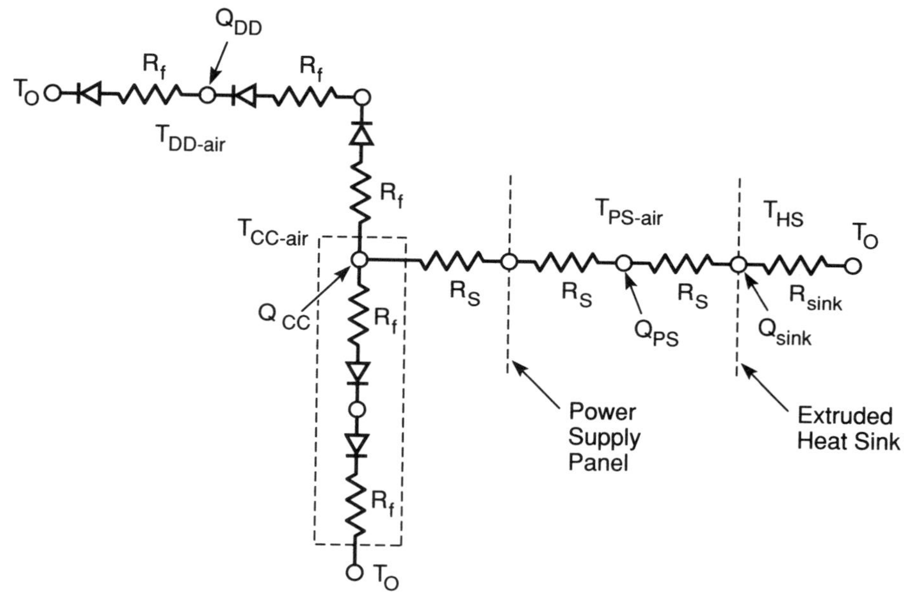

$T_{PS\text{-}air}$	= power supply air temp	R_f	= thermal fluid resistance (diode, "1-way")
T_{HS}	= heat sink temp	R_S	= surface resistance (convection and/or radiation)
$T_{CC\text{-}air}$	= card cage air temp	R_{sink}	= extruded heat sink resistance (convection and/or radiation)
$T_{DD\text{-}air}$	= disk drive air	Q_{DD}	= disk drive dissipation
T_O	= ambient air	Q_{sink}	= heat sink dissipation
Q_{CC}	= total card cage dissipation	Q_{PS}	= power supply dissipation

Figure 3.10 Thermal analog for generic workstation in Fig. 3.8 [Courtesy, Tektronix, Inc.].

3 TOOLS

3.1 Simple Closed Form Methods — Mostly Convection

The impressively low cost/performance ratio of modern computer platforms and popularization of the personal computer make it easy to ignore the value in very simple computational methods that require little more than an inexpensive pocket calculator. Simple methods can be the first level of analysis that provide justification for continued evaluation of a design concept, or the basis upon which the idea should be rejected without further exploration. Furthermore, anyone not experienced in either the programming or use of applications software is aware of just how easy it is to obtain incorrect answers with even the easiest-to-use software — basic methods provide a minimal level of insurance for the more complex computational procedures. Finally, the generality of many digital computer programs is such that the user may be required to calculate some of the input using basic methods.

Because the literature is quite complete with regard to the elucidation of simple techniques [see, for example, Ellison (1989), Kraus and Bar-Cohen (1983), Steinberg (1991), and Dean (1985)], the actual elaboration of non-digital computer solution methods will not be explored here except for two examples of systems that illustrate the aspect of a very preliminary design analysis. Before proceeding, it should be noted that in what follows numerous references are made to basic formulae in Ellison (1989), where a hybrid system of English and metric units are used. Where these formulae are quoted here, a conversion from hybrid to metric units has been incorporated.

Consider then, the rectangular enclosure in Fig. 3.11a with dimensions given as W = 12.7 cm, L=25.4 cm, H=12.7 cm, and a dissipation of about 10 watts. The most desirable enclosure design would be a non-vented, non-fan cooled box. Also suppose that the internal air temperature is required not to exceed 15° C above fairly normal room ambient temperatures.

A simple thermal circuit for the non-fan cooled box is shown in Fig. 3.11b. The resistances, R_{CI} and R_{CE}, represent internal and external natural convection heat transfer. Internal radiation exchange is ignored on the assumption that several circuit boards

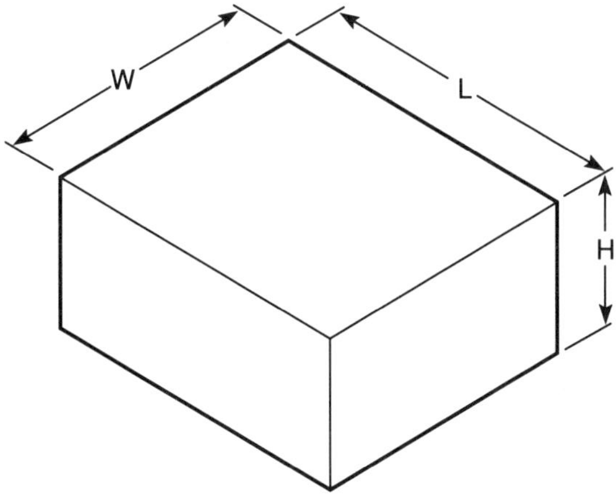

(a) Enclosure Geometry

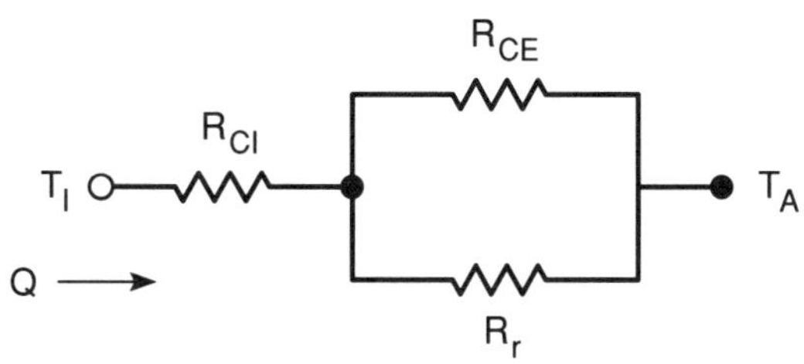

(b) Simple Thermal Circuit

Q = total heat dissipation

T_A = ambient room temperature

T_I = internal air temperature

R_{CI} = convection resistance, internal

R_{CE} = convection resistance, external

R_r = radiation resistance

Figure 3.11 Basic enclosure cooling consideration: (a) enclosure geometry, (b) simple thermal circuit [Courtesy of Tektronix, Inc.].

would effectively introduce an equivalent radiation shield between the majority of the components and the internal enclosure wall.

Making another approximation, that all convecting walls possess the same convection coefficient, significantly simplifies the problem. The convection resistances are therefore given as

$$R_{CI} = R_{CE} = 1/h_C A_S \tag{3.1}$$

An order of magnitude for h_C for this enclosure (average dimension $\overline{L} \approx 17$ cm) and a reasonable temperature rise above ambient of about 10° C is [Ellison (1989), page 36]

$$\begin{aligned} h_C &\approx 3.9 \times 10^{-4}(\Delta T/\overline{L})^{0.25} \\ &\approx 3.4 \times 10^{-4} \mathrm{W/cm^2} - {}^\circ \mathrm{C} \end{aligned}$$

Radiation resistance for a painted enclosure radiating to surrounding surfaces that are at approximately the same temperature as the ambient air may be represented by

$$R_r = 1/h_r A_S \tag{3.2}$$

Surfaces with modest temperatures [Ellison (1989), Fig. 3.16] have an h_r of about 6.2×10^{-4} W/cm² °C.

Neglecting the small temperature difference due to conduction through the box walls, the internal air temperature is estimated to be

$$T_I = \left[R_{CI} + \frac{R_{CE} R_r}{R_{CE} + R_r}\right] Q + T_A \tag{3.3}$$

using

$$\begin{aligned} A_S &= 2(WH + LH + WL) = [(12.7)^2 + 2(25.4)(12.7)] \\ &= 1612.9 \text{ cm}^2 \\ R_{CI} &= R_{CE} = 1/h_C A_S = 1/(3.4 \times 10^{-4})(1612.9) \\ &= 1.8^\circ \text{ C/W} \\ R_r &= 1/h_r A_S = 1/(6.2 \times 10^{-4})(1612.9) \\ &= 1.0^\circ \text{ C/W} \end{aligned}$$

and substituting R_{CI}, R_{CE}, R_r, and Q into Eq. (3.3)

$$T_I - T_A = \left[1.8 + \frac{(1.8)(1.0)}{1.8 + 1.0}\right](10) = 24.4^\circ \text{ C}$$

This 24.4° C is clearly too large and other cooling means must be considered. Natural ventilation air draft can be calculated (although not recommended) without the aid of a digital computer [see Ellison (1989), Chapter 6]. However, even ventilation might not be expected to sufficiently reduce the 24.4° C. A fan should be considered next.

A well-mixed fluid increases in temperature for a heat input, Q, [see Ellison (1989), page 56] by the amount

$$\Delta T = Q/\dot{M}C_p = 3.73Q/G \tag{3.4}$$

where G is the volumetric flow rate in liters per second. The required air flow to obtain the 15° C air temperature rise is therefore

$$G = 3.73Q/\Delta T = 3.73(10)/15 = 2.49\ L/\text{s}$$

The designer therefore knows the approximate flow range of the fan and may start a search for an appropriate device. It is at this point that a more detailed system design would be appropriate. Basic calculations indicated that a detailed analysis of a sealed or vented enclosure would be a waste of time and resources. Additionally, the preceding simple calculations could be used as a credibility validation of more exhaustive analysis methods.

3.2 Complex Closed Form Solutions

The Problem. The variety of closed form solutions is enormous and it is not our purpose to elaborate upon them here. The reader is referred to any of the many texts written for this purpose or a general heat transfer text [Lienhard (1987), Holman (1990), or Sucec (1985) for example]. However, a class of solutions has evolved that is particularly useful in electronics cooling: the three-dimensional flat plate with discrete surface heat sources. The utility of these solutions is demonstrated by articles that appear in the heat transfer literature quite regularly. For example, Hein and Lenzi (1969) provided a general approach which was followed by Ellison (1973, 1976), David (1977), Ramakrishna (1982), Ellison (1984), Pinto and Mikic (1986), Estes (1987), and Ellison (1991). All of these cited methods use a rather standard Fourier series or integral approach to represent the finite size source dimensions. They also exhibit some commonality in the mathematical development with expected differences according to the preferences of the respective authors. Some of the approaches are applicable to a variety of electronic systems, but may be limited, from a practical standpoint, when applied to packages such as integrated circuit chips with sources numbering in the hundreds or thousands. More recently, some novel techniques that appear to have promise in the IC area have been described by Palisoc and Lee (1988).

Figure 3.12 illustrates some simple application possibilities of the Fourier series method applied to three dimensions. A power transistor (a) attached to a flat plate chassis panel is probably the simplest practical example. It is not a trivial problem, however, as the analyst must consider the three-dimensional heat conduction spreading from the transistor-plate interface out into the remainder of the panel. The remaining examples, (b) through (d), have a similar conduction problem, but with the added complexity of multi-source interaction or "crosstalk" effects.

The transistor chip problem (c) has an additional complexity in that it is a two-fold heat spreading problem. There is a spreading resistance from the chip junction heat sources to the chip base, and another spreading resistance from the die-bond base out into the package substrate.

The fourth possibility, the finned heat sink (d), is not even a flat plate problem. There may be some vertical conductive heat transfer due to the fin mass. However, this extra conductive effect is not usually of great concern and may be neglected in most instances. The extended fin area is easily incorporated into an "effective" heat transfer coefficient [Ellison (1989), Section 7.4]. The standard multi-source spreading occurs from the base of the power transistors out to the fin base.

Rather detailed discussions of computational results from application of the method are provided later in this chapter in Component Examples, Section 4.2: Hybrid Package with Heat Spreader, Power Transistors on Panel, and Power Transistors on Finned Heat Sink. A summary of the mathematical formulation of a flat plate solution is covered in the next several subsections.

Partial Differential Equation and Boundary Conditions for the Multilayer, Flat Plate Problem. Fully effective utilization of any given technique is enhanced by a good understanding of the basic mathematics and methods used to solve the differential equation. The methods of Ellison (1973, 1984) and those contained in other

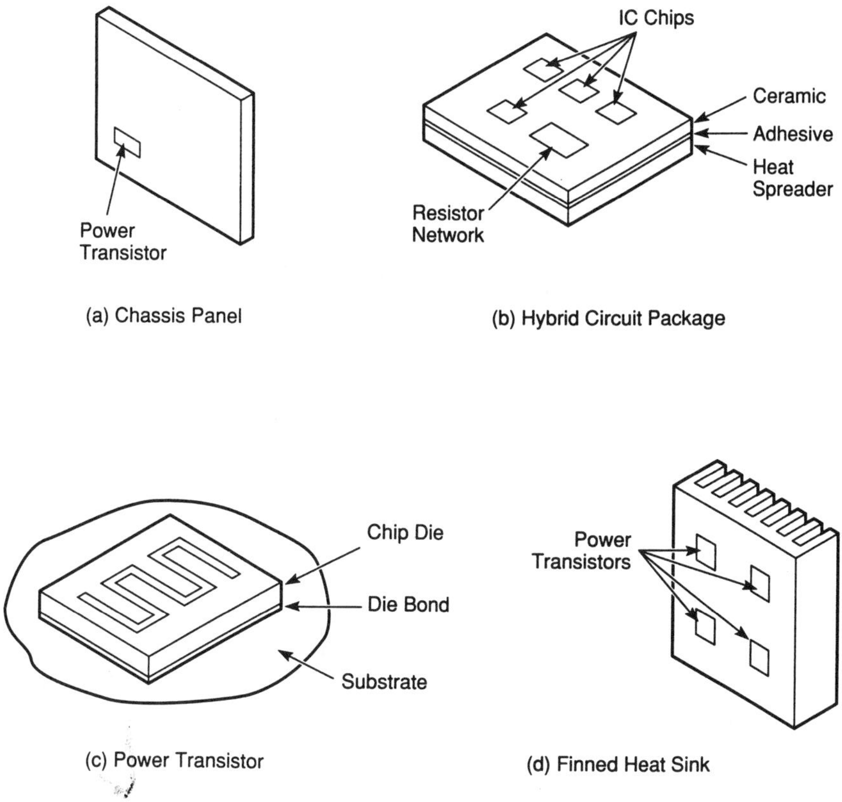

Figure 3.12 Application possibilities for flat plate closed form solutions [Courtesy of Tektronix, Inc.].

non-cited publications are reviewed here. A FORTRAN source code for a resulting computer program is obtainable from Appendix V in Ellison (1989).

The differential equation for steady-state heat conduction in a solid with orthotropic thermal conductivity and a heat source per unit volume Q_V is

$$k_x \frac{\partial^2 T}{\partial x^2} + k_y \frac{\partial^2 T}{\partial y^2} + k_z \frac{\partial^2 T}{\partial z^2} = -Q_V \tag{3.5}$$

where k_x, k_y, and k_z are constant, i.e., not a function of x, y, or z.

Boundary conditions are best described with reference to Fig. 3.13 which shows the orientation of the plate in a Cartesian coordinate system. Because the typical problem is expected to require

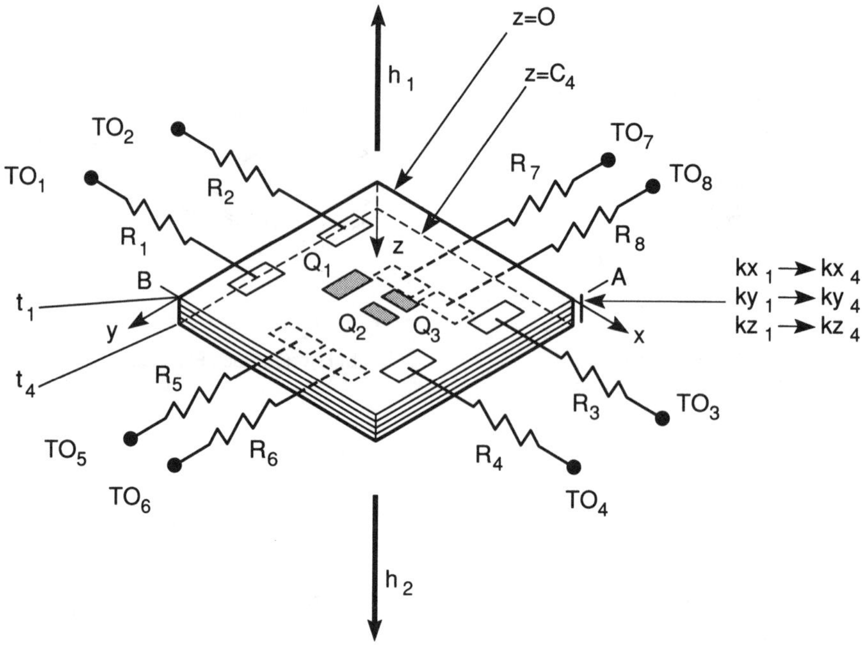

Figure 3.13 Geometry and revelant heat transfer quantities for heat sources and lumped parameter thermal resistances on a multilayer substrate [Ellison (1984)] [Reprinted with premission of the International Society for Hybrid Microelectronics, Reston, VA].

accounting for surface convection and radiation, Newtonian boundary conditions are selected for the two large plate surfaces $z = 0$ and $z = C_4$ (total thickness):

$$k_z \frac{\partial T}{\partial z} = h_1 T, \qquad z = 0 \tag{3.6}$$

$$k_z \frac{\partial T}{\partial z} = -h_2 T, \qquad z = C_4 \tag{3.7}$$

where h_1 and h_2 are the total heat transfer coefficients at the $z = 0$ and $z = C_4$ surfaces, respectively. The temperature T is defined as the plate temperature rise over the surrounding ambient above and below the plate.

Edge boundary conditions are conveniently selected as no-flux conditions:

$$k_x \frac{\partial T}{\partial x} = 0, \qquad x = 0, A \tag{3.8}$$

$$k_y \frac{\partial T}{\partial y} = 0, \qquad y = 0, B \tag{3.9}$$

The Form of the Solution for One Heat Source. The form of solution can be obtained by employing a method known as "separation of variables" and is in terms of a double Fourier cosine series where the x and y dependence is satisfied by the cosine terms and the z dependence is accounted for in the Fourier coefficient. In this case, the selected cosine terms are eigenfunctions which satisfy the no-flux boundary conditions at the edges:

$$T(r) = \sum_{\ell=0}^{\infty} \sum_{m=0}^{\infty} \varepsilon_\ell \varepsilon_m \psi_{\ell m}(z) \cos\left(\frac{\ell \pi x}{A}\right) \cos\left(\frac{m \pi y}{B}\right) \tag{3.10}$$

where

$$\begin{aligned} \varepsilon_\ell &= \frac{1}{2}, & \ell &= 0 \\ \varepsilon_\ell &= 1, & \ell &\neq 0 \\ \varepsilon_m &= \frac{1}{2}, & m &= 0 \\ \varepsilon_m &= 1, & m &\neq 0 \end{aligned}$$

The Representation of One Heat Source. It will be seen to be particularly convenient to represent a heat source in a form similar to the expansion for the temperature:

$$Q_V(r) = \sum_{\ell=0}^{\infty} \sum_{m=0}^{\infty} \varepsilon_\ell \varepsilon_m \phi_{\ell m}(z) \cos\left(\frac{\ell \pi x}{A}\right) \cos\left(\frac{m \pi y}{B}\right) \tag{3.11}$$

The heat source Q_V is assumed to generate uniform power over the plate coordinates, x to $(x + dx)$, y to $(y + dy)$, covered by the source. With this definition of Q_V, the Fourier source coefficients $\phi_{\ell m}$ may be derived. Multiply both sides of the Q_V expansion by

$$\cos\left(\frac{\ell' \pi x}{A}\right) \cos\left(\frac{m' \pi y}{B}\right)$$

where ℓ' and m' are integers in the range 0 to infinity. The result is integrated over the x, y dimensions of the substrate, A, B, respectively:

$$\int_0^A \int_0^B Q_V \cos\left(\frac{\ell' \pi x}{A}\right) \cos\left(\frac{m' \pi y}{B}\right) dx dy = \Lambda_1 + \Lambda_2 + \Lambda_3 + \Lambda_4$$

where

$$\Lambda_1 = \frac{1}{4}\phi_{00}\int_0^A \cos\left(\frac{\ell'\pi x}{A}\right)dx\int_0^B \cos\left(\frac{m\pi y}{B}\right)dy$$

$$\Lambda_2 = \frac{1}{2}\sum_{m=1}^{\infty}\phi_{0m}\int_0^A \cos\left(\frac{\ell'\pi x}{A}\right)dx\int_0^B \cos\left(\frac{m'\pi y}{B}\right)\cos\left(\frac{m\pi y}{B}\right)dy$$

$$\Lambda_3 = \frac{1}{2}\sum_{\ell=1}^{\infty}\phi_{\ell 0}\int_0^B \cos\left(\frac{m'\pi y}{B}\right)dy\int_0^A \cos\left(\frac{\ell'\pi x}{A}\right)\cos\left(\frac{\ell\pi y}{A}\right)dx$$

and

$$\Lambda_4 = \sum_{\ell=1}^{\infty}\sum_{m=1}^{\infty}\phi_{\ell m}\int_0^A \cos\left(\frac{\ell'\pi x}{B}\right)\cos\left(\frac{\ell\pi x}{B}\right)dx \bullet$$

$$\int_0^B \cos\left(\frac{m'\pi y}{B}\right)\cos\left(\frac{m\pi y}{B}\right)dy \qquad (3.12)$$

The preceding is simplified by observing that

$$\int_0^A \cos\left(\frac{\ell'\pi x}{A}\right)dx = A\delta_{\ell'0}$$

$$(3.13)$$

$$\int_0^A \cos\left(\frac{\ell'\pi x}{A}\right)\cos\left(\frac{\ell\pi x}{A}\right)dx = \frac{A}{2}\delta_{\ell'\ell}; \quad \ell' \neq 0, \ell \neq 0$$

where $\delta_{\ell'\ell}$ is the Kronecker delta function with the properties

$$\delta_{\ell'\ell} = 0 \qquad \ell' \neq \ell$$

$$\delta_{\ell'\ell} = 1 \qquad \ell' = \ell$$

The integration over Q_V becomes

$$\int_0^A\int_0^B Q_V \cos\left(\frac{\ell'\pi x}{A}\right)\cos\left(\frac{m'\pi y}{B}\right)dxdy = \frac{AB}{4}[\xi_1 + \xi_2 + \xi_3 + \xi_4] \qquad (3.14)$$

where

$$\xi_1 = \phi_{00}\delta_{\ell'0}\delta_{m'0}$$

$$\xi_2 = \sum_{m=1}^{\infty}\phi_{0m}\delta_{\ell'0}\delta_{m'm}$$

$$\xi_3 = \sum_{\ell=1}^{\infty}\phi_{\ell 0}\delta_{m'0}\delta_{\ell'\ell}$$

and

$$\xi_4 = \sum_{m=1}^{\infty} \sum_{\ell=1}^{\infty} \phi_{\ell m} \delta_{\ell' \ell} \delta_{m' m}$$

If the heat source is further restricted to be of zero thickness, the Dirac delta function, $\delta(z - z_0)$, may be utilized. For a heat flux $q(r)$

$$Q_V = q(r)\delta(z - z_0) \tag{3.15}$$

By respectively setting $l' = 0$, $m' = 0$; $l' \neq 0$, $m' = 0$; $l' = 0$, $m' \neq 0$, $\ell' \neq 0$, $m' = 0$, the source Fourier coefficients are obtained:

$$\begin{aligned} \phi_{00} &= \frac{4}{AB} \int_0^A \int_0^B Q_V dx dy \\ &= \frac{4}{AB} q(r)\delta(z - z_0)(x_2 - x_1)(y_2 - y_1) \end{aligned} \tag{3.16}$$

$$\begin{aligned} \phi_{\ell 0} &= \frac{4}{AB} \int_0^A \int_0^B Q_V \cos\left(\frac{\ell \pi x}{A}\right) dx dy \\ &= \frac{4}{AB} q(r)\delta(z - z_0)(y_2 - y_1) \int_{x_1}^{x_2} \cos\left(\frac{\ell \pi x}{A}\right) dx \\ &= \frac{4}{\pi \ell B} q(r)\delta(z - z_0)(y_2 - y_1) \left[\sin\left(\frac{\ell \pi x_2}{A}\right) - \sin\left(\frac{\ell \pi x_1}{A}\right)\right] \end{aligned} \tag{3.17}$$

Similarly,

$$\phi_{0m} = \frac{4}{\pi m A} q(r)\delta(z - z_0)(x_2 - x_1) \left[\sin\left(\frac{m \pi y_2}{B}\right) - \sin\left(\frac{m \pi y_1}{B}\right)\right] \tag{3.18}$$

and

$$\phi_{\ell m} = \frac{4}{\pi^2 \ell m} q(r)\delta(z - z_0)\Gamma_A \Gamma_B \tag{3.19}$$

where

$$\begin{aligned} \Gamma_A &= \sin\left(\frac{\ell \pi x_2}{A}\right) - \sin\left(\frac{\ell \pi x_1}{A}\right) \\ \Gamma_B &= \sin\left(\frac{m \pi y_2}{B}\right) - \sin\left(\frac{m \pi y_1}{B}\right) \end{aligned}$$

With these representations, the source functions are completely known.

The Reduction of the Differential Equation from Three Dimensions to One Dimension. Specification of the x and y terms in the solution as cosine functions implies that the remaining problem is to find the z-dependent portion of T. This is of course the Fourier coefficient $\psi_{\ell m}$.

One begins by substituting Eq. (3.10):

$$T(r) = \sum_{\ell=0}^{\infty} \sum_{m=0}^{\infty} \varepsilon_\ell \varepsilon_m \psi_{\ell m}(z) \cos\left(\frac{\ell \pi x}{A}\right) \cos\left(\frac{m \pi y}{B}\right)$$

and Eq. (3.11):

$$Q_V = \sum_{\ell=0}^{\infty} \sum_{m=0}^{\infty} \varepsilon_\ell \varepsilon_m \phi_{\ell m}(z) \cos\left(\frac{\ell \pi x}{A}\right) \cos\left(\frac{m \pi y}{B}\right)$$

into Eq. (3.5):

$$k_x \frac{\partial^2 T}{\partial x^2} + k_y \frac{\partial^2 T}{\partial x^2} + k_z \frac{\partial^2 T}{\partial x^2} = -Q_V$$

to obtain

$$\sum_{\ell=0}^{\infty} \sum_{m=0}^{\infty} \Omega_1 \Omega_2 = -\sum_{\ell=0}^{\infty} \sum_{m=0}^{\infty} \phi_{\ell m} \Omega_2$$

where

$$\Omega_1 = -\left[k_x \left(\frac{\ell \pi}{A}\right)^2 + k_y \left(\frac{m \pi}{B}\right)^2\right] \psi_{\ell m} + k_z \frac{d^2 \psi_{\ell m}}{dz^2}$$

and

$$\Omega_2 = \varepsilon_\ell \varepsilon_m \cos\left(\frac{\ell \pi x}{A}\right) \cos\left(\frac{m \pi y}{B}\right) \tag{3.20}$$

Next, the coefficients of like terms:

$$\varepsilon_\ell \varepsilon_m \cos\left(\frac{\ell \pi x}{A}\right) \cos\left(\frac{m \pi y}{B}\right)$$

are equated as required by Eq. (3.20). This results in

$$\frac{d^2 \psi_{\ell m}}{dz^2} - \gamma_{\ell m}^2 \psi_{\ell m} = -\frac{1}{k_z} \phi_{\ell m} \tag{3.21}$$

where

$$\gamma_{\ell m}^2 = \alpha^2 + \beta^2$$

$$\alpha^2 = \left(\frac{k_x}{k_z}\right)\left(\frac{\ell\pi}{A}\right)^2$$

$$\beta^2 = \left(\frac{k_y}{k_z}\right)\left(\frac{m\pi}{B}\right)^2$$

This shows that the multilayer conduction problem with convective/radiative boundary conditions is reduced to a one-dimensional problem in the z-coordinate.

Solution of $\psi_{\ell m}$ for the Multilayer Problem. There are a variety of methods for solving Eq. (3.21). For example, a Green's function technique has been applied for two- and four-layer versions of the problem by Ellison (1973, 1976, and 1984). Regardless of the method chosen, boundary conditions in addition to the Newtonian cooling at the $z = 0$, C_4 surfaces must be satisfied. These are the continuity of temperature and heat flux at the boundaries separating the various layers:

$$\psi_{\ell m}\Big|_{z=C_i^-} = \psi_{\ell m}\Big|_{z=C_i^+} \qquad i = 1, 2, 3 \tag{3.22}$$

$$k_z \frac{d\psi_{\ell m}}{dz}\Big|_{z=C_i^-} = k_z \frac{d\psi_{\ell m}}{dz}\Big|_{z=C_i^+} \qquad i = 1, 2, 3 \tag{3.23}$$

The Extension to Multiple Sources. Equations (3.16) through (3.19) showed that the $\phi_{\ell m}$ are proportional to source dissipation. Similarly,

$$\psi_{\ell m} \alpha Q \tag{3.24}$$

Ellison (1976) then shows that the infinite series solution for the substrate temperature T with the Fourier coefficients $\psi_{\ell m}$ is

$$T = AQ + T_A \tag{3.25}$$

where A is the Fourier series solution for a source of unit dissipation. In what follows, A is referred to as a single source influence coefficient.

The temperature $T(r_i)$ at any fixed location r_i on the substrate due to heat sources j is accounted for by superposition of single source solutions. Thus,

$$T(r_i) = \sum_{j=1}^{NS} A_{ij} Q_j + T_A \tag{3.26}$$

where A_{ij} is the influence coefficient for a temperature at r_i due to a unit source at r_j and NS is the total number of all sources Q_j.

Heat Removal by Lumped Thermal Resistances. It is often necessary to account for a heat loss from the structure that is not associated with the convective coefficients, h_1 and h_2. An example would be a hybrid circuit substrate with not only surface convection, but also conduction through an edge connector. This subsection illustrates how lumped thermal resistances are incorporated into the solution, thus accounting for heat transfer through such appurtenances as edge connectors.

The principle of superposition may also be applied here. Observe in Fig. 3.14 that heat, F_k, is removed from a surface region of the plate through a thermal resistance, R_k. If the temperature at the end of the resistance is specified as TO_k, a plate surface temperature, $T(r_k)$, is produced. Surface area is specified in the same manner as a source: the corner is located at x, y, and the area has edge dimensions Δx, Δy. Edge dimensions are presumed sufficiently small so that the temperature is uniform over the area.

The temperature rise from the resistance termination to the surface temperature is

$$T(r_k) - TO_k = F_k R_k \tag{3.27}$$

With reference to the ambient temperature T_A surrounding the plate:

$$T(r_k) - T_A = F_k R_k + (TO_k - T_A) \tag{3.28}$$

Using the representation for surface temperature $T(r_i)$ at any point r_i on the plate, resistance is treated in a manner similar to that for a source but using an influence coefficient A_{ik}. It is also recognized that the resistance governs the heat removal from the plate:

$$T(r_i) - T_A = \sum_{j=1}^{NS} A_{ij} Q_j - \sum_{k=1}^{NR} A_{ik} F_k \tag{3.29}$$

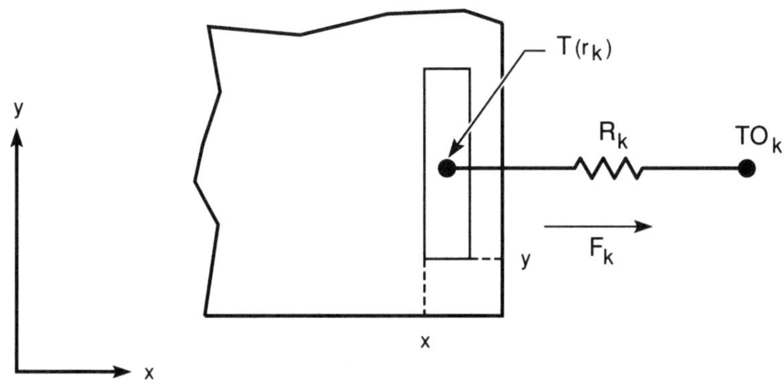

Figure 3.14 Heat removal (F_k) via thermal resistance R_k from plate surface area at x, y [Ellison (1990)] [Reprinted with permission of Computational Mechanics Publications, Southampton, England].

where there are NS sources and NR resistors. The temperature at any resistor location r_k is then

$$T(r_k) - T_A = \sum_{j=1}^{NS} A_{kj}Q_j - \sum_{\ell=1}^{NR} A_{k\ell}F_\ell \tag{3.30}$$

Combining the preceding equation with the temperature rise of the resistor above ambient produces

$$F_kR_k + (TO_k - T_A) = \sum_{j=1}^{NS} A_{kj}Q_j - \sum_{\ell=1}^{NR} A_{k\ell}F_\ell \tag{3.31}$$

and a little algebra provides the desired result:

$$F_kR_k + \sum_{\ell=1}^{NR} A_{k\ell}F_\ell = \sum_{j=1}^{NS} A_{kj}Q_j + (T_A - TO_k) \tag{3.32}$$

Examination of this result indicates that this is actually a set of simultaneous equations where the F_k may be specified as the unknowns. The heat source dissipations Q_j, ambient T_A, resistances R_k, resistance termination temperatures TO_k, and influence coefficients A_{ij} are all knowns. Of course, the determination of A_{ij} and A_{ik} requires considerable calculation effort.

Upon completion of the F_k calculation, the temperature $R(r_i)$ at any location r_i may be calculated from

$$T(r_i) = \sum_{j=1}^{NS} A_{ij}Q_j - \sum_{k=1}^{NR} A_{ik}F_k \tag{3.33}$$

Thus, the mathematical formulation of a closed form method can be quite formidable. However, once the computer program is written [Ellison (1989)], the true power of the method becomes apparent through its ease of use. The reader may wish to refer to the examples in the subsections to Section 4.2 at this point.

3.3 Network Method — Thermal Conduction/ Convection/Radiation

Engineers have long been familiar with the network method because of its easy progammability and, of even more importance, its value in making a clear translation of the physical problem to a mathematical representation. The fact that exact geometrical definitions for every piece of the problem are not needed is both a strength and a weakness. The strength is that difficult-to-calculate quantities such as thermal interface resistance may be approximated with an order of magnitude estimate to at least provide for this thermal mechanism. The weakness is that graphic presentation suffers, because geometric definition data is not available for combination with external results. The following treatment is not intended to be all-inclusive, but rather of an introductory nature.

Conduction. The theoretical introduction to the thermal network method necessarily begins with the definition of the more basic elements of a system that accounts for conduction, convection, and radiation. Consider the problem represented by a one-dimensional approximation of heat conduction in the arbitrary element in Fig. 3.15. Heat Q_k uniformly enters the cross-section at $x = x_1$ and exits uniformly at $x = x_2$. Fourier's law of heat conduction is applied to an arbitrary location x between x_1 and x_2 where the cross-sectional area that varies with x is designated as A_k:

$$Q_k = -kA_k \frac{dT}{dx} \tag{3.34}$$

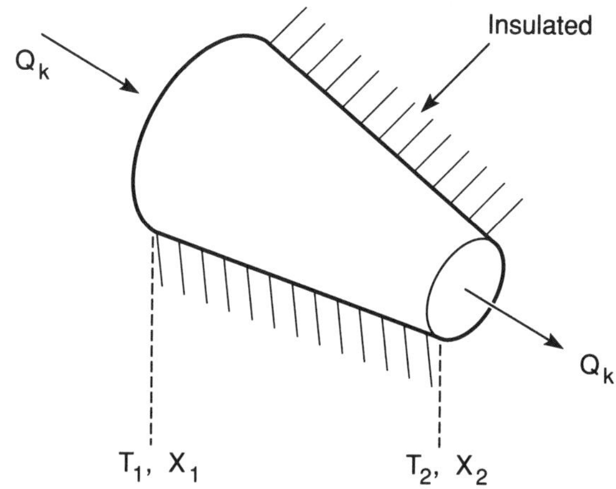

Figure 3.15 Conductive heat transfer in a one-dimensional insulated structure with end temperatures T_1, T_2 [Courtesy of Tektronix, Inc.].

Solving for dT via separation of the variables gives

$$dT = -\frac{Q_k}{kA_k}dx \tag{3.35}$$

This result is integrated over the length $L = x_2 - x_1$ to yield the temperature difference:

$$\int_{T_1}^{T_2} dT = -Q_k \int_{x_1}^{x_2} \frac{dx}{kA_k} \tag{3.36}$$

$$T_1 - T_2 = Q_k \int_{x_1}^{x_2} \frac{dx}{kA_k} = Q_k R_k \tag{3.37}$$

where, with an origin translation,

$$R_k = \int_0^L \frac{dx}{kA_k}, \qquad L = x_2 - x_1 \tag{3.38}$$

is defined, for this element, as the thermal resistance for conduction. Clearly, this general expression for the thermal resistance is applicable to a temperature-dependent thermal conductivity as well as the arbitrarily varying cross-section area, A_k. Very often, however, models may be developed where A_k is uniform and k is independent of temperature and x. In this instance the resistance becomes the familiar:

$$R_k = \frac{L}{kA_k} , \qquad \text{uniform } A_k , \qquad k \neq k(T,x) \tag{3.39}$$

Convection. The geometry pertinent to the basic convection element is an area A_S at uniform temperature T_S (Fig. 3.16). Heat Q_C is transferred by convection from this surface to a surrounding ambient at temperature T_A. The relationship describing this in a phenomenological sense is Newton's law of cooling:

$$Q_C = h_C A_S (T_S - T_A) \tag{3.40}$$

Solving for the temperature rise above ambient:

$$T_S - T_A = Q\frac{1}{h_C A_S} \tag{3.41}$$

or

$$T_S - T_A = Q R_C \tag{3.42}$$

where the thermal resistance for convection is

$$R_C = \frac{1}{h_C A_S} \tag{3.43}$$

Radiation. Many radiation problems in electronic cooling applications may be treated in a rather simplistic fashion. This requires adequately meeting the condition that the component under consideration is a diffusely reflecting and emitting surface in a non-reflecting

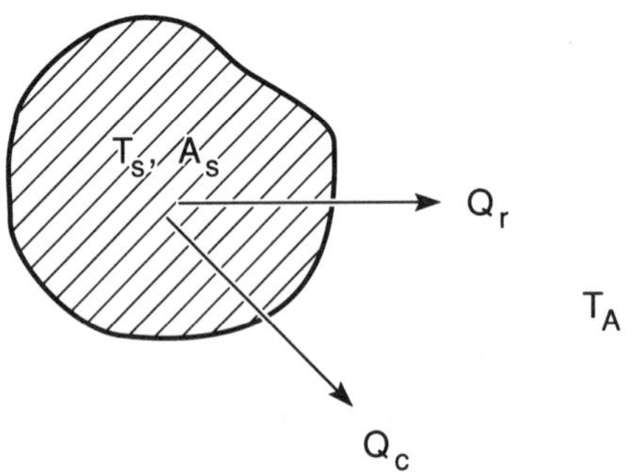

Figure 3.16 Convective, Q_C, and radiative, Q_r, heat transfer from surface A_S at temperature T_S to ambient at temperature TA [Courtesy of Tektronix, Inc.].

ambient. This condition is certainly met by an electronics enclosure in a highly absorbing room. In the case that the latter is highly reflecting, only a fraction of the reflected radiation will actually return to the component, due to its small size.

For those situations where the receiving surface of interest intercepts less than all of the source's transmitted radiation (the "shape factor" is less than unity) the complexities of multi-surface radiation must be addressed [see Ellison (1989)].

The net radiative heat transfer between a surface area A_S at a temperature T'_S (prime indicating absolute temperature scale) and an ambient at temperature T'_A where a shape factor of unity may be assumed is

$$\begin{aligned} Q_r &= \sigma\varepsilon A_S(T'^4_S - T'^4_A) && (3.44) \\ &= \sigma\varepsilon A_S(T'^3_S + T'^2_S T'_A + T'_S T'^2_A + T'^3_A)(T'_S - T'_A) && (3.45) \\ &= h_r A_S(T'_s - T'_A) && (3.46) \end{aligned}$$

where the temperature-dependent radiative heat transfer coefficient is

$$h_r = \varepsilon\sigma(T'^3_S + T'^2_S T'_A + T'_S T'^2_A + T'^3_A) \qquad (3.47)$$

The thermal resistance for radiation with the foregoing approximations is therefore

$$R_r = \frac{1}{h_r A_S} \qquad (3.48)$$

Conservation Principle — Nodal Energy Balance. The thermal network representation of any real problem is an assembly of nodes interconnected by thermal elements of the type previously described. The governing equations of the resultant circuit are derived by using the conservation of thermal energy. Figure 3.17 is a piece of a larger circuit system. The arbitrary "node" i is interconnected to an assembly of nearest neighbor nodes j. A source of internal heat generation at node i is indicated by Q_i.

Conservation of energy may be applied in steady-state conditions:

$$\text{HEAT INTO NODE } i = \text{HEAT OUT OF NODE } i \qquad (3.49)$$

$$Q_i = \sum_j K_{ij}(T_i - T_j) \qquad (3.50)$$

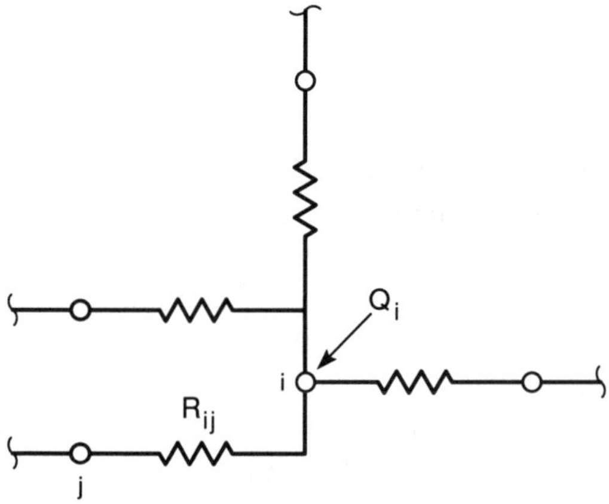

Figure 3.17 Portion of a steady-state thermal circuit showing node i interconnected to nodes j [Courtesy of Tektronix, Inc.].

where the use of conductance elements

$$K_{ij} = \frac{1}{R_{ij}} \tag{3.51}$$

is preferred to resistance elements.

The energy balance equation may be rewritten in different forms depending on the solution method desired. The Gauss-Seidel method is straightforward for programming and requires minimal computer memory. This format is obtained by solving for T_i:

$$T_i = \frac{\sum_j K_{ij}T_j + Q_i}{\sum_j K_{ij}} \tag{3.52}$$

A computer program utilizing the Gauss-Seidel method would of course work in an iterative fashion. A first guess of all nodal temperatures (except fixed temperatures) would be used to start. One iteration would consist of solving for each T_i using a combination of the first guess and computed temperatures for node numbers less than i. Non-linear problems such as those with temperature-dependent thermal conductivity, natural convection elements, or radiation are easily managed by merely updating the conductances, K_{ij}.

Computer programs seldom use the Gauss-Seidel method in exactly the form of Eq. (3.52). It is preferred practice to include a relaxation parameter that permits damping or acceleration of each iteration, depending on the nature of the problem [see Ellison (1989), page 23].

A matrix solution method requires writing the energy balance equation in a different format:

$$\left(\sum_j K_{ij}\right) T_i + \sum_j (-K_{ij})T_j = Q_i \tag{3.53}$$

This equation is actually descriptive of a set of simultaneous equations with unknowns T_i, T_j, but with the possibility of fixed source and/or sink temperatures. Terms involving fixed temperatures are placed in the source vector column. Non-linear temperature-dependent problems may be managed by repeatedly applying a linear equation solver, and updating the conductances at each iteration. There are of course a number of different solution techniques employed by software authors.

Two commonly used methods to indicate satisfactory solution are a check on temperature changes during the iteration and/or a check on the energy balance at each iteration. A rapidly converging solution will readily result in small temperature changes between any two iterations for any node when the equations have been appropriately solved. However, in the case of a slowly converging solution, the method of depending on small nodal temperature changes as an indication of a good solution is very risky. The obvious reason is that a slowly convergent solution will, by its very nature, have a small temperature change between any two iterations, even when the solution is still inaccurate.

For this reason, a check on the nodal energy balance, or "residual" r_i, is appropriate. The residual for each node is simply the unaccounted-for energy in the original node equation:

$$r_i = \sum_j K_{ij}(T_i - T_j) - Q_i \tag{3.54}$$

Clearly, when r_i is small for each node, an adequate solution of the equations has been obtained. In this sense, "small" is a relative term. Those systems of equations with nodal sources may utilize a comparison with the source dissipations as a measure of smallness.

A different approach is required to determine the transient thermal behavior of an electronics system. The derivation of a time-

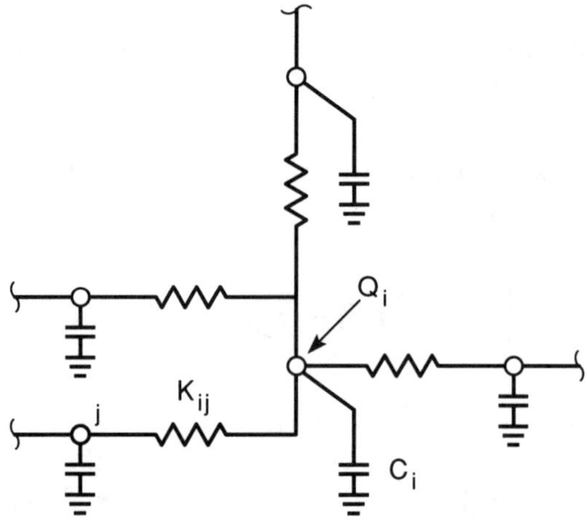

Figure 3.18 Portion of a time-dependent thermal circuit showing node i with capacitance C_i, interconnected to nodes j [Courtesy of Tektronix, Inc.].

dependent differential equation is quite straightforward when a forward finite difference method in time is used. One merely adds the appropriate storage term to the energy balance equation. The result for a nodal volume ΔV_i with specific heat C_p and density ρ is

$$Q_i - \sum_j K_{ij}(T_i - T_j) = \frac{\rho C_p \Delta V_i (T_i^{\Delta} - T_i)}{\Delta t} \tag{3.55}$$

where T_i is the temperature of node i at time t, T_i^{Δ} is the temperature at time $t + \Delta t$ and T_j is the temperature of node j at time t.

The solution for node i at time $t + \Delta t$ is

$$T_i^{\Delta} = T_i \left(1 - \frac{\Delta t}{C_i} \sum_j K_{ij}\right) + \frac{\Delta t}{C_i}\left(Q_i + \sum_j K_{ij} T_j\right) \tag{3.56}$$

where C_i is the capacitance of node i. Figure 3.18 indicates the appropriate circuit representation.

The convergence issue is one of appropriately sizing the time step Δt. Examination of the temperature solution shows that

$$\Delta t \leq \frac{C_i}{\sum_j K_{ij}} \tag{3.57}$$

The justification for this is that if Δt is sufficiently large, i.e., if

$$\Delta t > \frac{C_i}{\sum_j K_{ij}} \tag{3.58}$$

the factor

$$\left(1 - \frac{\Delta t}{C_i}\sum_j K_{ij}\right)$$

is negative. The implication would be the impossible condition, in a heating process, that the greater the temperature of node i at time t, the smaller the temperature for this node at time $t + \Delta t$. An additional implication is that heat could flow from a lower to a higher temperature. This would be in violation of the Second Law of Thermodynamics.

The various T_i are solved by starting out with a set of initial conditions, or starting temperatures. A new set of temperatures are calculated at the first step, $t = \Delta t$. Another set of temperatures are calculated at the new time $= t + \Delta t$. This procedure continues until terminated by the program user.

3.4 Network Method — Fluid Transport

Adequate design characteristics regarding cooling fluid rates may often be obtained by using a hydraulics approach to the electronics problem. While perhaps not as elegant or accurate as the full solution of the Navier Stokes equations, with or without turbulent flow models, the method does not typically require significant computer power or expensive software. It may even provide the basis for manual computations using an electronic calculator.

The foundations of the method rely on interrelating a form of the well-known Bernoulli equation with the definitions of fan pressure as specified by manufacturers of these systems. Bernoulli's equation applied to two different stations in a flow system with pressure losses is given as

$$\frac{p_1}{\rho g} + \frac{v_1^2}{2g} + z_1 = \frac{p_2}{\rho g} + \frac{v_2^2}{2g} + z_2 + \Delta p_{1-2} \tag{3.59}$$

where p = pressure, v = velocity, and z = elevation at stations 1 or 2 in the flow system. The last term on the right side of the

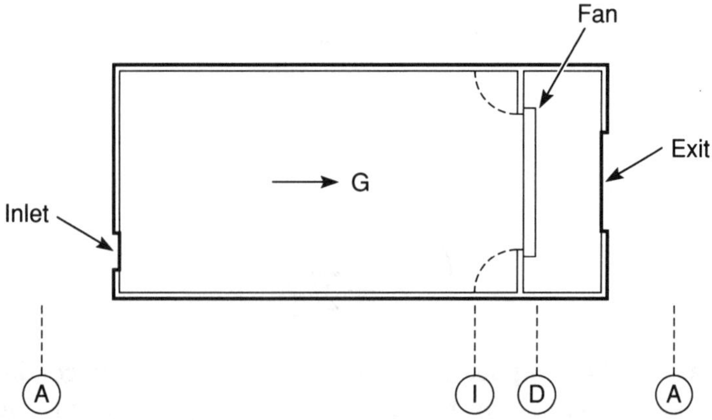

Figure 3.19 Fan-cooled enclosure with pre/post fan pressure losses [Courtesy of Tektronix, Inc.].

equation represents the frictional or irreversible loss in total pressure that occurs in all real systems.

Practical applications typically involve the use of "pressure head" rather than pressure. Fan manufacturers typically use units of millimeters or inches of water column height:

$$h_{s1} + h_{v1} = h_{s2} + h_{v2} + H_{1-2} \tag{3.60}$$

where h now represents the respective "static" and "velocity" pressure contributions to the total pressure. H is the total pressure head loss. Elevation terms have been eliminated because in most forced flow-cooled electronic systems, the two elevation terms are nearly identical. The sum of static and velocity heads is defined as total pressure head, thus

$$h_T = h_v + h_s \tag{3.61}$$

Figure 3.19 illustrates the elements of a typical electronics enclosure in which it is required to apply Bernoulli's equation. The equation is readily applied across the fan, where the bell shape at the fan entrance implies that entrance losses to the fan are contained in the fan performance specifications. Here

$$h_{TA} = h_{TI} + H_{A-I} \tag{3.62}$$

$$h_{vA} + h_{SA} = h_{TI} + H_{A-I} \tag{3.63}$$

where

h_{TA} = total head at A, the external ambient

h_{TI} = total head at I, the fan entrance (inlet)

H_{A-I} = total head loss between ambient and fan entrance

The total pressure at the external ambient is the sum of the external velocity and static heads. The external velocity head is zero and all static heads are determined with reference to the external static head. Then the total external pressure head may be set at zero. The result is

$$h_{TI} = -H_{A-I} \tag{3.64}$$

for the total pressure head at the fan entrance.

It may be shown in similar fashion that for the fan discharge

$$h_{TD} = h_{TA} + H_{D-A} \tag{3.65}$$

$$= h_{vA} + h_{SA} + H_{D-A} \tag{3.66}$$

and, again, with $h_{vA} = h_{SA} = 0$

$$h_{TD} = H_{D-A} \tag{3.67}$$

It is now necessary to define fan pressures according to ASHRAE Standard 51-75:

$$h_{fT} = \text{fan total pressure head} = h_{TD} - h_{TI} \tag{3.68}$$

$$h_{fS} = \text{fan static pressure head} = h_{fT} - h_{vD} \tag{3.69}$$

where

h_{vD} = velocity pressure head at the fan discharge

The desired result is obtained by combining the applied Bernoulli's equations with the fan pressure head definitions:

$$h_{fT} = H_{D-A} + H_{A-I} \tag{3.70}$$

or

$$h_{fT} = H - h_{vD} \tag{3.71}$$

where H is the sum of the total pressure head losses from inlet to exit, exclusive of the fan. Figure 3.20 is a graphic representation of this result.

In an actual application analysis such as would be required for the simple system shown in Fig. 3.9, H must be computed as the total pressure head loss as seen across the "nodes" associated with the fan inlet and exit. As indicated by Ellison (1989), very small problems may be solved using manual means but complex circuits consisting of ten to fifty nodes are readily encountered. Software for the steady-state analysis of such systems is based on the conservation of flow at each circuit node. This is expressed for node i with a flow source s_i and interconnecting air flow resistances R_{ij} with flow G_{ij} as

$$\sum_j G_{ij} = s_i \tag{3.72}$$

The total head loss across each element R_{ij} is

$$\Delta h_{ij} = R_{ij} G_{ij}^{N_{ij}} \tag{3.73}$$

where N_{ij} is exactly one for totally laminar flow (often characteristic of smooth ducts) and is exactly two for totally turbulent flow (turns and perforated plates, for example). The flow can be evaluated as

$$G_{ij} = \varepsilon_{ij} \left(\frac{|\Delta h_{ij}|}{R_{ij}} \right)^{1/N_{ij}}, \qquad \varepsilon_{ij} = \left\{ \begin{array}{ll} +1, & \Delta h_{ij} > 0 \\ -1, & \Delta h_{ij} < 0 \end{array} \right\} \tag{3.74}$$

and substituted into Eq. (3.72), which is the conservation equation:

$$\sum_j \varepsilon_{ij} \left(\frac{|\Delta h_{ij}|}{R_{ij}} \right)^{1/N_{ij}} = s_i \tag{3.75}$$

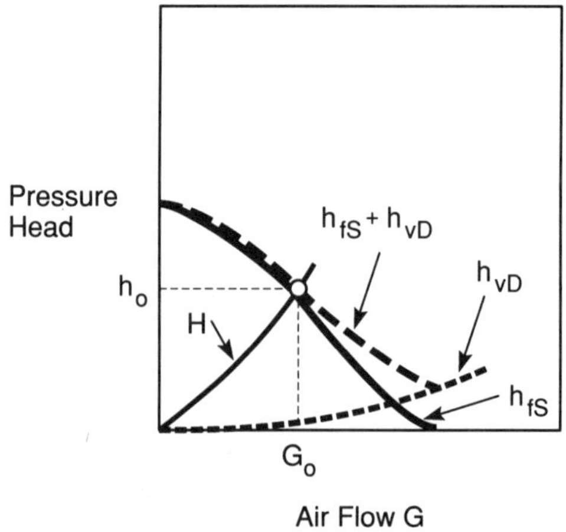

Figure 3.20 Pressure head/air flow curves for fan-cooled enclosure [Courtesy of Tektronix, Inc.].

The numerator and denominator on the left side may be multiplied by $|\Delta h_{ij}|$ to obtain:

$$\sum_j \Delta h_{ij} \frac{1}{R_{ij}^{1/N_{ij}} |\Delta h_{ij}|^{(N_{ij}-1)/N_{ij}}} = s_i \tag{3.76}$$

which expands to

$$h_i \left[\sum_j \frac{1}{R_{ij}^{1/N_{ij}} |\Delta h_{ij}|^{(N_{ij}-1)/N_{ij}}} \right]$$

$$+ \sum_j \left[\frac{1}{R_{ij}^{1/N_{ij}} |\Delta h_{ij}|^{(N_{ij}-1)/N_{ij}}} \right] h_j = s_i \tag{3.77}$$

Note that this equation defines a set of simultaneous non-linear algebraic equations in unknown h_i and h_j. A variety of equation-solving methods are applicable. The Gauss-Seidel technique, however, is not recommended because the non-linearity is sufficiently great as to make the convergence of many problems quite difficult. A Gauss-Jordan substitution method, or variations of such, iteratively applied as the matrix coefficients are updated, has been found to converge well for all but the most poorly conceived problems.

Application of a solution technique to the simultaneous equations is an iterative procedure. A first guess of the fan flow is applied to the source term s_i. The set of equations is then solved. An extrapolation, preferably quadratic, of this solution to the fan curve is next applied. The flow determined by the intersection of the extrapolated solution and the fan curve is used as the next choice of source s_i. The solution-extrapolation procedure is continued until the extrapolation and the previous source term are sufficiently identical. If all circuit elements depend quadratically on air flow, that is, all $N_{ij} = 2.0$, a single quadratic extrapolation is sufficient. If all N_{ij} are not identically 2.0, a weighted average of a linear and quadratic extrapolation is required.

3.5 Finite Difference Method — Conduction/Convection

This section is intended to demonstrate, for the sake of completeness, that the finite difference and thermal network methods are nearly identical. The difference in the two techniques lies in the method of

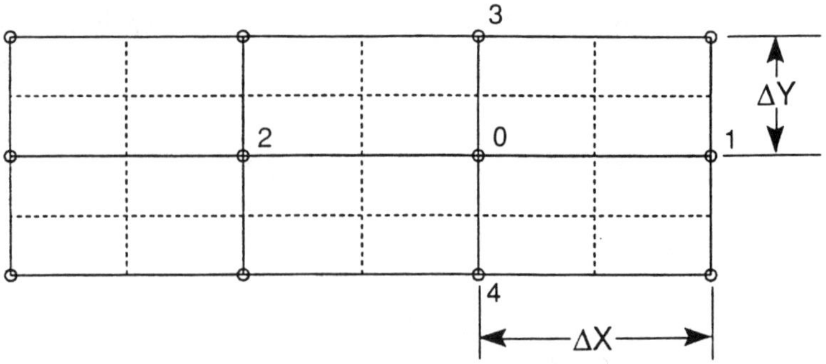

Figure 3.21 Grid/node layout for two-dimensional finite difference method [Courtesy of Tektronix, Inc.].

data input to the computer program. A true finite difference program would require the user to specify the grid mesh in terms of the number of elements in each of the coordinate directions, whereas a thermal network analyzer would take input in terms of conductances. The following paragraphs develop the basic steady-state thermal equations by putting the partial differential equation for heat conduction into a finite difference form. It is then demonstrated that the equations are equivalent to a thermal network system.

A portion of a two-dimensional grid in Fig. 3.21 indicates a mesh in the x and y dimensions. The plane is assumed to have a uniform thickness, Δz. The partial differential equation for this problem with a heat generation per unit volume of Q_V and temperature independent thermal conductivity k is

$$\frac{\partial^2 T}{\partial x^2} + \frac{\partial^2 T}{\partial y^2} = -\frac{Q_V}{k} \tag{3.78}$$

The finite difference approximation is developed around "node-0."

The first forward finite difference of node-0 in the x-direction is

$$\frac{\partial T}{\partial x} \cong \frac{T_1 - T_0}{\Delta x} \tag{3.79}$$

The first backward difference of node-0 in the x-direction is

$$\frac{\partial T}{\partial x} \cong \frac{T_0 - T_2}{\Delta x} \tag{3.80}$$

With similar differences in the y-direction the central differences of the derivatives for the x and y-directions are

$$\frac{\partial^2 T}{\partial x^2} \cong \frac{(\partial T/\partial x)_{0-1} - (\partial T/\partial x)_{2-0}}{\Delta x} \cong \frac{(T_1 + T_2 - 2T_0)}{\Delta x^2} \tag{3.81}$$

$$\frac{\partial^2 T}{\partial y^2} \cong \frac{(\partial T/\partial y)_{0-3} - (\partial T/\partial y)_{4-0}}{\Delta y} \cong \frac{(T_3 + T_4 - 2T_0)}{\Delta y^2} \tag{3.82}$$

respectively.

Substituting these central differences and the heat dissipation

$$Q_0 \cong Q_V \Delta x \Delta y \Delta z \tag{3.83}$$

into the partial differential equation and rearranging, gives

$$\left(\frac{k\Delta y\Delta z}{\Delta x}\right)(T_1 - T_0) + \left(\frac{k\Delta y\Delta z}{\Delta x}\right)(T_2 - T_0) + \left(\frac{k\Delta x\Delta z}{\Delta y}\right)(T_3 - T_0) + \left(\frac{k\Delta x\Delta z}{\Delta y}\right)(T_4 - T_0) = -Q_0 \tag{3.84}$$

Each factor of the temperature differences is readily noted to be identical to a thermal conductance, thus the following representation is possible:

$$Q_0 = K_{0-1}(T_0 - T_1) + K_{0-2}(T_0 - T_2) + K_{0-3}(T_0 - T_3) + K_{0-4}(T_0 - T_4) \tag{3.85}$$

If a surface boundary condition is set at a node such as node-1, the source dissipation term is transformed:

$$Q_0 \rightarrow Q_0 - h\Delta y\Delta z(T_0 - T_A) \tag{3.86}$$

so that the final system of equations becomes

$$Q_i = \sum_j K_{ij}(T_i - T_j) \tag{3.87}$$

where both conduction terms, typically of the form:

$$K_{ij} = \frac{k\Delta y\Delta z}{\Delta x} \tag{3.88}$$

and convection terms of the form:

$$K_{ij} = h\Delta y\Delta z \tag{3.89}$$

apply where appropriate. Clearly then, the finite difference method results in a set of equations identical to those of the thermal network.

3.6 Finite Element Analysis — Conduction/Convection

The finite element technique is a very powerful and commonly used method. However, existing and potential users are often unaware of the mathematical foundations of the method. This section provides a rudimentary introduction to the technique. The exposition is given in such a manner as to provide the reader with an appreciation of the basic difference between finite element methods (often referred to as FEM) and finite difference methods (FDM) or the thermal network method. There are numerous texts that may be referenced for a more detailed account [see, for example, Segerlind (1976), Reddy (1984), Burnett (1987), and Hsu (1990)].

The example of one-dimensional heat transfer in a solid conducting bar of uniform cross-section and thermal conductivity will be used (Fig. 3.22a). The bar is insulated over its length. A heat flux, q_0, is injected at $x = 0$ and at the end at $x = L$, heat flows by convection to an ambient temperature, T_∞. The end surface area A_c has an associated heat transfer coefficient h.

The variational calculus method will be used, first to show that the differential equation for heat conduction with boundary conditions can be derived from the method. An alternate path in the variational development will be shown as leading to the formulation

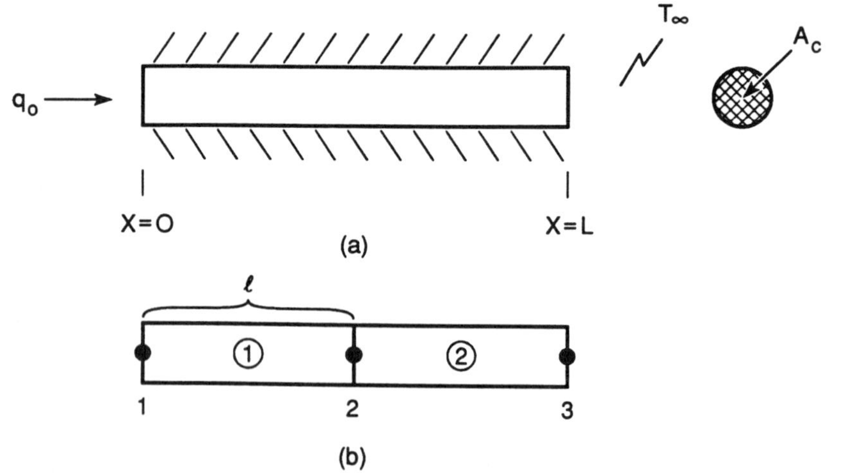

Figure 3.22 Finite element representation of heat transfer in a bar. (a) problem geometry (b) two-element model [Courtesy of Tektronix, Inc.].

of a set of equations that can be solved for the temperature distribution in the specified geometry.

The "functional" F is proposed as the sum of volume and surface integrals where T and T_x are temperature and temperature gradients, respectively:

$$F = \int_V f_V(T, T_x, x)dv + \int_S f_S(T, T_x, x)ds \tag{3.90}$$

In the case of the present problem, the volume and surface functions f_V and f_S are "proposed" to be

$$f_V = \frac{1}{2}k\left(\frac{dT}{dx}\right)^2 - Q_V T \tag{3.91}$$

$$f_S = qT + \frac{1}{2}h(T - T_\infty)^2 \tag{3.92}$$

for Q_V, the heat dissipation per unit volume and q, the heat dissipation per unit area.

Justification of this choice will be the proof that the forcing of F, defined by Eq. (3.90), to be a stationary condition (minimum or maximum) will lead to the correct differential equation and boundary conditions for steady-state conduction.

F is made stationary by setting the variation of F to zero and following through with the variational calculus steps. Thus,

$$\delta F = 0 \tag{3.93}$$

$$\delta F = \int_V \delta f_V dv + \int_S \delta f_S ds \tag{3.94}$$

$$= A_C \int_0^L \left(\frac{\partial f_V}{\partial T_x}\delta T_x + \frac{\partial f_V}{\partial T}\delta T\right) dx + \int_S \left(\frac{\partial f_S}{\partial T}\delta T + \frac{\partial f_S}{\partial T_x}\delta T_x\right) ds \tag{3.95}$$

$$= A_C \int_0^L \left(k\frac{dT}{dx}\delta T_x - Q_V \delta T\right) dx + \int_S [q + h(T - T_\infty)]\delta T ds \tag{3.96}$$

$$= A_C \int_0^L \left[k\frac{dT}{dx}\frac{d}{dx}(\delta T) - Q_V \delta T\right] dx + \int_S [q + h(T - T_\infty)]\delta T ds \tag{3.97}$$

The surface integral is immediately simplified by recognizing that q and h are zero except at $x = 0$ and $x = L$. Integrating the first term in the first integral by parts yields

$$
\begin{aligned}
\delta F &= A_C \int_0^L \left(-k\frac{d^2T}{dx^2} - Q_V \right) \delta T dx + k\frac{dT}{dx} A_C \delta T \Big|_0^L \\
&\quad + \left\{ \left[q + h(T - T_\infty) \right] A_C \delta T \right\} \Big|_0^L \qquad (3.98) \\
&= -A_C \int_0^L \left(k\frac{d^2T}{dx^2} + Q_V \right) \delta T dx \\
&\quad + A_C \delta T \left[k\frac{dT}{dx} + q + h(T - T_\infty) \right]_0^L \qquad (3.99)
\end{aligned}
$$

Because variational calculus permits the variation in T to be arbitrary over the path, and, if the variation in T is not required to be zero at the ends of the bar, the following conditions are obtained for stationary F:

$$\frac{d^2T}{dx^2} = -\frac{Q_V}{k}, \qquad (3.100)$$

$$\left[k\frac{dT}{dx} + q + h(T - T_\infty) \right]_{x=0,L} = 0 \qquad (3.101)$$

The first condition is the differential equation for heat transfer in one dimension, and the second is the set of boundary conditions.

For the problem at hand, boundary conditions are developed to be

$$h = 0, \quad -k\frac{dT}{dx} = q_0 \qquad x = 0 \qquad (3.102)$$

$$q = 0, \quad -k\frac{dT}{dx} = h(T - T_\infty) \qquad x = L \qquad (3.103)$$

The desired result has been achieved: the given functional, when required to be stationary, results in the differential equation and boundary conditions for the problem. The next step is to show that the stationary requirement on this same functional may be used in a somewhat different fashion to result in a set of "finite element" equations that could be solved for the temperature distribution in the bar.

If, in this case, it is correctly presumed that the internal heat dissipation Q_V is zero, the functional is again

$$F = \int_V \frac{1}{2} k \left(\frac{dT}{dx}\right)^2 dx + \int_S \left[qT + \frac{1}{2} h (T - T_\infty)^2\right] ds \qquad (3.104)$$

The first step that actually represents a finite element aspect of the physical problem is that the bar is now broken down into discrete pieces of length ℓ (Fig. 3.22b). In this example, only two elements will be used for the sake of simplicity. The standard method is to represent the temperature distribution in each element by approximating "polynomials" (linear here) $N_1^{(1)}$, $N_2^{(1)}$, $N_2^{(2)}$, and $N_3^{(2)}$. The subscripts are associated with node numbers and the superscripts are associated with element numbers. The temperatures in elements (1) and (2) are represented by two different approximating polynomials:

$$T^{(1)} = N_1^{(1)} T_1 + N_2^{(1)} T_2 \qquad (3.105)$$

$$N_1^{(1)} = \frac{x_2 - x}{\ell} \qquad (3.106)$$

$$N_2^{(1)} = \frac{x - x_1}{\ell} \qquad (3.107)$$

$$T^{(2)} = N_2^{(2)} T_2 + N_3^{(2)} T_3 \qquad (3.108)$$

$$N_2^{(2)} = \frac{x_3 - x}{\ell} \qquad (3.109)$$

$$N_3^{(2)} = \frac{x - x_2}{\ell} \qquad (3.110)$$

where x_1, x_2, and x_3 are the coordinates of nodes 1, 2, and 3.

The integration required for the functional is now performed over each individual element using $T^{(1)}$ and $T^{(2)}$ for the appropriate x. Omitting the algebra, the result is

$$F = \frac{K^{(1)}}{2}(T_2 - T_1)^2 + \frac{K^{(2)}}{2}(T_3 - T_2)^2 + q_0 A_C T_1 + \frac{hA_C}{2}(T_3 - T_\infty)^2 \qquad (3.111)$$

where

$$K^{(1)} = K^{(2)} = kA_C/\ell \qquad (3.112)$$

Finally, the step is taken that requires the functional F to be stationary. It is proposed that such a condition will transpire if, and only if, F is minimum with respect to each independent variable, in this case, the nodal temperatures:

$$\frac{\partial F}{\partial T_1} = -K^{(1)}(T_2 - T_1) - q_0 A_C = 0 \tag{3.113}$$

$$\frac{\partial F}{\partial T_2} = K^{(1)}(T_2 - T_1) - K^{(2)}(T_3 - T_2) = 0 \tag{3.114}$$

$$\frac{\partial F}{\partial T_3} = K^{(2)}(T_3 - T_2) + hA_C(T_3 - T_\infty) = 0 \tag{3.115}$$

The preceding may be reformulated as a set of simultaneous equations with temperatures T as the unknowns:

$$\begin{bmatrix} K^{(1)} & -K^{(1)} & 0 \\ -K^{(1)} & (K^{(1)} + K^{(2)}) & -K^{(2)} \\ 0 & -K^{(2)} & (K^{(2)} + hA_C) \end{bmatrix} \begin{bmatrix} T_1 \\ T_2 \\ T_3 \end{bmatrix} = \begin{bmatrix} q_0 A_c \\ 0 \\ hA_C T_\infty \end{bmatrix} \tag{3.116}$$

Using more conventional matrix algebra notation

$$[K]\,[T] = [s] \tag{3.117}$$

The matrices $[K]$ and $[s]$ are conductance and source matrices, respectively. However, more consistent with common terminology used in the FEM (and originating in structural analysis), the matrices are referred to as the global stiffness and force matrices. Clearly, the only remaining requirement would be to input numerical geometric and physical constants and then select a numerical procedure to solve for the temperatures. Further discussion of these and the appropriate programming procedures will not be presented here.

No discussion of the finite element method would be complete without at least mentioning the "boundary element method" (BEM). BEM has become a powerful alternative to FEM. The principal feature of BEM is that only the surface (boundary for 2-D) of a volume must be discretized. The simplification of the initial modeling process as well as later model modification is therefore of great significance as a productivity issue. The reader is referred to works such as Brebbia and Dominquez (1989) for further reading on this subject.

3.7 Experimental Methods

Experimental methods play an important role in all aspects of heat transfer, including the electronics cooling sub-discipline. In an era of rapidly increasing computer power per unit cost and sophisticated computer graphics displays, the need for experimental studies is often overlooked. It might even be suggested that in the intense competition for limited funds, the "high cost" of computer graphics displays with animation and other impressive effects leads to difficulty for experimentalists in generating the same level of interest in their projects. It is not unusual for an engineer with predominantly computational interests to never consider the issue of correlation between calculation and experiment.

No matter what the interest or time constraints of the engineer, some effort must be devoted to ascertaining the accuracy of any given analysis procedure. If, after a series of analyses and tests, it is concluded that the analysis fails to yield agreement with test results, the model may possibly be modified in a judicious manner to more accurately reflect reality.

Finally, there are critical parameters that are difficult to calculate from first principles with sufficient accuracy. In particular, computational fluid dynamics (CFD) has made remarkable progress in recent years, as well as significant inroads into the industrial arena. On the other hand, as shown by Anderson and Moffat (1990), the experimental determination of convection correlations for components on printed circuit boards is a very active research field.

The laboratory of the practicing thermal analyst need not be particularly elegant. A stock of thermocouples and a digital thermometer may be all that is required. Equipment designers working with forced air cooled systems would do well to equip themselves with a setup that collects and measures volumetric air flow (see Fig. 3.23). Such a system is useful in measuring actual operating flow to or from a packaged system. The same system may be used to interface with a test section designed to provide experimental data for various device types. Figure 3.24 illustrates a Plexiglas test section designed for measuring the thermal resistance of small extruded heat sinks mounted on PCB-type panels. The housing is fabricated to permit simulation of board separation and various flow rates. The resultant experimental data may be plotted so as to be used as design guidelines or even as element data in a thermal network model.

Figure 3.23 Air flow test system for measuring flow rates of 2 to 475 L/sec [Courtesy of Airflow Measurement Systems].

4 THERMAL ANALYSIS APPLICATIONS

4.1 Introduction

The following examples are intended to illustrate most of the analysis methods discussed in the previous pages. Complexity is sacrificed to permit an adequate description of the problem. The reader will be able to better ascertain the relative merits of the various analysis techniques after reviewing this section.

Four different component examples are considered: (1) the analysis of a simple flat pack using the thermal network method is reviewed; (2) a rather complex hybrid circuit package consisting of an alumina substrate/adhesive/aluminum heat spreader is modeled using the Fourier series closed form technique; (3) two power transistors on a heat sinking flat chassis panel provide an excellent opportunity to compare results from the Fourier series closed form method, the thermal network method, and the finite element method and (4) a final component problem demonstrating procedures that allow one to analyze a finned heat sink extrusion using the flat plate Fourier series method.

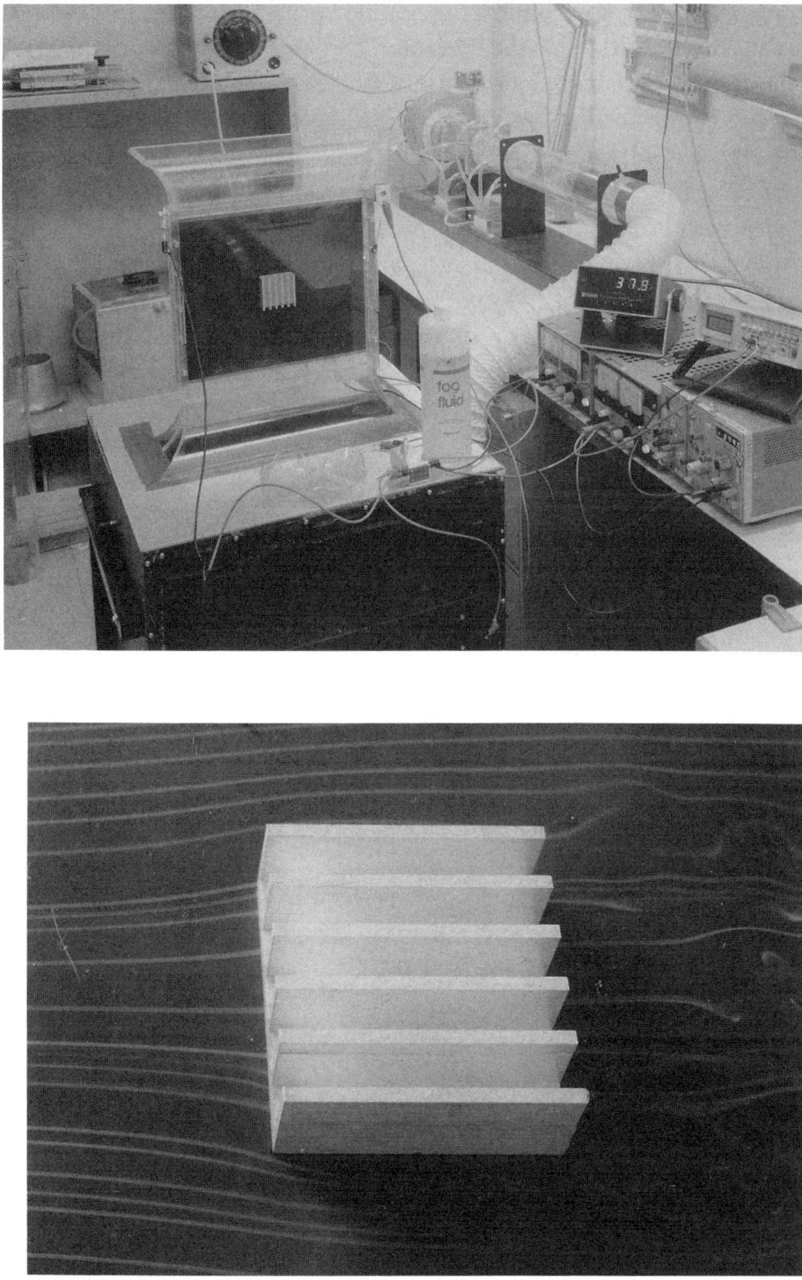

Figure 3.24 Experimental apparatus for evaluating the thermal characteristics of a forced air cooled heat sink on a board [Courtesy of Tektronix].

Various electronics packaging configurations are used to illustrate systems applications of the thermal and air flow network analysis methods developed in Sections 3.3 and 3.4. Three different enclosures are considered: (1) a small single-chamber, single circuit board box with and without venting; (2) a multi-chambered enclosure resembling a low packaging density, natural convection/radiation/air draft cooled, laboratory type electronics system and (3) a forced air-cooled enclosure with several internal heat sink extrusions. Results from examples (1) and (3) are compared with experimental data.

Finally, the applications section is concluded with a comparison of calculated and experimental results for a printed circuit board populated with 56 dual-in-line resistor packages.

4.2 Component Examples

16-Lead Ceramic Flat Pack (Thermal Network). The external geometry associated with this example of a typical flat pack is illustrated in Fig. 3.25. This package consists of a sandwich structure with eight leads on each of the two long dimensions. The base is a solid, uniformly thick (0.051 cm) ceramic slab. The next layer is a rectangular annulus placed on the top of the base, creating an internal cavity for the chip. A lead frame/glass layer follows the ceramic annulus, creating additional internal cavity height. The package is topped off with a ceramic lid extending out to the full dimensions of the component.

Heat transfer from a flat pack is via (1) the leads to the PCB, (2) radiation and air conduction from the package base to the PCB and (3) radiation and convection from the package surface (bottom excluded) to the local ambient air.

A portion of the package model is illustrated in Fig. 3.26, showing a planar view (a) with nodal structure as seen from the top of the package. A few of the thermal conductances interconnecting nodes are also shown. The base and lid utilize the full nodal structure shown, and the annular region contains nodes only around the periphery (best shown in the end view). Major heat transfer paths (b) are also partially shown.

The general problem usually requires thermal resistance or conductance calculation prior to data entry. The conductance for conduction is the inverse of Eq. (3.39):

$$K_{ij} = kA_k/L \tag{3.118}$$

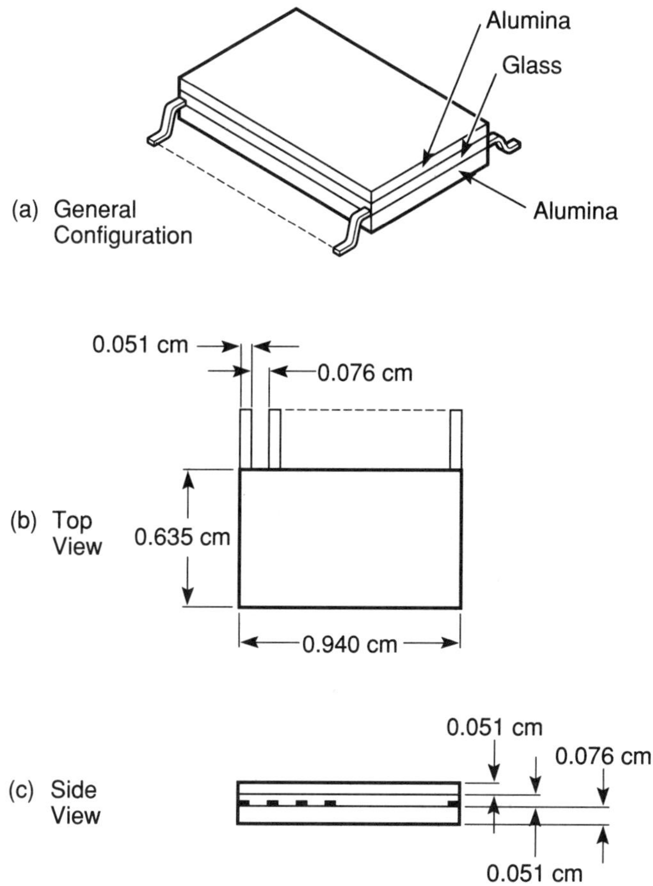

Figure 3.25 Typical 16-lead ceramic flat pack: (a) general configuration, (b) top view, (c) side view [Courtesy of Tektronix, Inc.].

A_k is the cross-sectional area for the heat flow path, which, for a rectangular node, is the product of the node thickness and width and L is the distance between nodes. The conductance for a single lead from package to board also utilizes Eq. (3.118), in which case A_k is the cross-sectional area of the lead and L is the lead length.

Convection and radiation conductances are handled according to the requirements of the program. Typically, a program would require convection/conductance input $K_{ij} = h_C A_S$. The minimal radiation input would consist of surface emissivity and nodal surface area.

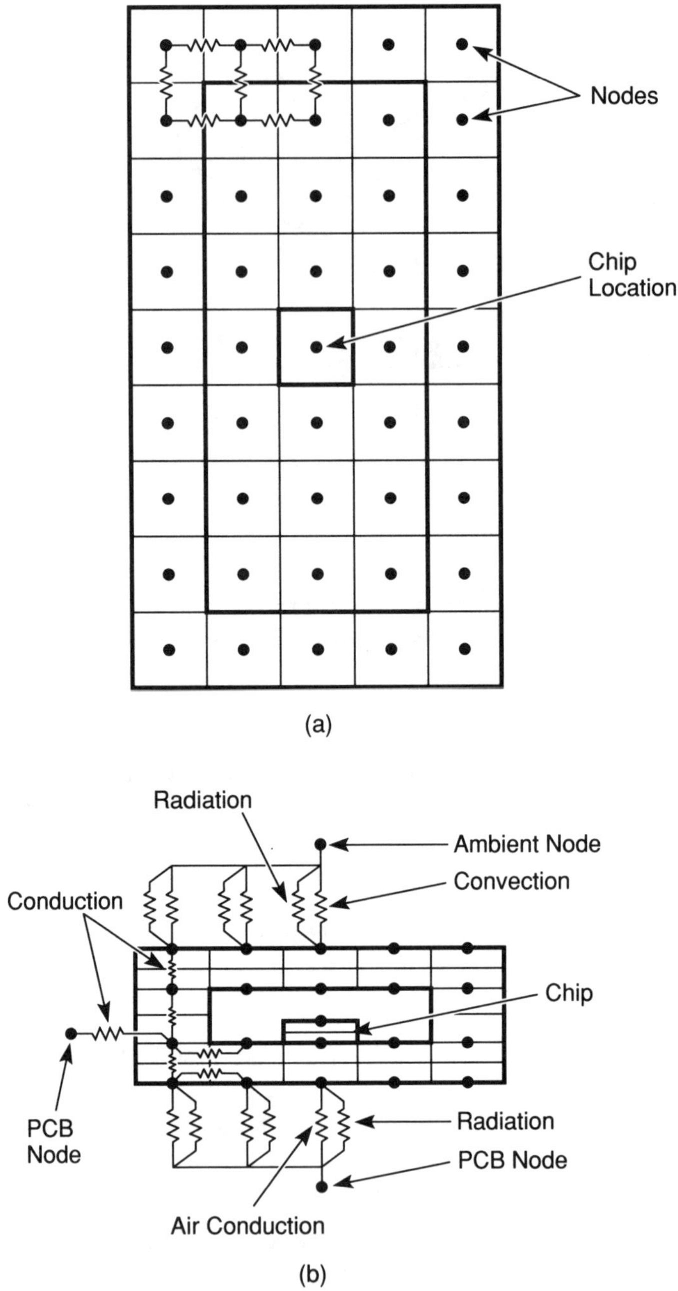

Figure 3.26 Thermal network model of 16-lead ceramic flat-pack. (a) top view, (b) end view [Courtesy of Tektronix, Inc.].

Any one of a number of commercially available thermal network analyzers may be used to solve this problem. The details of data entry are not relevant here except to say that these types of programs typically require the use of text editors for data entry. Special-purpose modules are sometimes available for special problem classes with specific geometry.

Minimum output from a thermal network analyzer consists of a table of temperatures for each node in the structure. Also desirable is the heat transfer between the various nodes.

Hybrid Package with Heat Spreader (Closed Form). This example illustrates the usefulness of the closed form analytical method outlined in Section 3.2. The hybrid geometry is illustrated in Fig. 3.27. The circuit board has components mounted on both sides. Heat dissipating IC chips are represented by four sources with dimensions of 0.127 cm × 0.127 cm, and two with dimensions of 0.508 cm × 0.508 cm. These sources simulate heat leaving the die attach base. Actual chip junction surface temperature differences must be separately calculated and added to the results obtained from the present analysis.

The package consists of a sandwich structure formed from the ceramic substrate-epoxy adhesive-aluminum heat spreader with thicknesses of 0.102, 0.005, 0.152 cm and thermal conductivities of 0.39, 0.0157, 1.58 W/cm −°C, respectively. The package is inverted such that the hybrid cover faces the circuit board. Interconnection between the hybrid and PCB is accomplished via twenty leads on each of the two long edges.

The thermal model of the real structure assumes that the major heat flow paths involve substrate conduction to the heat spreader and leads, convection from the heat spreader to the local air, lead conduction to the PCB, and PCB conduction/convection to the local air.

The data preparation and program input for closed form methods is usually quite simple. In this case, much of the required input consists of plate dimensions, layer thickness, thermal conductivities, and heat source location and dissipation. The heat transfer coefficients for the 5.33 cm × 2.54 cm surfaces of the plate are provided to represent forced convection from the heat spreader for a moderate air velocity ($h = 0.0016$ W/cm^2 −°C) and neglect heat loss from the

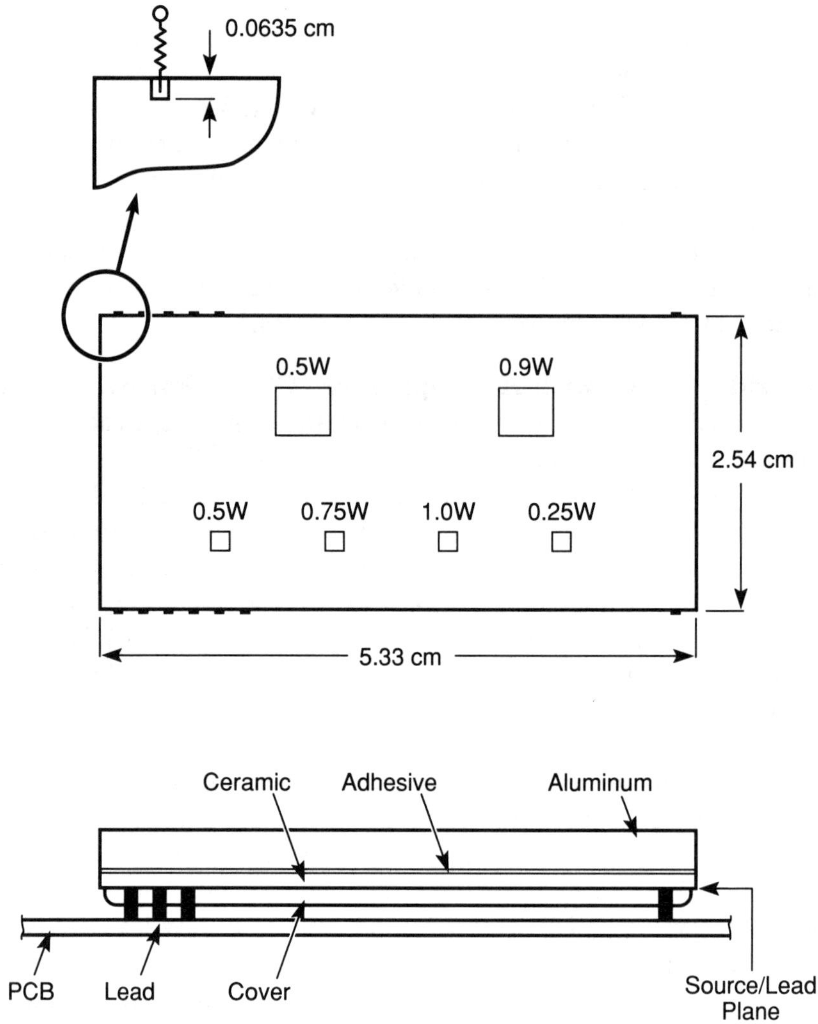

Figure 3.27 Geometry of hybrid package with heat spreader [Courtesy of Tektronix, Inc.].

source plane ($h = 1.0 \times 10^{-10}$ W/cm^2 $-$°C). The latter is justified by assuming that components on both sides of the board permit the board center line to be viewed as a plane of symmetry.

Accurate heat loss modeling via the package leads and PCB structure is complex with any analysis method. A very approximate thermal resistance for the conduction through a single lead, board conduction, and board convection resistance of 1000° C/W will be used for this problem. The side-brazed leads are modeled in the program by thermal resistances that remove heat from a 0.0635 cm × 0.0635 cm region on the ceramic surface (shown in Fig. 3.27).

As indicated in Section 3.2, the Fourier series x, y representation of surface heat sources and thermal resistances requires consideration of the number of terms at which the series should be truncated. The smallest area, whether it be a source or resistor region, is used to determine the number of terms. The resistors represent the smallest area in this problem. An analysis of the convergence issue is accomplished by computing the thermal resistance (temperature rise for one watt) for a 0.0635 cm × 0.0635 cm heat source versus the number of terms for each direction. The results for this analysis are shown in Fig. 3.28. The graph shows that 100 terms per direction is adequate, although 150 terms were selected for the full problem calculation. The reader is reminded that since the series expansion is in both the x and y directions, the actual number of terms is $150 \times 150 = 22,500$.

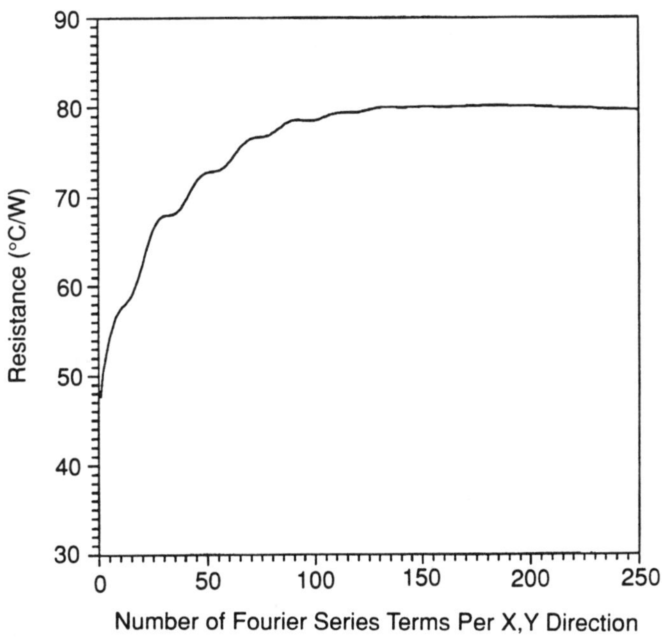

Figure 3.28 Convergence analysis of hybrid package with heat spreader [Courtesy of Tektronix, Inc.].

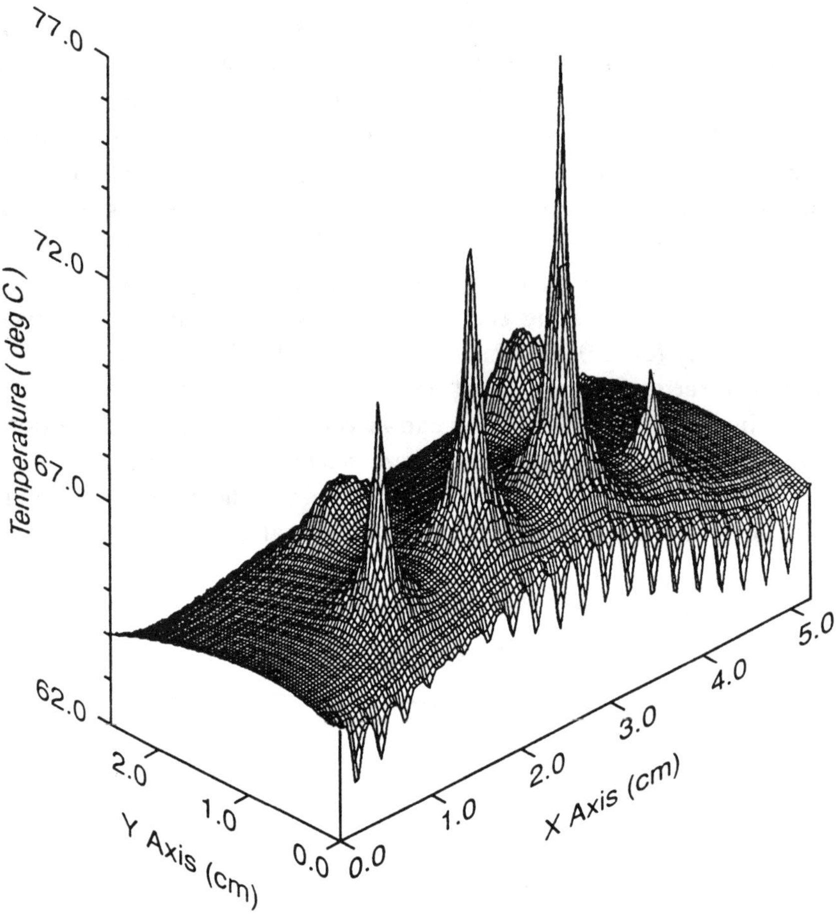

Figure 3.29 3-D mapping of hybrid results [Courtesy of Tektronix, Inc.].

The final solution is displayed in Fig. 3.29 as a three-dimensional contour map of substrate surface temperature difference above the ambient. A commercial software package, SURFER, developed by Golden Software [12] was used to generate the display from an x, y grid on the substrate surface.[1] The grid consisted of a coarse 40×21 point mesh with a finer mesh over the four small heat sources. This finer grid of 5×5 at the small sources extended over an area nine times the size of each source. The slight roughness of the temperature surface at the sources may be due to a combination of the effect of Fourier series truncation at less than an optimum number of

[1]Software references are designated by numerals within brackets.

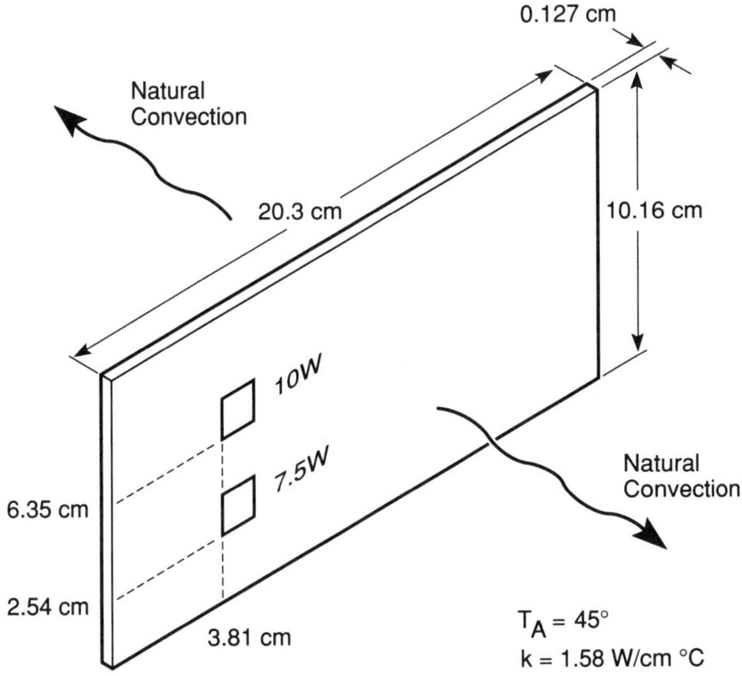

Figure 3.30 Power transistors on aluminum chassis panel [Courtesy of Tektronix, Inc.].

terms for all sources and resistors, Fourier series problems with the discontinuity in Q_V at each source edge (Gibb's phenomenon) plus the effects of the gridding operations in the graphics display package. The interaction effects of the heat sources are rather nicely indicated by this presentation technique.

Power Transistors on Panel (Closed Form, Thermal Network, FEM). This problem illustrates the application of three different solution methods: closed form, thermal network, and finite element method, to a simple but not uncommon problem encountered by an electronics packaging engineer. A flat, vertically-oriented, natural convection-cooled chassis panel is used to dissipate heat from two TO-220 power transistor packages (Fig. 3.30). The heat dissipation from the base of the transistor packages is represented by two 1.27 cm × 1.27 cm surface heat sources.

The average natural convection heat transfer coefficient may be calculated from a relationship recommended by Ellison (1989):

$$h_C = 4.73 \times 10^{-4} (\Delta T/H)^{0.35} \text{ W/cm}^2 - ^\circ\text{C} \tag{3.119}$$

for a plate height, H (cm), and average surface temperature rise above ambient ΔT.

The total heat dissipation (17.5 W) from the plate with total area A_S is then Eq. (3.40):

$$Q_C = A_S h_C \Delta T$$

Substitution of Eq. (3.119) into Eq. (3.40) permits solution for ΔT. The average temperature rise may be determined as

$$\Delta T = 51^\circ\text{C}$$

This value of ΔT is then substituted back into Eq. (3.119). The convective heat transfer coefficient is easily computed to be

$$h_C = 8.37 \times 10^{-4} \text{ W/cm}^2 - ^\circ\text{C}$$

The closed form model solution is again based on the Fourier series development of Section 3.2. The required program input consists of the plate thickness, length, width, thermal conductivity, convective heat transfer coefficients, and the location, dimensions, and dissipation of the two heat sources. Solution of the problem was accomplished with the TAMS program [18] found in Ellison (1989).

The thermal network model for the problem is illustrated in Fig. 3.31. A large number of grid combinations is possible and a straightforward and adequate single layer nodal layout is shown in the illustration. The single layer of nodes through the plate thickness neglects thermal gradients in this direction. Convection from plate to ambient is represented by convection conductances to nodes representing the ambient temperature.

The values of $k = 1.58$ W/cm $-^\circ$C and $h_C = 8.37 \times 10^{-4}$ W/cm^2$-^\circ$C that are used in the closed form solution model are also used here. The conduction conductances are calculated from

$$K_{ij} = kA_k/L = kWt/L \tag{3.120}$$

for a conduction path cross-sectional area equal to path width times the path thickness. In the case of interconnecting square nodes,

$W = L$, and $K_{ij} = kt = (1.58)(0.127) = 0.201$ W/°C. Conductances between non-square nodes are calculated using the appropriate L and W.

The convection conductance for each surface element is

$$K_{ij} = h_c A_s = h_c W L \tag{3.121}$$

For a square element with $W = L = 0.635$ cm,

$$K_{ij} = (8.37 \times 10^{-4}\ \text{W/cm}^2 - ^\circ\text{C})(0.635)^2 = .000337\ \text{W/}^\circ\text{C}$$

The solution of the thermal network problem is readily accomplished by any one of several commercial codes. Programming of this particular problem without the frills usually found in more general-purpose programs is quite simple. However, the TNETFA program [18] provided by Ellison (1989) was used for the actual problem.

A general-purpose code was used to solve the finite element model for the problem. Thermal constants identical to the other problem solutions were used. A finite element mesh identical to Fig. 3.31 was used except that the nodes were at the corners of each rectangle.

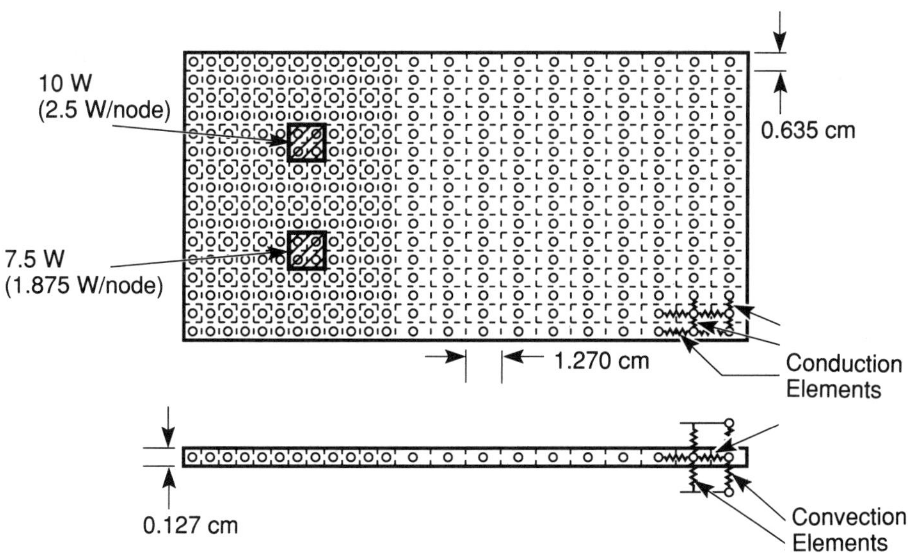

Figure 3.31 Thermal network geometry for 10 × 20 cm chassis panel [Courtesy of Tektronix, Inc.].

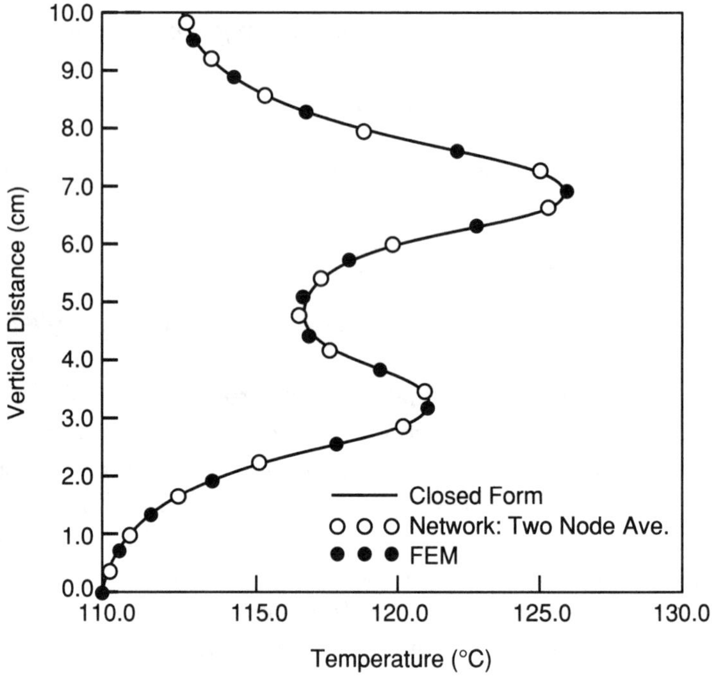

Figure 3.32 Comparison of profiles through sources [Courtesy of Tektronix, Inc.].

In addition, the particular code used, ALGOR Supersap [1], permits only one convection surface for each eight-noded brick element. For this reason, a two-element layer model was used. A personal computer version of the program was used to solve the problem.

A plot of the results for all three solution techniques is shown in Fig. 3.32 in terms of a temperature profile taken in the vertical direction. The closed form solution profile was plotted from a 400-point calculation. The curve shown has no mathematical smoothing and is therefore the exact solution for a uniform h. The solid curve is a profile directly through the source centers (x = 4.45 cm, y = variable).

Thermal network problems naturally compute temperatures at node points. Therefore, the profile temperatures are directly available from this solution only through the node points. The profile representing $x = 4.45$ is an average of the two nodal temperatures on either side of $x = 4.45$.

The finite element results were available as contour lines, contour bands, or temperatures directly at the element corners. The latter provides the most exact extraction of results from the output. The FEM profile is therefore taken along the line of nodes at exactly $x = 4.45$. Clearly the correlation between the methods is excellent.

Power Transistors on Finned Heat Sink. The geometry for the device under consideration is displayed in Fig. 3.33. Four TO-220 transistors dissipate a total of 44 watts and are arranged in a column format. This problem is typical of those encountered by power supply designers.

The most common approach to this problem would be to use a finite element or thermal network analyzer. A detailed analysis would encompass the total structure including the 0.508 cm thick base and the ten 0.254 cm thick fins. Convective and radiative heat transfer coefficients (with correction due to emissivity and fin shielding effects) would be applied to the fin and relevant base surfaces for heat transfer to the 20° C surrounding ambient. This straightforward application of a finite numerical analysis approach is not necessarily the most practical. Generation of the grid for the complete structure is a somewhat time-consuming process. In addition, it must be remembered that the accuracy of the convective heat transfer coefficient and, to a lesser extent, the surface emissivity and thermal conductivity of the extrusion, provide a basic limitation to the accuracy of the final results. Consideration of appropriate time-saving approximation issues is therefore appropriate.

Even if the analyst has access to limited tools, various levels of approximation may be useful. A manual calculation of the average heat sink temperature may be within 30% or better of a more extensive analysis, depending on the relative contributions of conduction, convection, and radiation. An easily achieved higher level of approximation would be to construct a model that does not directly include the fin structure in the conduction analysis. This model would be essentially that of a flat plate. The extended surface area of the fin structure would be accommodated by appropriately enhancing the total heat transfer coefficient of the finned side of the otherwise flat plate. Fin base-tip temperature gradient effects are incorporated via well known simple relationships. The following methods illustrate the application of this approximation scheme. A closed form method [TAMS program [18], see Ellison (1989)] is selected as the most expeditious technique for the purposes intended. The only calculation requiring explanation then becomes the total heat transfer coefficients.

The heat sink in Fig. 3.33 is assumed to transfer heat from both the source and finned side to ambient. It will be presumed that the

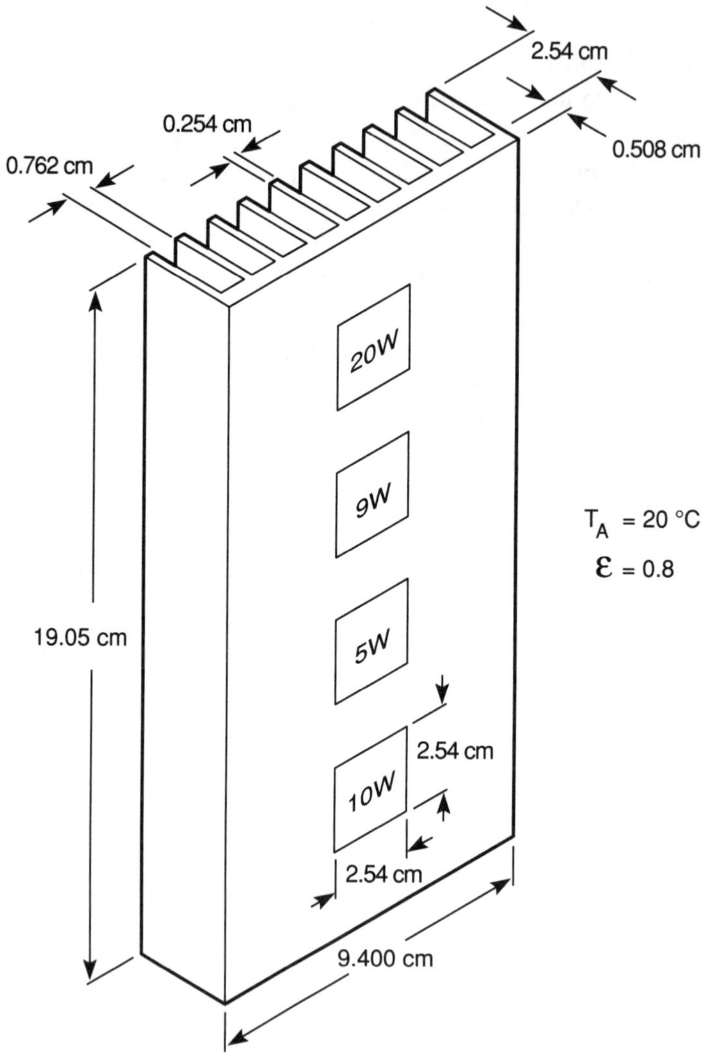

Figure 3.33 Extruded heat sink with square heat sources representing four TO-3 power transistors [Courtesy of Tektronix, Inc.].

desired goal is to determine the total power dissipation that may be placed on the flat side of the sink (in the form of four TO-4 transistor packages). This is 44 watts with the illustrated distribution. It is also desired that no heat sink transistor site exceed about 80° C. If such is not the case, the heat sink would presumably have to be re-designed or the power dissipation reduced.

The source side total heat transfer coefficient h_T is calculated first. An appropriate convective heat transfer coefficient for a vertical flat plate is [see Ellison (1989), page 36, E2.13]

$$h_H = 4.70 \times 10^{-4}(\Delta T/H)^{0.25} \text{ W/cm}^2 - {}^\circ\text{C} \tag{3.122}$$

for a plate of height H in centimeters. Using $\Delta T = 50^\circ$ C and $H = 19.05$ cm,

$$h_H = 5.98 \times 10^{-4} \text{ W/cm}^2 - {}^\circ\text{C}$$

The radiation heat transfer coefficient h_r is given by Eq. (3.47) with a shape factor of unity

$$h_r = \varepsilon 5.67 \times 10^{-12}(T_s'^3 + T_s'^2 T_A' + T_s' T_A'^2 + T_A'^3) \text{ W/cm}^2 - {}^\circ\text{C} \tag{3.123}$$

where $T' = T + 273$. Using $T_s = 70^\circ$ C and $T_A = 20^\circ$ C,

$$h_r = 7.34 \times 10^{-4}\varepsilon \text{ W/cm}^2 - {}^\circ\text{C}$$

The source side thermal conductance K_S is now calculated using not only the planar area equal to the width times the height, but also the left and right edge base area contributions, the latter admittedly a very small contribution:

$$\begin{aligned} K_S &= h_{ST} A_S \\ &= (h_H + h_r)(W + 2t)H \qquad (3.124) \\ &= (5.98 \times 10^{-4} + 0.8 \times 7.34 \times 10^{-4})(9.4 + 2 \times 0.5)(19.1) \\ &= (0.00119)(198.64) \\ &= 0.236 \text{ W/}^\circ\text{C} \end{aligned}$$

An enhanced total heat transfer coefficient h_{Se} is readily computed using K_S and the heat sink base area:

$$\begin{aligned} h_{Se} &= K_S/WH = 0.236/(9.4 \times 19.1) \\ &= 0.0013 \text{ W/cm}^2 \ {}^\circ\text{C} \qquad (3.125) \end{aligned}$$

If the base area edge contributions ($2tH$) were omitted from Eq. (3.124), the non-enhanced total $h = h_H + h_r = 0.0012$ would be only slightly less than the enhanced value (0.0013).

The calculation of the total h for the finned side is complicated by air choking and fin radiation shielding effects in the extrusion channels. The average convection coefficient h_C in the channels and the radiation h_r contribute to the finned side conductance in the expected manner:

$$K_f = [h_C + (\mathcal{F}/\varepsilon)h_r]A_i + (h_H + h_r)A_0 \tag{3.126}$$

for an interior fin channel area A_i and an exterior fin area A_0. $\mathcal{F}$ is a shielding, emmissivity factor dependent on the fin geometry via the geometric configuration factor F from each fin channel to ambient. The factor, $\mathcal{F}$, is also dependent on the heat sink emissivity.

Both h_C and $\mathcal{F}$ are readily estimated using methods developed by Ellison (1989) and will not be considered further here other than in the presentation of the results. The coefficient, h_C, is found to be 70% of the full vertical flat plate value, thus

$$h_C = 0.7(5.98 \times 10^{-4}) = 4.19 \times 10^{-4}\ \mathrm{W/cm^2-{}^\circ C}$$

The finned shielding, emissivity factor, $\mathcal{F} = 0.15$, is a significant reduction from the 0.8 emissivity.

The total fin channel area, A_i, including the between-fin base and fin tip area for nine channels is

$$\begin{aligned} A_i &= 9(19.05)[2(2.54) + (0.762)] + 10(19.05)(0.254) \\ &= 1050.0\ \mathrm{cm^2} \end{aligned}$$

The area, A_0, of the two outer fins is

$$A_0 = 2(19.05)(2.54) = 96.8\ \mathrm{cm^2}$$

The finned side conductance is therefore found by substituting into Eq. (3.126)

$$\begin{aligned} K_f &= [4.19 \times 10^{-4} + (0.15/0.8)(0.8)(7.34 \times 10^{-4})](1050.0) \\ &\quad +[5.98 \times 10^{-4} + (0.8)(7.34 \times 10^{-4})](96.8) \\ &= 0.670\ \mathrm{W/{}^\circ C} \end{aligned}$$

and the enhanced heat transfer coefficient h_{fe} for the finned side is

$$h_{fe} = K_f/WH = 0.67/(9.4)(19.1) = 0.00373\ \mathrm{W/cm^2\ {}^\circ C} \tag{3.127}$$

A check on the average temperature rise of the sink for 44 watts is computed using the total sink conductance:

$$k = k_f + k_S = 0.670 + 0.236 = 0.906 \text{ W/ °C} \tag{3.128}$$

The temperature rise is

$$\Delta T = Q/k = 44/0.906 = 48.6\text{°C} \tag{3.129}$$

which is very nearly the original goal.

Finally, the issue of a fin base-tip temperature gradient that could invalidate the above conductance calculations has not yet been considered. The calculation of "efficiency" is well covered in basic heat transfer treatises as well as those specializing in electronic equipment cooling. An adequate relationship for fin efficiency for a fin with negligible tip area is

$$\eta = \frac{1}{\sqrt{\frac{2L^2h}{kt_f}}} \tanh \sqrt{\frac{2L^2h}{kt_f}} \tag{3.130}$$

for fin length $L = 2.54$ cm, fin thickness $t_f = 0.254$ cm, sink thermal conductivity $k = 1.97$ W/cm $-$°C, and a fin average h of

$$\begin{aligned} h &= h_C + (\mathcal{F}/\epsilon)h_r = 4.19 \times 10^{-4} + (0.15/0.8)(0.8)(7.34 \times 10^{-4}) \qquad (3.131) \\ &= 0.000529 \text{ W/cm}^2 \text{ °C} \end{aligned}$$

Use of Eq. (3.130) shows that $\eta = 0.996$, i.e., a negligible fin base-tip temperature gradient. The preceding conductance and enhanced heat transfer coefficient calculations are acceptable as they stand.

The results of the closed form Fourier series calculations using the values of h_{fe} and h_{Se} are shown in Fig. 3.34 for the source (transistor) side of the equivalent flat plate problem. The temperature variation across the sink surface is about $\pm 10\%$. This is sufficiently small to justify the original approximation of a single average 50° C rise for the entire heat sink.

In summary, the significant simplifications of reducing the computational effect of an extended fin structure conduction/convection/radiation problem to using an enhanced total heat transfer coefficient, plus the avoidance of setting up the heat sink base grid for a finite numerical approach, is a time/cost-effective approach for a frequently encountered application.

4.3 Printed Circuit Board Example

A planar view of a printed circuit board with fifty-six dual-in-line resistor packages soldered into the board is illustrated in Fig. 3.35. The board is epoxy-glass with a single layer of nearly continuous one-ounce Cu (0.00356 cm). The board was actually the middle board of a five-board card cage where the boards were placed with a pitch of 2.54 cm. The cases (top surface) of a single column of DIPs were instrumented with thermocouples as indicated in Fig. 3.35. Each resistor DIP dissipated 0.33 W. The inlet air temperature was approximately 20° C.

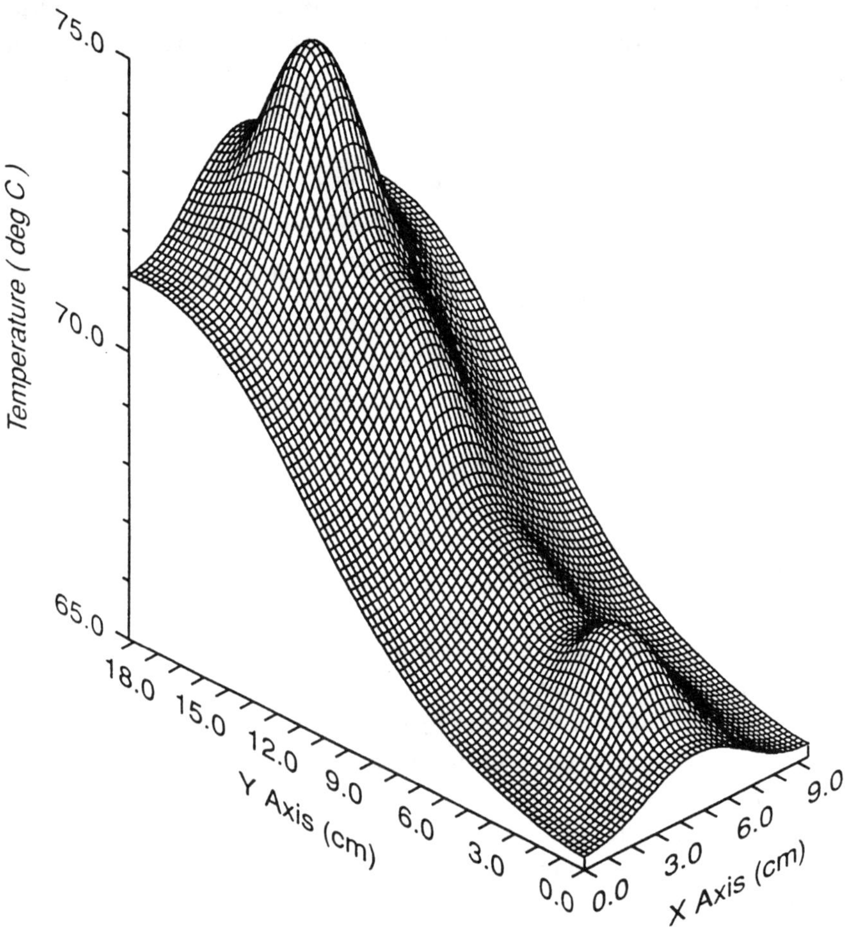

Figure 3.34 Extruded heat sink thermal results for source plane [Courtesy of Tektronix, Inc.].

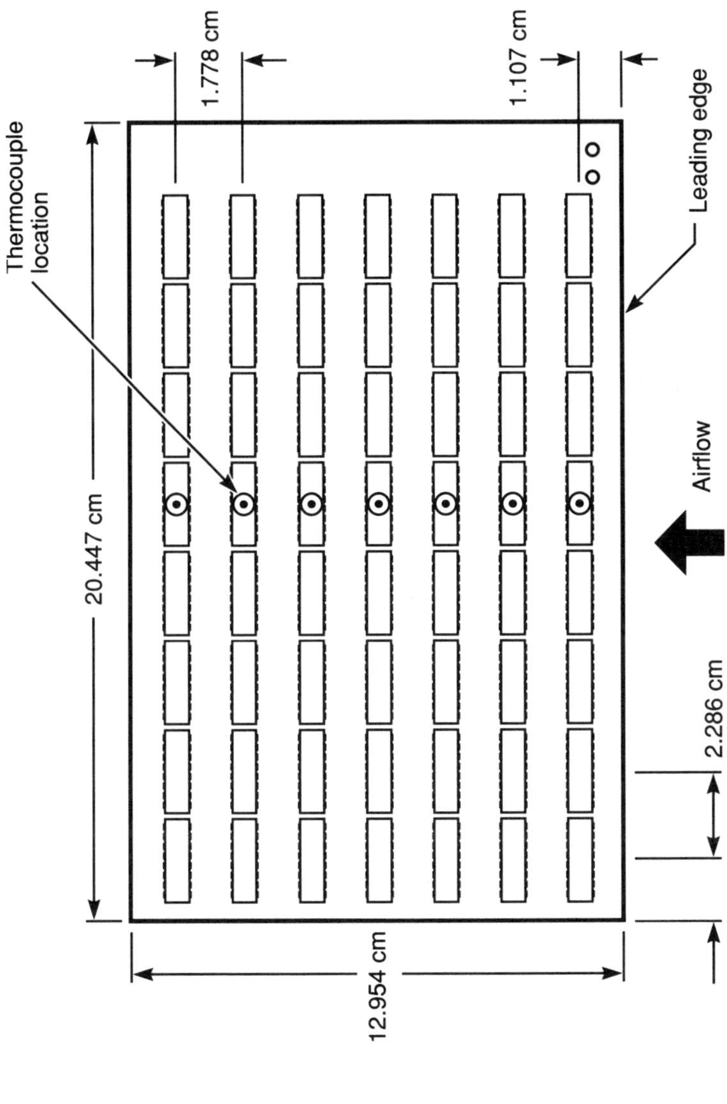

Figure 3.35 Printed circuit board with DIP resistor packs [Based on Ellison (1991)]. [Reprinted with the permission of the Institute of Electrical and Electronics Engineers, Inc.].

The major features of a thermal network for the board are shown in Fig. 3.36. The resistor layout in the "planar view" represents the board conduction model. Because only one layer of nodes is used for the board, each conduction element in this rather simple representation is actually a parallel combination of epoxy-glass and copper conduction resistances calculated from the basic $K_{ij} = k(wt/L)$ for a node material with thermal conductivity k and a nodal path width, thickness and length of w, t, and L, respectively. Temperature gradients through the board thickness are neglected.

Interaction of the board with other elements of the model are illustrated in the "edge view" of Fig. 3.36. This drawing shows four different node layers: component side air, DIP cases, circuit board, and non-component side air. The elements interconnecting the DIPcases and the board are actually each representative of three different elements (only one is shown in the figure): lead conduction, air-gap conduction, and DIP-to-board radiation. The lead resistance for each DIP was calculated for 16 leads in parallel:

$$R_L = \frac{\ell}{16kA_k} \tag{3.132}$$

for a lead length, conductivity, and cross-sectional area of ℓ, k, A_k, respectively.

The air gap resistance R_G is similarly calculated:

$$R_G = \frac{t}{kA} \tag{3.133}$$

for a gap thickness t(0.102 cm), an air thermal conductivity k, and a conduction cross-sectional area A.

The air gap radiation resistances $R_R = 1/(\mathcal{F}/\epsilon_1)h_r A_1$ were calculated using $\mathcal{F}$ as a result for two-surface radiation network theory:

$$\mathcal{F} = \frac{1}{\dfrac{1-\varepsilon_1}{\varepsilon_1} + \left(\dfrac{1-\varepsilon_2}{\varepsilon_2}\right)\left(\dfrac{A_1}{A_2}\right) + \dfrac{1}{F_{1-2}}} \tag{3.134}$$

where the board and DIP emissivities $\varepsilon_1, \varepsilon_2 = 0.8$, the radiating DIP and board surface areas A_1, A_2 are taken as equal, and the geometric configuration factor or shape factor is, $F_{1-2} = 1.0$.

The DIP to ambient convection resistance $R = 1/h_C A$ for a DIP surface area A (top plus all four edges) used an h_C correlation by Wills (1983).

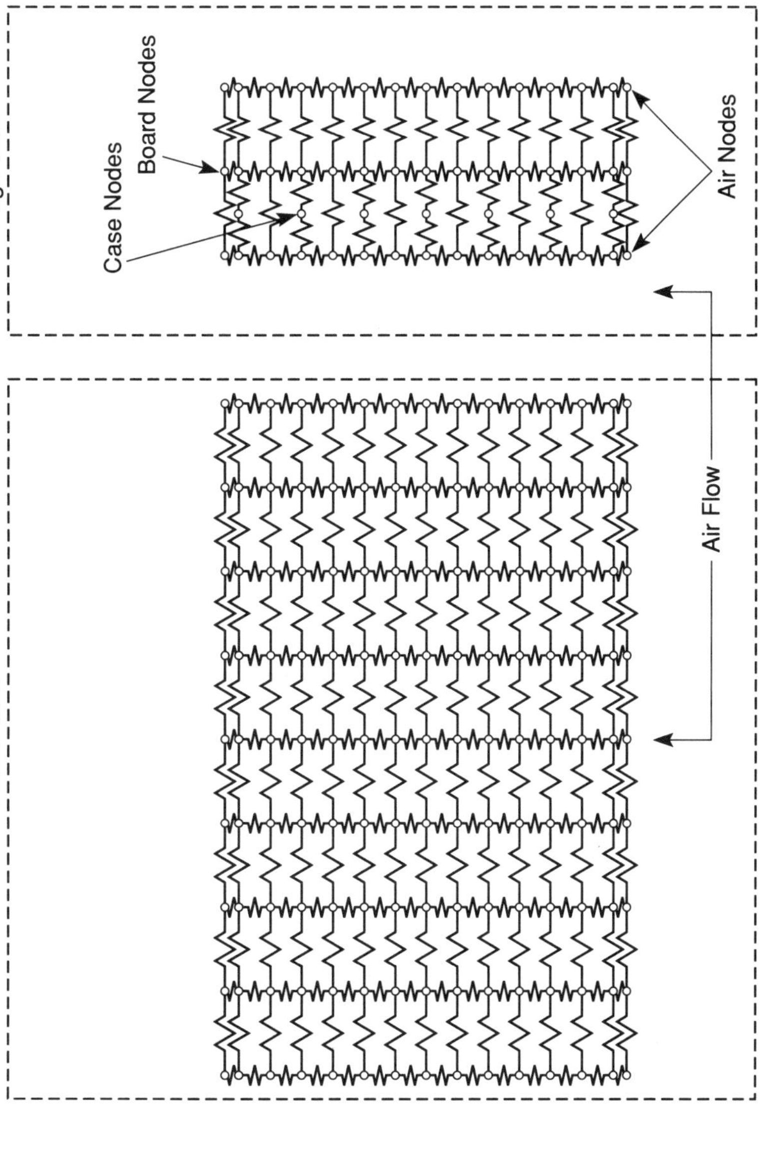

Figure 3.36 Thermal network model of printed circuit board.

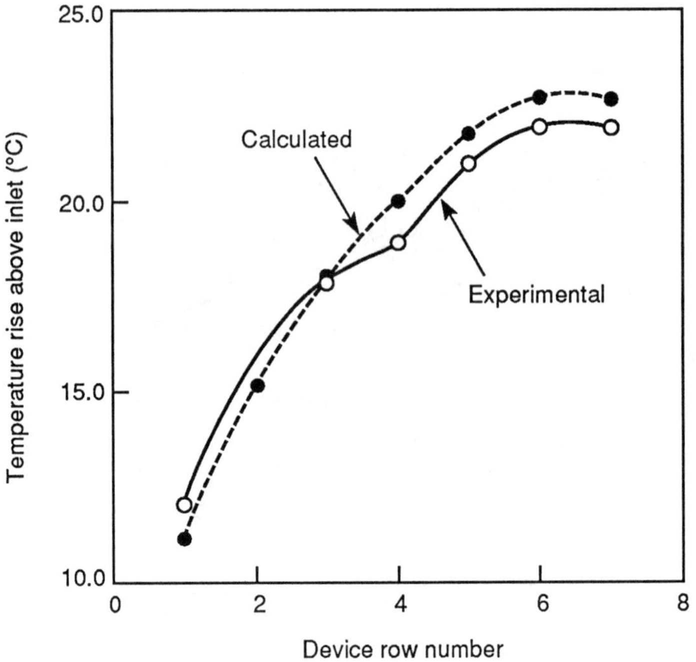

Figure 3.37 Printed circuit board with DIP resistor packs — plot of case temperature for experimental and thermal network results. The board channel air velocity equals 1.52 m/s and the board pitch equals 2.54 cm.

$$h_C = 5.3\left(\frac{b}{d}\right) + 7.6\frac{(\rho v)^{0.8}}{(Nd)^{0.2}}\left[\left(\frac{b}{d}\right) - 1\right]^{0.13} \tag{3.135}$$

where h_C has dimensions of W/m^2 −°C, b is the component length in m, d is the component pitch in m, ρ is the air density in kg/m^3, v is the air velocity in m/s, and N is the component row number.

Both the bare board space on the component side and all of the area on the non-component side of the PCB used a flate plate h_c correlation from Ellison [Eq. 2.23 and Fig. 2.12, (1989)].

A comparison of calculated thermal network results and measured case temperatures is shown in Fig. 3.37 for an air flow of 1.52 m/s between the DIP cases and the opposing board surface. The agreement is within about 10% for all temperatures illustrated. However, caution is required in assuming that the displayed accuracy of the calculation is representative of the prediction accuracy to be expected in PCB studies. The choice of the component and board h_C for this experiment was fortunate indeed. Extensive measurements for other test board conditions (board spacing, component pitch, and air velocity) indicated much poorer agreement with calculations. However, as the knowledge base of convective heat transfer coefficients improves, one hopes that the time is almost ripe for the illustrated prediction and measurement correlation to be typical of actual component on board designs.

4.4 Enclosure Examples

Simple Enclosure—Results Compared with Experiment. This example provides a good representation of natural convection/radiation/ventilation modeling using the thermal network method. Although the selected example does not have the complex geometry of more difficult design cases, the modeling principles apply equally.

A card/enclosure geometry is illustrated in Fig. 3.38. The essential elements of the circuit board structure consist of a nearly solid ground plane sandwiched in a 0.152 cm thick epoxy-glass board. Fifty-six equally dissipating, 16-lead DIPs are mounted on the circuit board.

The box was analyzed and tested in both a sealed and ventilated mode. In both cooling orientations, the 15.16 cm dimension was the enclosure height. When the vented mode was used, the inlet perforation pattern was at the bottom of the side panel facing the non-component side of the board. Conversely, the exit perforation pattern was on the top of the side panel facing the component side of the circuit board. The enclosure interior was unpainted and non-anodized with an estimated emissivity of less than 0.1. The painted exterior had an emissivity estimated to be 0.8.

Two network models are associated with the problem: a thermal network (Fig. 3.39) and an air flow network (Fig. 3.40). The thermal network is used with both the non-vented and vented cases. The air flow network is required only for the vented problem.

Thermal resistances are identified as to type:

- R_C — convection
- R_r — radiation

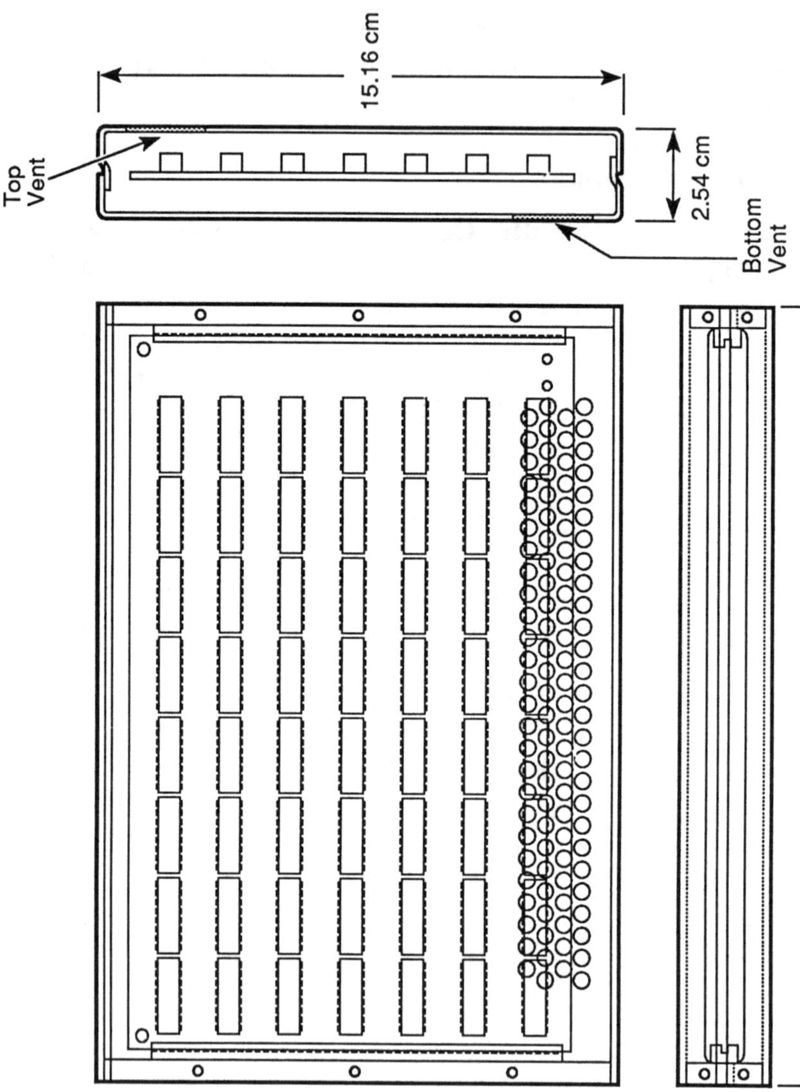

Figure 3.38 Top and side views of box (single-side vented) and card assembly [Courtesy of Tektronix, Inc.].

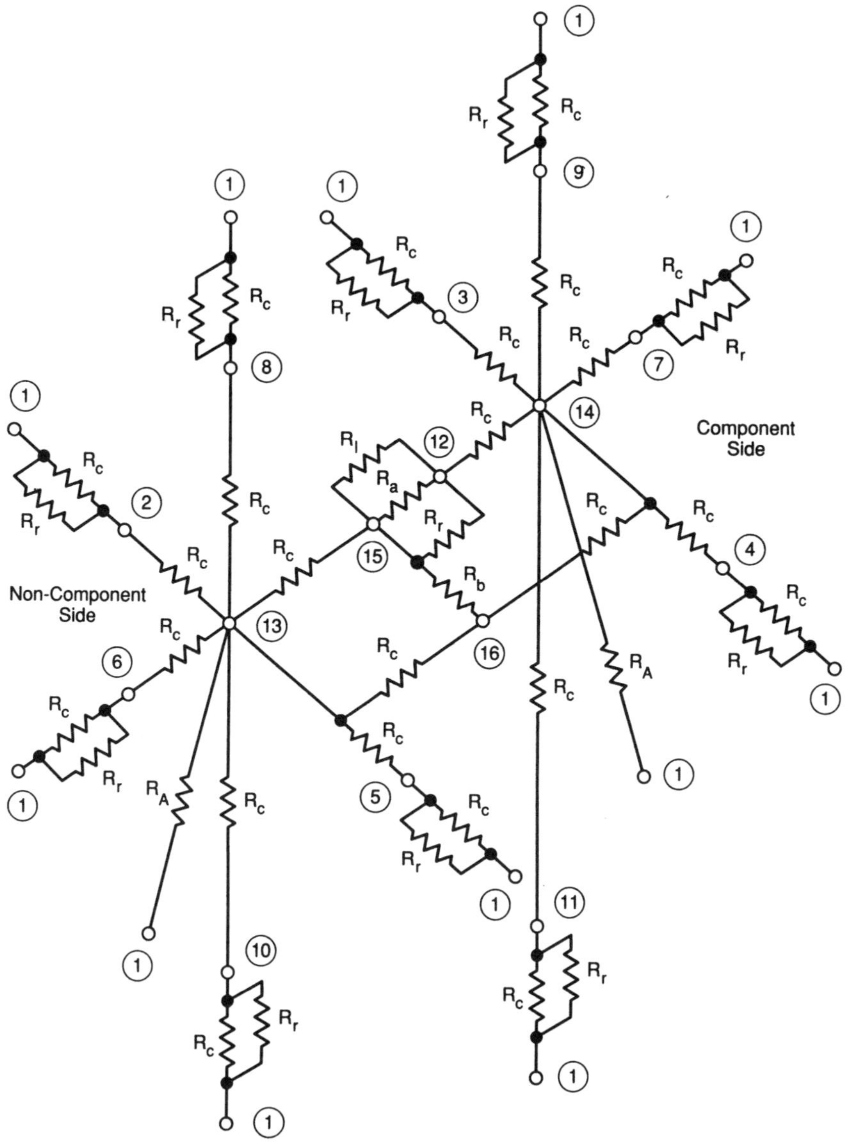

Figure 3.39 Thermal network model for simple-vented enclosure [Courtesy of Tektronix, Inc.].

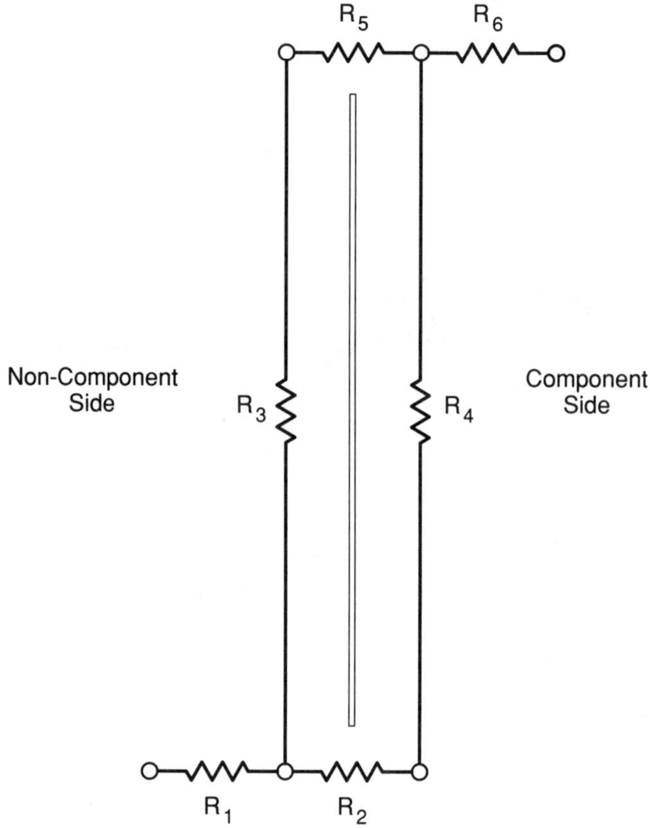

Figure 3.40 Pressure/air flow circuit for simple-vented enclosure [Courtesy of Tektronix, Inc.].

- R_A — one way air flow
- R_b — board conduction from components to non-component region of board
- R_l — conduction from components to board via leads
- R_a — conduction from components through air gaps to board

Internal radiation exchange from components and board to enclosure walls is assumed negligible because of the low emissivity of the interior surface of the various panels.

The thermal circuit nodes are identified as:

- Node 1: ambient external to enclosure

- Nodes 5–7: side and end panels
- Nodes 8–11: top and bottom panels
- Node 12: component cases
- Nodes 13–14: internal air on non-component and component sides of the board, respectively
- Node 15: board locations beneath components
- Node 16: non-component locations of board

The simple representation of the components and board as a four-element circuit is possible because all devices are identical. Because the board contained 56 devices, the total component convection resistance used an area 56 times the single component area; the total component lead to board resistance used a resistance of (single lead resistance)/(56 × 16); the component to board radiation resistance used a value of (single component to board resistance)/56. In the latter case, a two surface resistance relationship was used:

$$R_r = 1/\left\{\left[\frac{1-\varepsilon_1}{\varepsilon_2} + \left(\frac{1-\varepsilon_2}{\varepsilon_2}\right)\frac{A_1}{A_2} + \frac{1}{F_{1-2}}\right]\left(\frac{1}{\varepsilon_1}\right) h_r A_1\right\} \tag{3.136}$$

Any thermal network analyzer may be used to solve this problem. The thermal analyst's greatest problem is probably the selection of the convective heat transfer coefficient. This problem was actually solved using the TNETFA program [18] with horizontal and vertical flat plate convective heat transfer coefficients for the enclosure panels, both front and rear board surfaces, and the component surfaces consisting of the component top and four sides [see Ellison (1989)]. The utilization of all four component surfaces was based on the assumption that the low flow velocities and somewhat large component separation in the vertical direction would encourage air into the cavities between components.

Most of the characteristic lengths used for the heat transfer coefficient were selected as expected. The enclosure vertical panels used the box height; the top and bottom panels used panel area/perimeter. The board and component characteristic lengths were more difficult to select and final judgement of the selection for such a simple representation of a truly complex air flow/heat transfer problem (on a microscopic scale) must be based on a theoretical/experimental correlation. However these calculations use the board height for both the board and component characteristic lengths.

As previously noted, the ventilated enclosure also requires solution of the air flow/pressure network shown in Fig. 3.40 where the air flow resistors as considered in Section 3.4 are:

- R_1 — inlet vent
- R_2 - slot (between board edge and bottom panel)
- R_3 - channel between vertical enclosure wall and non-component side of board
- R_4 — channel between vertical enclosure wall and component side of board
- R_5 — slot (between board edge and bottom panel)
- R_6 — exit vent

The relationships for the air flow resistors are taken from Ellison (1989):

$$\text{Vents } R = 11.4/A_f^2 \tag{3.137}$$

$$\text{Slots } R = 8.6/A_f^2 \tag{3.138}$$

$$\text{Circuit board } R = 2.5L/A^2 \tag{3.139}$$

where A_f (cm^2) is the the total free area of the perforated vent regions and the open area of the slot; L (cm) is the board length parallel to the air flow direction; A is the product of the board surface to side panel distance and the board width. The channel resistance R_3 was estimated based on the ratio of open area for R_3 and R_4 to be $R_3 = 0.33R_4$.

The procedure used to incorporate air draft into the thermal model was based on an iterative method explained by Ellison (1989) (Chapter 6). This method is based on the requirement of the solving program, TNETFA [18], to use a volumetric air flow rate as input for the one-way air flow resistors R_A in Fig. 3.40. The first step was to guess the air flows in R_A, execute TNETFA [18], then determine the effective heat transfer Q_A through R_A. Ellison (1989) then shows how to calculate the two-channel air flows due to Q_A. This provides

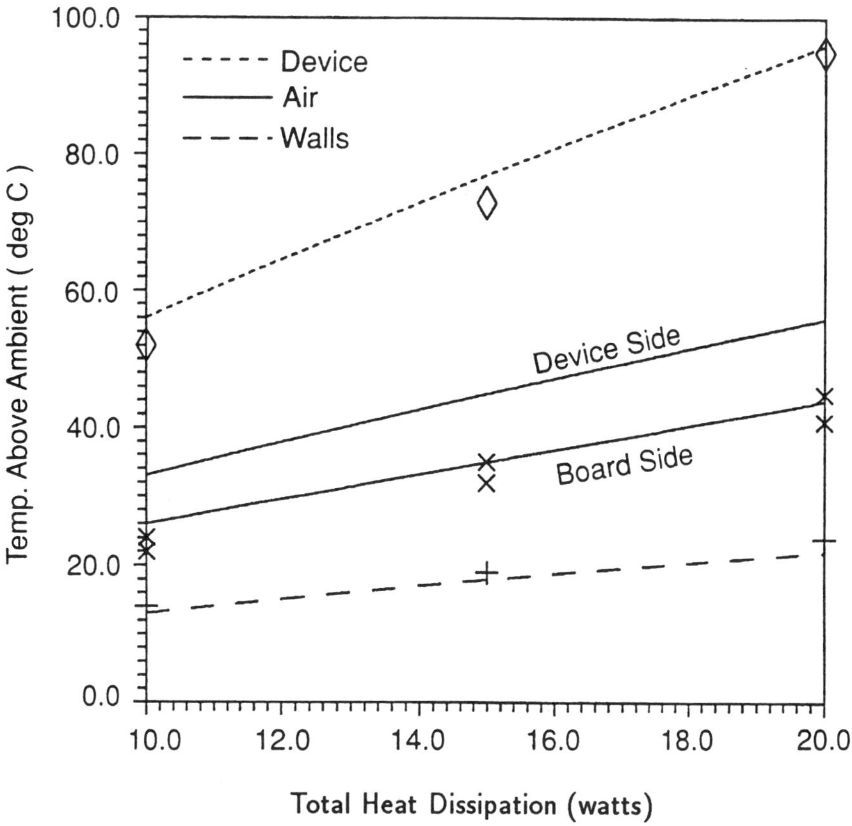

Figure 3.41 Measured and calculated results for simple non-vented enclosure. Symbols represent measured data [Courtesy of Tektronix, Inc.].

a better estimate of the required TNETFA air draft input. The procedure seldom requires more than about three iterations of this type to produce a convergent solution.

Computational results as well as measurements are plotted in Figs. 3.41 and 3.42 for the sealed and vented enclosure problems, respectively. The computed temperatures were, of course, evaluated at the nodal points in the network. The computed wall temperatures shown on the plot were obtained by averaging the two vertical wall temperatures, which were nearly identical in all cases.

A study of Figs. 3.41 and 3.42 clearly shows a decrease in both computed and measured component internal air, and wall temperatures when the box is changed from sealed to vented. The measured air temperatures for component and board side are quite similar in the sealed box, with only slightly higher air temperature for the component side. However, the vented box results show large air temperature differences between the component and board sides. It would seem that the component side air draft is sufficiently greater than the other side to lower the DIP case to air thermal resistance causing greater heat dissipation into the component side air stream. This interpretation presumes the DIP case (and board surface) resistance is decreased more than the increased draft can offset the fluid enthalpy change.

The measurements also require some explanation. The test box used in the experiments utilized DIP resistor packs as components. The component temperature is taken from the center DIP case surface facing the vertical wall. The measured wall temperatures are

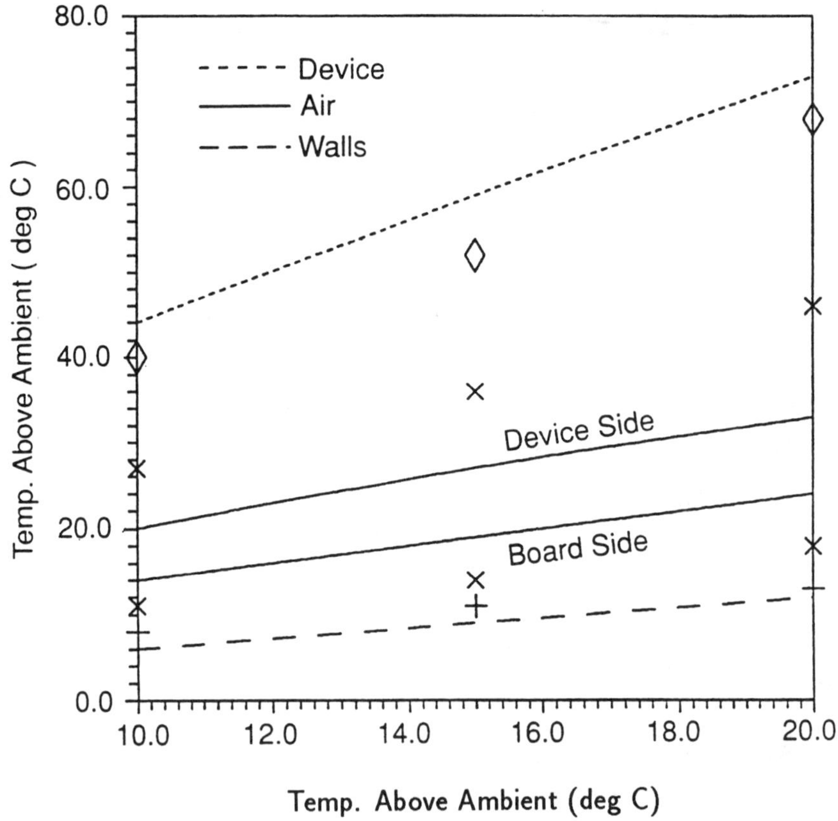

Figure 3.42 Measured and calculated results for simple vented enclosure. Symbols represent measured data [Courtesy of Tektronix, Inc.].

averages of the two vertical wall temperatures as measured at the wall centers. The air temperatures were taken by thermocouples placed midway between the component and board surfaces and the vertical enclosure walls, but at the top of the air channel. Because the warm air rises to the top of the enclosure, these air temperatures are approximately indicative of a mixed mean ambient.

The accuracy of the method is quite good considering the complexity of the actual hardware and the simplicity of the analysis method. With regard to the latter, however, the reader must be aware that the modeling and selection of convective heat transfer coefficients play a crucial role in the validity of the computed results.

Multi-Chambered, Vented Enclosure. Many enclosure designs have an interior structure with solid or nearly solid vertical and/or horizontal panels that prohibit the internal air from circulating throughout the box. The model construction methodology and analysis procedure for this type of system is usually a natural extension of the methods used in the preceding problem. Analysis results will not be provided, but differences from the simple, single card problem are discussed. The example for this is taken from Ellison (1989) and is illustrated in Figs. 3.43 and 3.44.

The complete structure is shown in Fig. 3.43. The design consists of a three-chamber system with the chambers as A, B, and C. Chambers A and B contain a set of vertical circuit cards whereas chamber C houses only a few miscellaneous components plus a cathode ray tube.

The cabinet is cooled by natural convection and radiation from the external surfaces as well as buoyancy induced air draft through the system. Chamber C also has a finned extrusion on the rear panel with power transistors attached.

The anticipated overall air flow paths are quite simple. Air enters an inlet pattern in the bottom of B, passes through a set of slots in the horizontal panel separating A and B, and then exits through a set of perforations in the side of chamber A. There are no vent holes between chambers A, B, and C. Air draft in chamber C is maintained by the bottom and top side perforation patterns.

Part of the thermal model (chamber A) is illustrated in Fig. 3.44. Various external convection and radiation resistances as well as the internal convection resistances are shown. Internal radiation is neglected because the internal cabinet surfaces have a low emissivity.

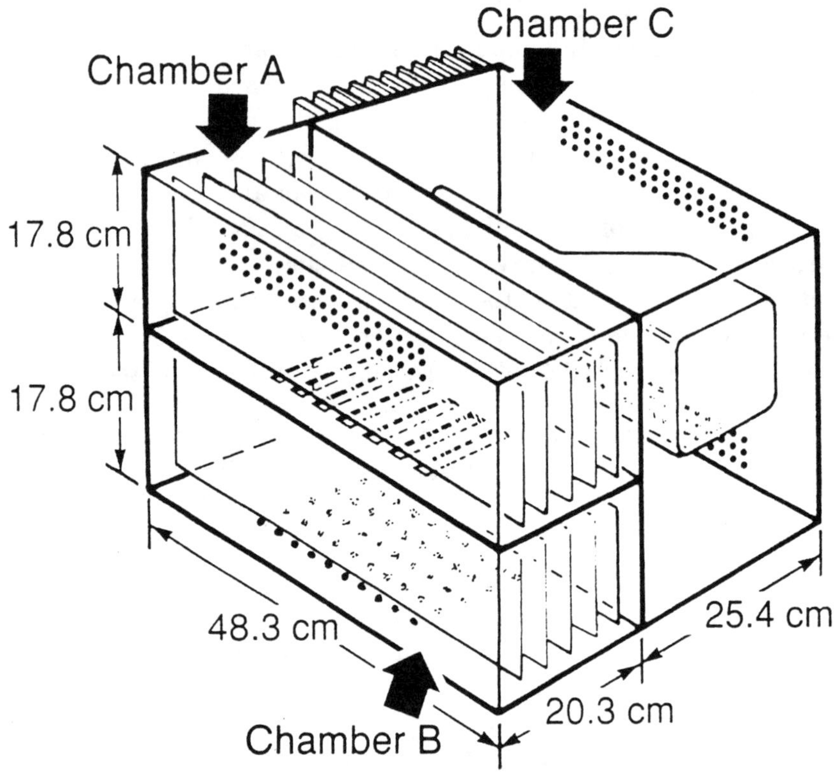

Figure 3.43 Application example of bench-top enclosure [Ellison (1989)] [Reprinted with permission of Robert E. Krieger Publishing Co., Malabar, FL].

In addition, a set of several circuit boards may be expected to act as radiation shields, particularly for the more central boards.

Particularly noteworthy nodal issues are (1) node 6 is common to both chambers A and C and node 8 is common to chambers A and B; (2) the rear panel node on chamber C (not illustrated) must connect with the ambient via a thermal resistance that incorporates the temperature dependencies of convection and radiation from a finned surface. The latter issue requires a thermal network analyzer that contains an appropriate non-linear solution feature (usually standard in commercial codes).

The incorporation of the air draft effects may be accomplished exactly as in the preceding simple enclosure problem. The major difference is that the present problem has two separate air flow circuits requiring solution along with the thermal problem. Ellison (1989) describes this procedure in considerable detail.

Forced Air Cooled Enclosure — Results Compared with Experiment. This section summarizes the published analysis of Ellison and Patelzick (1986) concerning a 1-kilowatt, forced air cooled power supply. As shown in Fig. 4.45, the major air flow paths consist of flow entering a grilled front panel, passing through an air filter, and through two identical fans. The flow is then distributed into seven identical "PCB/heat sink modules" and one "control board" channel. The air diverges from these eight channels into a common chamber from where it exits the enclosure via a slotted grill panel.

An additional structural detail of one of the PCB/heat sink modules is shown in Fig. 3.46 where the mockup nature of the resultant test data is illustrated by discrete resistors on the PCB. The drawing is shown reversed from Fig. 3.45; thus air should be considered to flow from the left over the PCB, where it splits into a heat sink branch and also a PCB branch beneath the heat sink. The lower

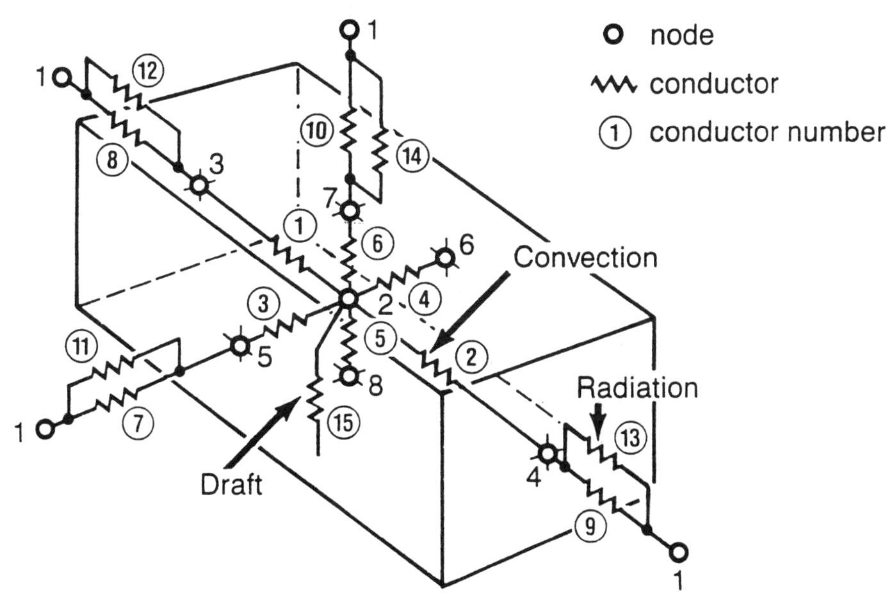

Figure 3.44 Application example of thermal-network model for Chamber A of bench-top enclosure [Ellison (1984)] [Reprinted with permission of Robert E. Krieger Publishing Co., Malabar, FL].

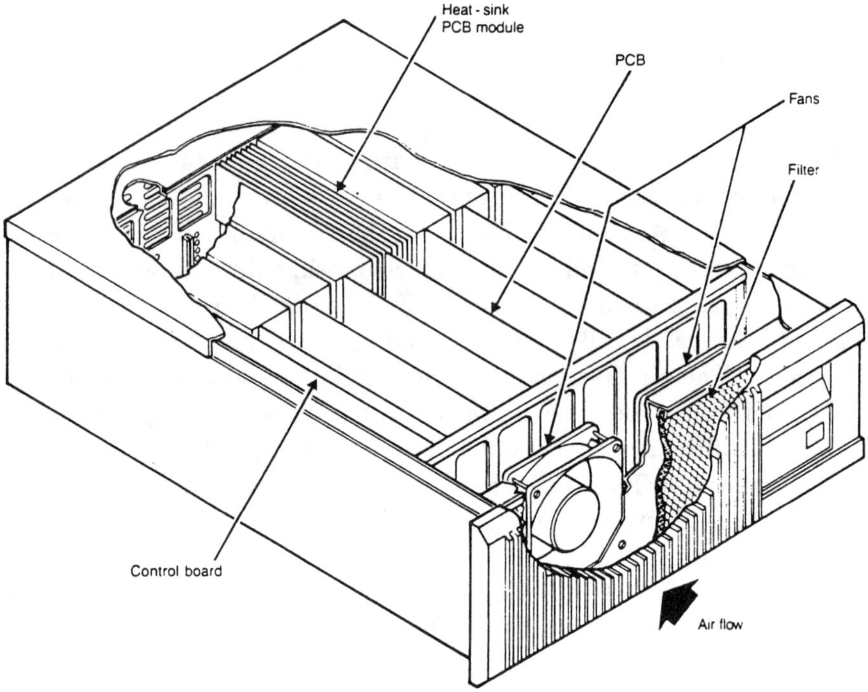

Figure 3.45 Forced air cooled enclosure with seven heat sink/PCB modules and two fans [Reprinted with permission of the International Electronics Packaging Society, Inc.].

PCB section contains a perforated plate at the air flow exit to assist in appropriately balancing the air flow between the two branches. Four power transistors are attached to the flange of each heat sink.

The fluid network method of Section 3.4 was applied to this problem. As previously explained, the operating point of the system is obtained by finding the intersection of the sum of the fan static pressure plus the velocity pressure at the fan discharge, with the total system pressure head loss. The air flow circuit of Fig. 3.47 was constructed to determine the system pressure loss. Various elements in the circuit are air flow resistors identified according to the correponding portion of the instrument. The elements are labeled for only one of the PCB/heat sink modules. The estimate of various air flow resistance elements was based on element recommendations provided by Ellison (1989).

Actual solution of the equations could be accomplished using any simultaneous equation solver appropriately incorporated into a code that correctly updates the "stiffness" or "conductance" matrix to account for the non-linear nature of the problem. The actual intersection of fan and system pressures may be found by writing a code that repeatedly solves the pressure/flow equations until the intersection point is found [see Ellison and Patelzick (1986)].

A less desirable, but not impossible, method is to use a computer code that only solves the equations for one flow source (fans) value at a time. The program user would run the code for multiple source values, plot the total pressure head loss/air flow results for each

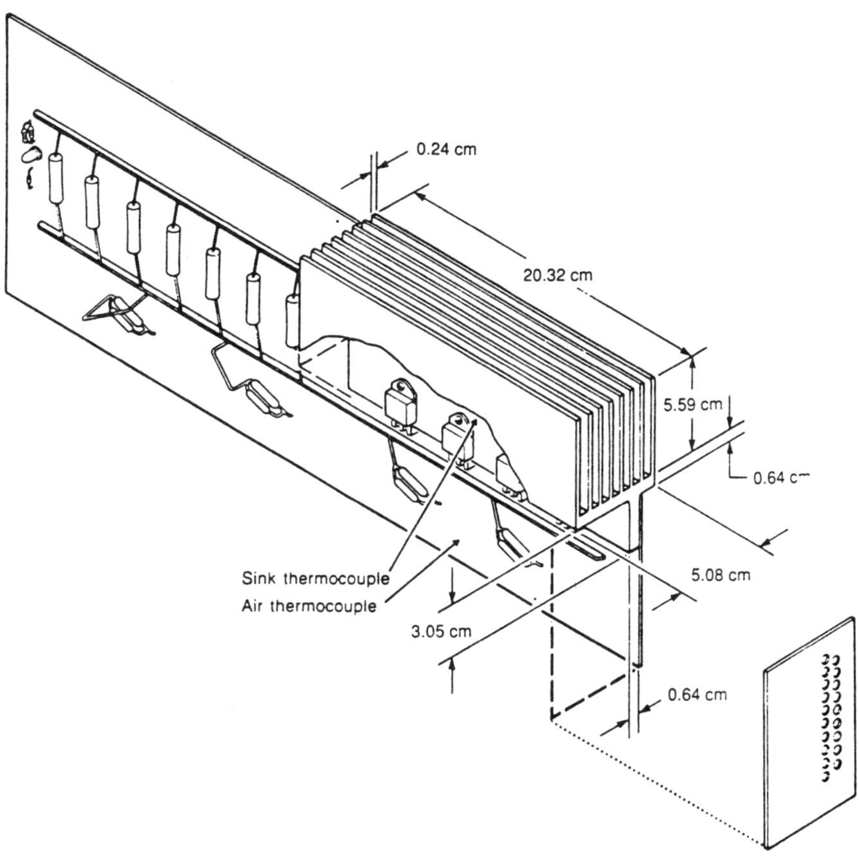

Figure 3.46 Heat sink/PCB module [Reprinted with permission of the International Electronics Packaging Society, Inc.].

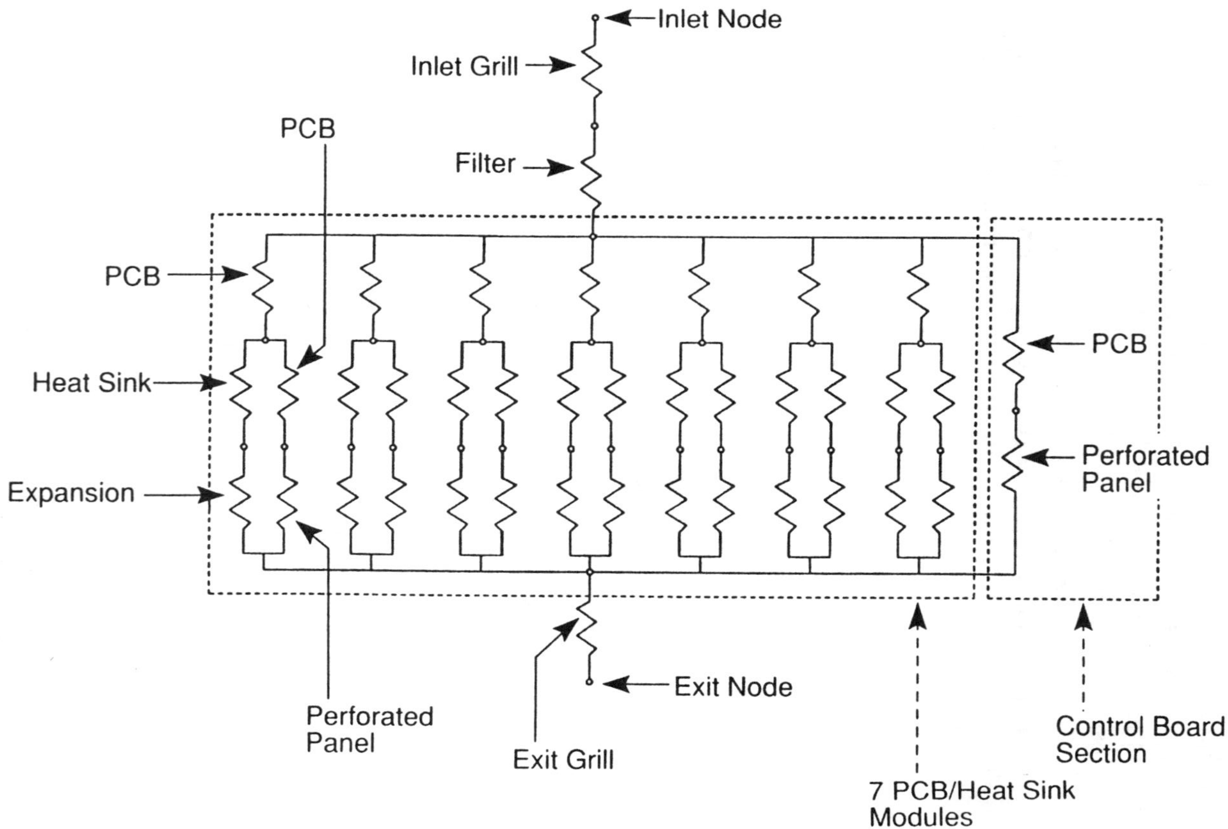

Figure 3.47 Air flow circuit for forced air cooled power supply [Courtesy of Tektronix, Inc.].

source value, and graphically determine the fan-system intersection. The code would then be executed one last time using the operating point air flow. Program output must then include the computed air flow in each branch air flow resistor.

The thermal portion of the problem requires set up and solution of an appropriate thermal network. Thus the resistive elements in Fig. 3.48 are thermal resistors. The heat dissipations of the various sections of the instrument are shown at the most appropriate nodal locations. All air flow resistors shown in Fig. 3.48 are "one way" thermal elements with the exception of the elements identified as "heat sink" with 93 watt source nodes. These heat sink resistances are actually convection resistances calculated for each finned heat sink. The resistance values are dependent on the geometry and air velocity in each fin channel. Each fin channel may be treated as a rectangular duct. The geometry of these ducts is such that one should consider entry length effects. A mean heat transfer coefficient for a circular tube, but corrected for the aspect ratio of the rectangular duct, has been satisfactorily used in several applications as well as that discussed here. The interested reader is referred to pp. 62–65 of Ellison (1989) for additional detail and relationships required for the calculation.

The reader should observe that a significant feature of the thermal circuit is that the heat sink convection resistance is connected to a node that is at the heat sink air flow exit. This means that this resistance is defined as "referenced to the exit." Attachment of the convection resistance to the heat sink inlet would incorrectly preheat the air entering the heat sink. The mathematical solution would be correct, but this placement would be an incorrect utilization of the thermal network method. Details of the methodology used for the heat sink model are discussed specifically for this problem by Ellison and Patelzick (1986) and Ellison (1989).

An important part of the analysis objective for the thermal system was to predict temperatures in the vicinity of the power transistor mounting surface for each of the heat sinks. However, because the system thermal network model provided only a single nodal temperature for each heat sink, additional effort was required to obtain an answer at a location in closer proximity to the transistors. A variety of methods are available. Ellison and Patelzick (1986) chose to use a closed form tool of the type reviewed in Section 3.2.

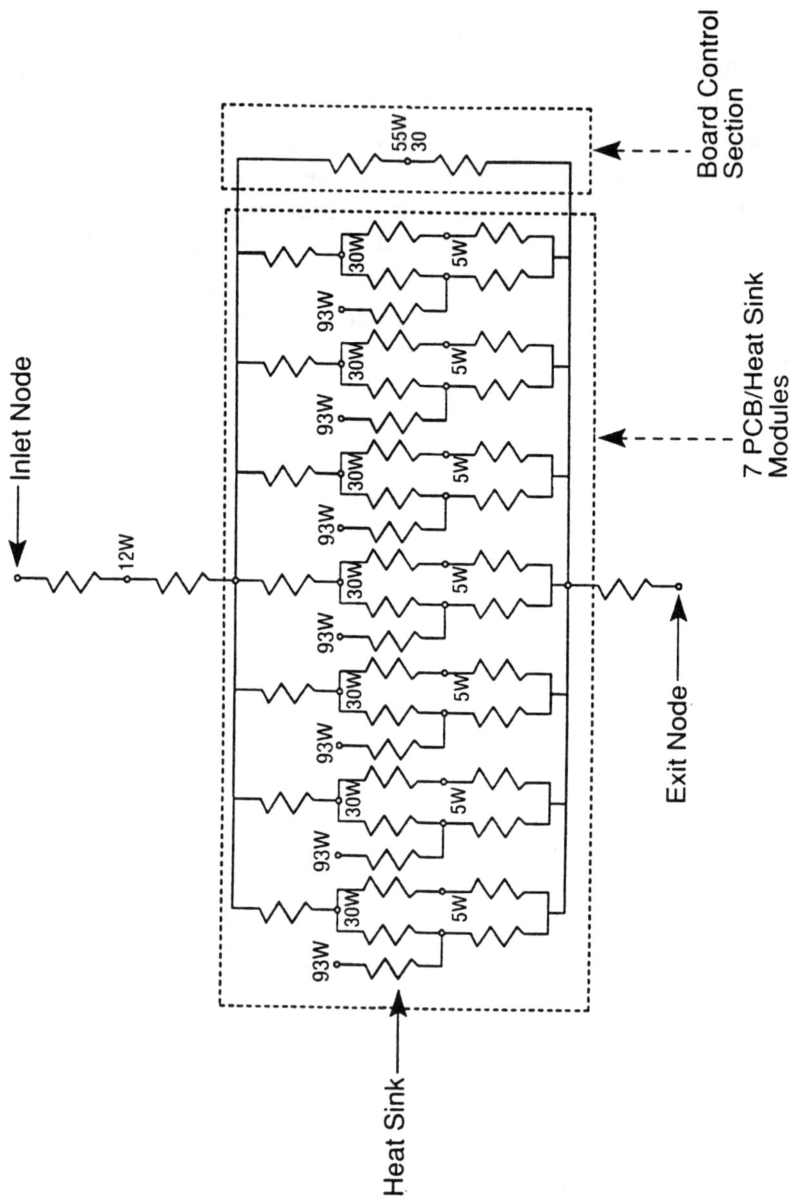

Figure 3.48 Thermal circuit for forced air cooled power supply [Courtesy of Tektronix, Inc.].

Models intended to give the desired results are shown in Fig. 3.49. A model is illustrated that uses a strip heat source to simulate the total heat from the four power transistors leaving the mounting flange and entering the heat sink base. The thermal network results were used to calculate an enhanced heat transfer coefficient from the base of a flat plate to a local heat sink ambient air temperature:

$$h_e = Q_{\text{sink}}/[T_{\text{sink}} - T_{\text{air}})(LW)_{\text{sink}}] \tag{3.140}$$

where L and W are the length and width, respectively, of the heat sink. The temperature is calculated within the strip heat source which represents the flange base.

The flange model in Fig. 3.49b uses zero values of lumped thermal resistances (see Section 3.2) that terminate in temperatures *TO* calculated previously as the flange base temperature. Finally, this flange model produces the desired mounting surface temperatures for the power transistors.

An experimental comparison of the computed results with a few measurements is shown in Table 3.1. The PCB/heat sink module air temperature was measured using thermocouples placed in the air stream near the trailing edge of each PCB. Each heat sink was also instrumented with a single thermocouple at the edge of a transistor mounting base. A single computed heat sink temperature is given because the model did not distinguish differences between the seven heat sinks.

The heat sink temperature measurements and calculations shown in Table 3.1 are in substantial agreement. But, the correlation between measured and computed air temperatures is not as good. This is due in large part to the fact that the calculation assumes "well-mixed air" when, in actuality, the air has a substantial temperature gradient. The elevated temperatures for modules 5 and 7 may have been due to irregular placement of the thermocouples with regard to heat dissipating component location.

4.5 Compute Times for Selected Examples

There is some risk attached to quoting compute times for software benchmarks (Table 3.2). It should be recognized that other variables may be of equal or greater importance. In particular, the problem set-up time may be more important than execution speed. For example, a short set-up/long compute time combination permits one

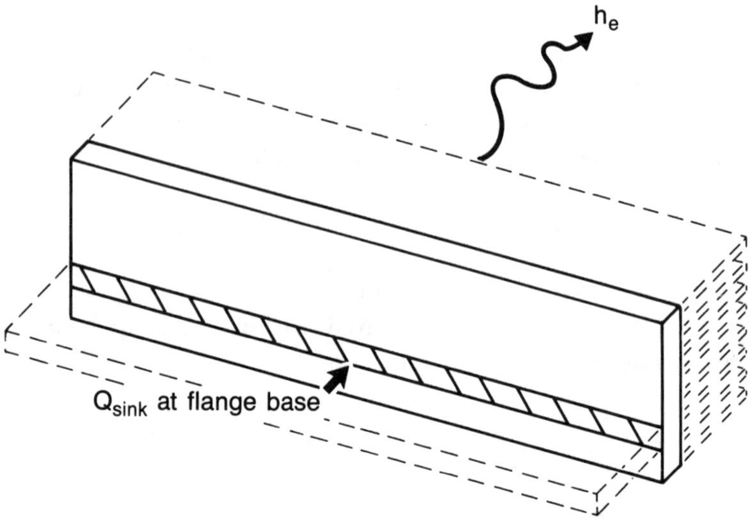

(a)

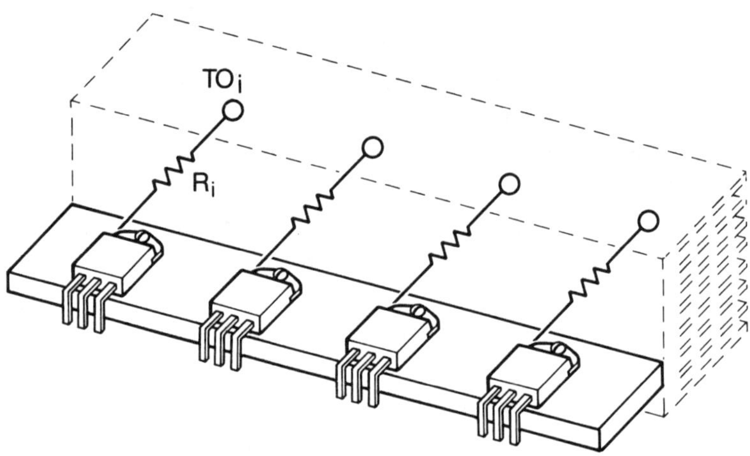

(b)

Figure 3.49 (a) Heat spreading from flange base, (b) heat spreading from transistor package bases to flange base [Reprinted with permission of the International Electronics Packaging Society, Inc.].

Table 3.1 Comparison of Results for Forced Air Cooled Enclosure: Temperature Rise Above Ambient.

Location	Module	Measured (°C)	Computed (°C)
Heat sink	1	32.4	33.5
	2	36.0	33.5
	3	35.0	33.5
	4	34.2	33.5
	5	35.5	33.5
	6	35.4	33.5
	7	35.7	33.5
PCB air	1	4.6	4.5
	2	5.4	4.5
	3	7.5	4.5
	4	5,6	4.5
	5	10.4	4.5
	6	6.0	4.5
	7	10.5	4.5

to perform some other task while the computer program is "crunching." Of even greater significance is the issue that a problem that is easy to set up might well contain fewer hidden errors, as well as encourage problem modification for parameter studies.

Table 3.2 must not be taken too literally. Other compute platforms and peripheral storage options would, of course, give different results. The data in Table 3.2 were obtained using a machine with an Intel 80386 processor, an Intel 80387 math coprocesser, and a hard disk with a vendor-specified 15 millisecond average access time.

5 AN OVERVIEW — COMPUTATIONAL SOFTWARE

This section is not intended to be an exhaustive study of all available software, but rather to serve as a focal point for a discussion of the author's view as to the applicability of the various categories to some

Table 3.2 Comparison of Compute Times for Selected Example Problems.

Problem		Program Execution Time		
Description	Section	TAMS [18]	TNETFA [18]	ALGOR [3]
Hybrid	3.2	5 min w/o gridding (1). 3 hr, 32 min w/gridding (2).	NA	NA
Panel	3.2	29 sec	9 sec	20–200 sec decoding, 21 min solving (3).
Heat sink	3.2	1 min, 18 sec	NA	NA
Enclosure	3.4	NA	3 sec	NA
Board	3.3	NA	46 sec	NA

(1) Output consists of substrate temperatures at all 6 sources and 40 lead areas.
(2) Output consists of (1) and temperatures at 994 grid points. The grid points were computed in three separate executions using 447, 447, and 100 grid points, respectively.
(3) Decoding consists of the computer translating the geometric model to meshed model input ready for solution. The variation in time required by the decoder depends on the skill and experience of the analyst in model preparation.

of the present day electronics cooling problems. The author makes no claim as to the quality of the systems listed in Table 3.3, nor does he necessarily have personal experience with each and every product. He also gives his apologies regarding those products that are not incorporated in the list, with the emphasis that no one product was intentionally excluded.

Table 3.3 Sampling of Commercial Codes

Category	Solution Style	Product Name	Personal Computer (1)	Workstation Mini/ Mainframe (2)	Ref.
Solid solvers	3-D Closed Form Fourier series	TAMS	√		18
	Network/FDM	TNETFA	√		18
	Network/FDM	TAK II	√		19
	Network/FDM	SINDA		√	16
	Network/FDM	TAEE	√		17
	Network/FDM	TMG		√	20
	FEM and/or FDM	TDS		√	21
	FEM	ANSYS	√	√	3
	FEM	ALGOR	√	√	1
	FEM	NISA II	√	√	13
	FEM	COSMOS/M	√	√	7
	BEM	BEASY		√	5
Fluid solvers	FEM	FIDAP	√	√	8
	FVM	FLUENT	√	√	11
	FEM	FLOTRAN	√	√	10
	FVM	PHOENICS	√	√	15
	FVM	FLOTHERM	√	√	9
	Network	SINDA/ FLUINT		√	16
	Network/FDM	TMG		√	20
PCB	FEM	AUTOTHERM		√	4
	FEM	PCB Explorer		√	14
	FDM	BETAsoft	√		6
	FDM	THERMAL		√	2

(1) Indicates availablity on platforms based on one of the Intel 8086, 8088, 80286, 80386, or 80486 processors.

(2) Indicates availability on some form of non-Intel CPU based workstation, including minicomputers and or mainframe computer.

The software products listed in Table 3.3 are arranged according to the categories of "solid solvers," "fluid solvers," and "PCB solvers." Each of the three categories contains at least two different "solution styles." Some of the programs may have applicability to more than one category. For example, fluid solvers may usually be used for conduction solving. The programs are listed according to this author's understanding of the program's principal use, but the reader is referred to the various vendors for more specific and up-to-date information.

An attempt has been made to classify the programs as to "compute platforms." The reader should not take the classification too literally, not only because the classification scheme will quite soon become outdated, but also because the distinction between a high-end personal computer and a workstation or mini-computer is somewhat blurred. Furthermore, some codes that are available over the full range of hardware systems may be reduced in capability for personal computer versions with regard to the maximum allowed number of nodes and elements. Generally, however, it is safe to assume that thermal network software that appears as only a personal computer version may well solve problems of less than a thousand nodes. Finite element codes on the personal computer may be capable of solving problems up to several thousand nodes and elements. The interested reader should consult with the software vendors for the latest details concerning program capability.

Problems encountered in the solid solver category would typically involve component and heat sink geometries. The principal advantage of closed form solution methods, when applicable, is that geometry meshing is not required. Although the commercial availability of such software is limited, many individual researchers have developed their own solutions. Moreover, in addition to the usefulness of such codes for design problems, they are invaluable for providing benchmark solutions when evaluating a new finite element or thermal network code. The reported closed form solutions often utilize different solution techniques, but more important to the user, the possible boundary conditions may or may not be appropriate. Computer programs designed to analyze single and multichip packages, or hybrids cooled by convection/radiation, should accept heat transfer coefficient(s) for the major surface(s).

The thermal network method is the second solution style of solid solvers listed in Table 3.3. This method is one of the earliest used by thermal analysts and therefore there are a number of commercial codes from which to choose. An advantage of these codes is that they are easy to learn, modeling is somewhat intuitive, and they do not require exact specification of geometric boundaries via nodes or some type of grid point specification. The programs also may have a degree of flexibility not available in the highly structured finite element method (FEM) codes. For example, the SINDA program [16], which is so familiar to nearly a generation of thermal analysts in the United States (and perhaps elsewhere too), permits the incorporation of FORTRAN statements to describe complex physical processes for which other codes may not be able to account. Perhaps the greatest disadvantage of the typical network program is that graphical input/output interfaces are not generally available, although some organizations have written special purpose input file generators that meet specific needs such as PCB modeling. Historically, thermal networks have, and still are, receiving extensive use for solid conduction problems, although the effort required for very complex geometries can be extensive. For such configurations, the user should consider an alternate technique, principally the finite element method.

The last solution style to be considered as a solid solver comes in the form of FEM solutions. The user has a variety of codes from which to choose, only a few of which are listed in Table 3.3. The disadvantages of this type of code (although the vendors might be quick to disagree) are (1) an extensive learning curve as well as regular use is required, (2) the codes may not have the flexibility of thermal network systems, (3) the modeling is not intuitive, and (4) the geometry must be rather exactly specified. A very attractive feature of FEM is that most commercial codes have quite elegant, interactive, graphic input/output displays. There is considerable variation in interface design amongst the vendors, but this permits the potential user to select a style that perhaps suits one's own idiosyncrasies. Another area in which the FEM vendors may vary in their offering is the completeness of the element library. These programs are excellent for complex, conduction-dominated geometry.

The second problem category is the fluid solvers, often referred to as "computational fluid dynamics" (CFD) solutions. These solvers are finite element (FEM), finite difference (FDM), or finite volume (FVM) solutions of the Navier-Stokes equations, with or without turbulence models. There are several commercial codes from which to select, and the question as to which one to use could be a significant evaluation project for the potential user. Some of these codes are receiving quite a lot of attention. The FLOTHERM program [9] is specifically designed for the analysis of electronics enclosures and subassemblies and therefore has an appropriately designed input/output display system. Schnipke et al (1989) used the FLOTRAN program to analyze the conjugate air flow/conduction problem for a single finned IC package. Using a coarse three-dimensional mesh, they obtained agreement within 30 percent of measurements. Agonafer and Moffat (1989) rather successfully predicted the convective heat transfer coefficient for an array of rectangular modules resembling a circuit board channel using the PHOENICS program [15]. Their results were within about eight percent of measurements, with the exception of the downstream region where the discrepancy was 30%. Linton and Agonafer (1989) developed quite a detailed model of a personal computer system, including printed circuit boards and disk drive housings, and compared several computed and measured temperatures with rather good agreement. However, even with these positive modeling experiences, the consensus in the industry seems to be that there is considerable work to be done in improving the codes to the point where they can be reliably used. CFD is a major topic in its own and is therefore not addressed in any detail in this section.

The second solution style for fluid solvers is the network method. Application of this technique was addressed earlier in both the theoretical summary and application sections. Admittedly, the network representation of a fluid problem is not as elegant as the CFD approach, nor is it nearly as costly from the point of view of software purchase price or time required to both set up and actually wait for a solution to be generated. It does have much appeal in the early conceptual design stages of rather complex air flow systems, and is also somewhat intuitive in modeling requirements. The accuracy of the method is not extremely great, but is certainly adequate for many engineering projects. For example, the total forced air flow for a fan-cooled system may be predicted to about $\pm 15\%$. The air flow in various internal chambers is often predicted less accurately, depending on the problem complexity. The difficulty of accurately measuring internal flows in complex geometry makes it difficult to estimate the actual accuracy that should be expected. Comparison of computed and measured air temperatures in electronics enclosures is difficult because in low flow regions, the actual flow is often unsteady, and in higher flow regions the temperature gradients are large. Unfortunately, air temperatures computed with network solvers are "mixed mean" values. Nevertheless, the experienced analyst is often obtaining accuracies in temperature rises of about ± 30. It should be possible to improve the accuracy of network air flow and thermal predictions by the development of better resistance elements. The implementation of experimental component convection and reference temperature models such as those of Anderson and Moffat (1990) would improve the network method accuracy.

The final category of software listed in Table 3.3 is that of the special-purpose PCB problem. The total problem addressed in this electronics cooling category is extremely complex because it encompasses very complex air flow patterns, convective heat transfer, and conduction in a variety of tortuous metallization patterns as well as the poor conduction capability of the typical epoxy-glass type substrate. Several PCB thermal programs make some attempt to address the total problem. The principal feature of these programs is not their accuracy, but rather that they are usually intended to be interfaced with a PCB layout program, usually via an appropriate file transfer scheme. This is nearly mandatory when one considers the hundreds of devices that may be anticipated on the board — certainly not a problem that a designer wishes to address on a component-by-component basis using a typical CAD system interface. Again, as with many of the general-purpose finite element programs, considerable effort has gone into making the display of results as attractive as possible, thus enhancing the user's ability to interpret them. All the listed PCB programs appear to have basic two-surface radiation as well as built-in convective heat transfer coefficient correlations. The latter may be expected to be rudimentary and not particularly well suited to all PCB problems. Furthermore, the thermal analyst will probably be expected to provide his/her own heat transfer coefficients to effectively analyze these problems. Most of the listed software also permits users to utilize their own preferred convection correlations if so required. The conduction modeling does not necessarily directly incorporate the details of metallization runs and power planes. The conduction material properties are addressed with an average thermal conductivity that one hopes will adequately approximate the problem.

The preceding paragraphs have touched on the range of applicability of a variety of software products for the electronics thermal analysis problem. The issue of accuracy is also of great concern, or at least should be, to the prospective user. Unfortunately, it is not easy to directly respond to the accuracy of the solution of any given problem, particularly in the real world of electronics cooling. All problem categories, with the exception of the PCB, permit to some degree, testing for accuracy by comparing with textbook solutions. The general-purpose code vendors seem to be quite conscientious in this area as demonstrated by the fact that they usually make available a manual of verification problems, against which they regularly test code changes. The thermal network solutions are quite easy to test against standard textbook solutions.

The accuracy of any code comparison with textbook solutions is, unfortunately, almost irrelevant with regard to obtaining design analysis results for electronics cooling problems. Agreement between textbook and computer code solutions is only an effective check of basic solution algorithms and possible code bugs. Serious thermal analysts surely understand that they are individually responsible for selecting the appropriate convective heat transfer coefficients, interface resistances, and making modeling decisions.

The utilization of special-purpose codes such as the PCB solvers is another issue. Although vendors supply built-in convection correlations, the user must be aware that this is a very difficult problem at best — convection from rectangular modules on plates is still an active area of university research at both domestic and foreign institutions. Because of a desire to reduce design costs and time-to-market, there is pressure from industrial management to purchase software solutions to design methodologies. The process of multiple efforts to lay out a printed circuit board that is both functional and meets thermal requirements can have a significant impact on product time-to-market as well as final cost to the customer. Although the automated thermal solution for PCBs ultimately promises to be very fruitful to industry, there is still much work to be done before this type of software can be expected to regularly provide results that correlate well with experiment.

In summary, the simplest of heat transfer electronics problems is complex indeed, considering the aspects of air flow, heat transfer by convection, and solid conduction geometry. Users of a program must use care in the selection and application of software solutions for any given problem. It is also an established fact that a good deal of excellent design engineering may be accomplished without resorting to computer analysis, or at the very least by using modest and inexpensive codes.

6 SUMMARY

A variety of computational methods has been examined with the express intent of evaluating the applicability and accuracy of electronics cooling applications. Such a survey covered in great detail would indeed be a monumental task and it has therefore been necessary to be somewhat brief and even omit topics that some people will believe to be essential. Whatever the views may be, the author has found the methods herein to be of significant use in producing reliable, cost-effective designs.

The use of basic design methodologies during the conceptual design phase cannot be over-emphasized. The application of closed form and network methods is not a particularly time-consuming process. Even complex forced air flow systems require surprisingly few elements and nodes for evaluation of the adequacy of a conceptual design. Many of these calculations do not even require the use of digital computer solutions. When the design has evolved such that it appears satisfactory to all parties involved (industrial designers, systems architects, manufacturing personnel, as well as mechanical packaging engineers), a more detailed design of the system is reasonable. At that point, a computer program may be valuable because of the size of the problem.

Detailed analysis of the fluid portion of a system design will probably require solution using a pressure/fluid network methodology. The non-linearity of the simultaneous equations requires a digital computer for all but the simplest of systems. The fluid solution will serve as input to the thermal portion of the problem because of the flow-dependent characteristics of the fluid temperature and convective heat transfer from surfaces to fluid. Application of CFD solution methods to complete systems is becoming more common.

Detailed component thermal analysis certainly requires digital computer solution methods and the analyst has a variety of tools from which to choose. The closed form is a natural choice for many problems, with its ease of application and accuracy when the geometry of the component package is appropriate. The thermal network method may also be used, but when the geometry becomes complex and large temperature gradients are anticipated, the finite element method is certainly an expected choice. When thermal stress considerations are anticipated, the thermal model and computed temperature distribution are easily adapted to the stress analysis.

The printed circuit board, as discussed previously, is an important problem, and commercial code vendors are providing some interesting computational tools. However, the complexity of the air flow and convective heat transfer problems is so great that potential users must be aware that they must be willing to bear the responsibility of determining the appropriate input that will result in truly useful temperature predictions.

NOMENCLATURE

A	a length dimension, also an area, also a single source influence coefficient, °C/W
B	a length dimension
b	component length in Eq. (3.135)
C	specific heat, J/kg −°C, also a thermal capacitance, Joules/°C, also a length dimension
d	component pitch in Eq. (3.135)
F	a particular heat dissipation, also a functional for variational problem
f	volume or surface function for variational problem
$\mathcal{F}$	radiation shielding factor, dimensionless
G	air flow, liters/s
g	acceleration of gravity
H	total head loss mm or inches of H_2O
h	head, mm or inches of H_2O, also heat transfer coefficient, W/cm^2−°C or W/m^2−°C
K	internode conductance, W/°C
k	thermal conductivity, W/m −°C
L	a length dimension
$\dot{M}$	mass flow rate
N	an exponent to account for laminar or turbulent flow, also indicates component row number, also a finite element polynomial

NR	number of thermal resistances in complex, closed form solution
NS	number of sources in complex, closed form solution
p	pressure, N/m^2
Q	heat flow, W
q	heat flux, W per unit area
R	thermal resistance, °C/W
r	a length dimension
s	air flow source, liters/s
T	temperature, °C
TO	temperature at termination of thermal resistance, °C
t	thickness of heat flow path, cm, also time, s
V	volume, m^3
v	velocity, m/s

Greek

α	parameter defined after Eq. (3.21)
β	parameter defined after Eq. (3.21)
ε	a constant in Eqs. (3.10), (3,11), and (3.74), also emissivity
ρ	density, kg/m^3
σ	Stefan-Boltzmann constant
ϕ	Fourier coefficient
ψ	Fourier coefficient

Subscripts

A	refers to ambient, also refers to one way air flow
a	refers to air gap
b	refers to board conduction
C	indicates a cross-sectional area, also indicates a convection mode
CC	refers to card cage
CE	refers to external convection
CI	refers to internal convection
D	indicates discharge
DD	refers to disk drive
e	indicates enhanced h

f	indicates free area, also indicates fluid, also indicates fin
H	indicates vertical flat plate of height H
I	indicates inlet
i	an index in a summation, also indicates interior
JC	refers to junction to case
j	an index in a summation
k	indicates the conduction mode, also an index in a summation
ℓ	an index in a summation
m	an idex in a summation
o	indicates exterior, also indicates ambient temperature or pressure
p	refers to a constant pressure condition
r	indicates radiation
S	indicates surface
s	indicates static head
T	indicates a total head
V	refers to unit volume
v	indicates a velocity head
x	refers to x-coordinate, also indicates derivative with respect to x
y	refers to y-coordinate
z	refers to z-coordinate
∞	indicates an ambient

REFERENCES

Agonafer, D and Moffat, D F (1989). Numerical modeling of forced convection heat transfer for modules mounted on circuit boards, *Num Sim of Conv in Elec Equip Cooling*, ASME Pub **HTD-121**, 1–5.

American Society of Heating, Refrigerating, and Air Conditioning Engineers, Standard 51-75.

Anderson, A M and Moffat, R J (1990). A heat transfer correlation for arbitrary geometries in electronic equipment, presented at *1990 ASME Winter Annual Meeting*, Dallas, TX.

Brebbia, C A and Dominguez, J (1989). *Boundary Elements, An Introductory Course*, Computational Mechanics Pubs, Boston, co-published with McGraw-Hill Book Co, New York.

Burnett, D S (1987). *Finite Element Analysis*, Addison-Wesley Pub Co.

David, R F (1977). Computerized thermal analysis of hybrid circuits, presented at the *1977 Elec Comp Conf*, Arlington, VA.

Dean, D J (1985). *Thermal Design of Electronic Circuit Boards and Packages*, Electrochemical Pubs Limited, 8 Barns St, Ayr, Scotland.

Ellison, G N (1991). Extensions of a closed form method for subtrate thermal analyzers to include thermal resistances from source-to-substrate and source-to-ambient, *7th Ann IEEE Semiconductor Thermal and Temp Mgt Symp*, Feb.

Ellison, G N (1990). TAMS: A thermal analyzer for multilayer structures, *Electrosoft*, **1**, Jun, 85–97.

Ellison, G N (1989). *Thermal Computations for Electronic Equipment*, Van Nostrand Reinhold, 1984. Reprinted by Robert E. Kreiger Pub Co, Malabar, FL.

Ellison, G N and Patelzick, D L (1986). The thermal design of a forced air cooled power supply, *Proc 6th Intl Elect Packaging Conf*, 829–838.

Ellison, G N (1984). A review of a thermal analysis computer program for micro-electronic devices, *Thermal Mgt Concepts in Microelec Packaging*, ISHM Tech Monograph Series 6984-003.

Ellison, G N (1976). The thermal design of an LSI single chip package, *IEEE Trans Parts, Hybrids, and Packaging*, **PHP-12**, Dec, 371–378.

Ellison, G N (1973). The effect of some composite structures on the thermal resistance of substrates and integrated circuit chips, *IEEE Trans Electron Dev*, **ED-20**, Mar, 233–238.

Estes, R (1987). Thermal design considerations for COB applications, presented at the *7th Ann Intl Elec Packaging Conf.*

Hein, V L and Lenzi, V K (1969). Thermal analysis of substrates and integrated circuits, presented at the *1969 Elec Comp Conf.*

Holman, J P (1990). *Heat Transfer*, 7th Ed, McGraw-Hill Book Co, New York.

Hsu, T R (1990). *The Finite Element Method in Thermomechanics*, Allen and Unwin, Boston, MA.

Lienhard, J H (1987). *A Heat Transfer Textbook*, 2nd Ed, Prentice-Hall Inc, Englewood Cliffs, NJ.

Linton, R A and Agonafer, D (1989). Thermal model of a PC, *Num Sim of Conv in Elec Equip Cooling–1989*, ASME Pub **HTD-121**, 69–72.

Kraus, A D and Bar-Cohen, A (1983). *Thermal Analysis and Control of Electronic Equipment*, McGraw-Hill, New York, NY.

Palisoc, A L and Lee, C C (1988). Thermal design of integrated circuit devices, *4th Ann IEEE Semiconductor Thermal and Temp Measurement Conf.*

Pinto, E J and Mikic, B B (1986). Temperature prediction on substrates and integrated circuit chips, presented at the *1986 AIAA/ASME Thermophys and Heat Trans Conf.*

Ramakrishna, K (1982). Thermal analysis of composite multilayer structures with multiple heat sources, presented at the *2nd Intl Elec Packaging Conf.*

Reddy, J N (1984). *An Introduction to the Finite Element Method*, McGraw-Hill, New York, NY.

Schnipke, R J (1989). A fluid flow and heat transfer analysis for evaluating the effectiveness of an IC package heat sink, *5th Ann IEEE Semiconductor Thermal and Temp Measurement Symp*, Feb, 81–87.

Segerlind, L J (1976). *Applied Finite Element Analysis*, John Wiley & Sons, Inc, New York, NY.

Steinberg, D S (1991). *Cooling Techniques for Electronic Equipment*, Wiley-Interscience Pubs, New York, NY.

Sucec, J (1985). *Heat Transfer*, Wm C Brown Publishers, Dubuque, IA.

Wills, M (1983). Thermal analysis of air-cooled PCBs, *Electronic Production*, May, 11–18.

SOFTWARE REFERENCES (with latest release date)

1. ALGOR Supersap, 1991. ALGOR Interactive Systems, Inc, 260 Alpha Drive, Pittsburgh, PA 15238.

2. Thermax, 1991. Cadence Design Systems, Inc, Systems Div, 555 River Oaks Parkway, San Jose, CA 95134.

3. ANSYS, 1991. Swanson Analysis Systems, Inc, Johnson Road, P.O. Box 65, Houston, PA 15342-0065.

4. AUTOTHERM, 1991. Mentor Graphics Corp, 8005 SW Boeckman, Wilsonville, OR 97070.

5. BEASY, 1991. Computational Mechanics, Inc, 25 Bridge Street, Billerca, MA 01821.

6. BETAsoft, 1991. Dynamic Soft Analysis Corp, 213 Guyasuta Road, Pittsburgh, PA 15215.

7. COSMOS/M, 1991. Structural Research and Analysis Corp, 1661 Lincoln Boulevard, Suite 100, Santa Monica, CA 90404.

8. FIDAP, 1991. Fluid Dynamics International Corp, Suite 400, 1600 Orrington Avenue, Evanston, IL 60201.

9. FLOTHERM, 1991. FLOMERICS, Inc, Lyman Office Center, Suite 202, 33 Lyman Street, Westborough, MA 01581.

10. FLOTRAN, 1991. Compuflo Corp, 1575 State Farm Boulevard, Charlottesville, VA 22901-8611.

11. FLUENT, 1991. Fluent, Inc, Centerra Resource Park, 1010 Cavendish Court, Lebanon, NH 03766.

12. SURFER, 1991. Golden Software, Inc, 807 14th Street, P.O. Box 281, Golden, CO 80402.

13. NISA II, 1991. Engineering Mechanics Research Corp, P.O. Box 696, Troy, MI 48099.

14. PCB Explorer, 1991. Pacific Numerix, 1200 Prospect Street, Suite 300, La Jolla, CA 92037.

15. PHEONICS, 1991. CHAM of North America, Inc, 4960 Corporate Drive, Suite 100A, Huntsville, AL 35805.

16. SINDA/85FLUINT, 1990. COSMIC, the University of Georgia, 382 E Broad Street, Athens, GA 30602.

17. TAEE, 1991. Intercept Software, 3425 S Bascom Avenue, Campbell, CA 95008.

18. TAMS, TNETFA, 1991. Thermal Computations, Inc, 3876 SE Bliss Court, Hillsboro, OR 97123.

19. TAKII, 1991. K&K Associates, P.O. Box 2882, Lyons, CO 80540.

20. TMG, 1991. Maya Heat Transfer Technologies, Ltd, 5180 Queen Mary Road, 380, Montreal, Quebec, Canada, H3W 3E7.

21. TDS, 1991. Sampson Technical Consultants, Inc, P.O. Box 4828, Auburn, CA 95604.

Chapter 4

THERMAL STRESS FAILURES OF SURFACE MOUNTED COMPONENTS

Dave S. Steinberg
Steinberg & Associates

1 INTRODUCTION

Electronic equipment failure rates have decreased significantly over the past decade because manufacturing process controls and cooling techniques have improved. Electronic components are performing at temperatures lower than ever before and component junction temperatures in sophisticated electronic assemblies very seldom exceed a maximum limit of 100° C under any operating condition. However, the failure rates in electronic assemblies are still too high. These failure rates must be reduced further in order to provide a quality, highly reliable product at a reasonable price, that can compete in an extremely competitive world market.

Experience has shown that many of the failures that occur in electronic systems can be directly related to high strains that result from temperature changes and from a mismatch in the thermal coefficients of expansion (TCE) of the various materials that are used in these systems. In the printed circuit board (PCB) the TCE mismatch can produce dimensional changes in the materials which typically result in high forces and stresses in the electronic component lead wires, solder joints and plated through holes (PTH). These stresses often lead to early failures.

Dimensional changes can also occur when there are significant differences in the thermal mass capacitances of two identical materials with exactly the same TCEs. Temperature changes due to heating or cooling are inversely proportional to the mass of the component. They will occur more rapidly in a small mass, while temperature changes will occur more slowly in a very large mass. The small mass will expand and contract more rapidly than the large mass creating an expansion difference which can again produce high forces and stresses, and possible failures, in the various structural elements of the electronic members.

Temperature changes can derive from a wide variety of conditions such as power cycling (turning the power "on" and "off"), season changes from summer to winter, solar heating, heat under the hood of an automobile, altitude changes in an aircraft, and heat developed by a chemical process. These temperature changes lead to thermally induced expansion forces and stresses which can be reduced, and the reliability of the electronic equipment can be improved, by developing a better understanding of the physical properties of the various materials used in the assembly of these electronic systems.

2 SURFACE MOUNTED ELECTRONIC COMPONENTS

Surface mounting of electronic components has evolved from the demand to reduce the size and cost of electronic equipment from commercial television sets to rugged and sophisticated military navigation and control systems. Surface mounted devices can be adapted easily to automated assembly using precision pick and place machines which show great promise to reduce long term manufacturing and assembly costs. Almost every type of electronic component can be obtained in a surface mounted configuration or can be adapted for surface mounted applications. This includes leadless as well as leaded devices.

Leadless surface mounted components have become very popular because of their small size which has resulted in smaller enclosures with more power and capability. Leadless devices are packaged in rectangular cases that are only about one third the size of a dual in-line package (DIP) with the same electrical functions. These devices are typically packaged in plastic cases and ceramic cases that surround a silicon chip. Sometimes the plastic cases are provided with lead wires, so they are called plastic leaded chip carriers (PLCCs). When the ceramic cases are not provided with lead wires they are called leadless ceramic chip carriers (LCCCs). Ceramic devices are hermetically sealed, with a ceramic or a kovar lid, and are usually used in military applications because the severe thermal, vibration, shock and high humidity conditions require the use of rugged electronic components.

Ceramics (aluminum oxide) have a low TCE of about 6 parts per million per °C (PPM/°C), so LCCC devices must be mounted with great care to avoid premature failures. Three major problems are involved in the mounting of these devices:

1. Large differences in the TCE between the LCCC and the PCB must be avoided, controlled, or compensated for in some way or high solder joint shear stresses may result in early failures. The TCE of a typical epoxy fiberglass PCB is about 15 PPM/°C along the x and y-axes and for polyimide fiberglass the value is about 13 PPM/°C. More exotic materials such as ceramics, kevlar and quartz with much lower TCEs, but with much higher costs, are being used in some applications to improve the solder joint fatigue life.

2. Improper mounting of the LCCC component on the PCB can result in a very high junction temperature which can damage the component and affect the electrical performance of the system.

3. Solder joints can be damaged due to PCB flexure during bench handling, or during insertion and removal of the PCB from a chassis assembly, which may lead to early failures.

Large LCCC devices with 20 or more electrical contacts must be examined very carefully before they are mounted on plain epoxy fiberglass or polyimide glass PCBs because thermal expansion differences can produce large relative displacements and stresses in the solder joints. This is shown in Fig. 4.1 and it is seen that these conditions can result in early solder joint failures.

Special precautions can be taken to reduce solder joint stresses by laminating one or more layers of stiff copper-invar-copper (C-I-C) sheets to a multilayer PCB to reduce its TCE. This assures that there is a closer match with the LCCC, as shown in Fig. 4.2. As shown by Hastings (1988), the addition of C-I-C sheets, usually about 0.006 inches thick, will reduce the TCE in the plane of the PCB (along the x and y-axes) and reduce the mismatch between the PCB and the LCCC. However, when the thermal expansion of the PCB is restrained along the x and y axes, the thermal expansion along the z axis (normal to the plane of the PCB) will increase. This will increase the forces and stresses developed in the copper plated

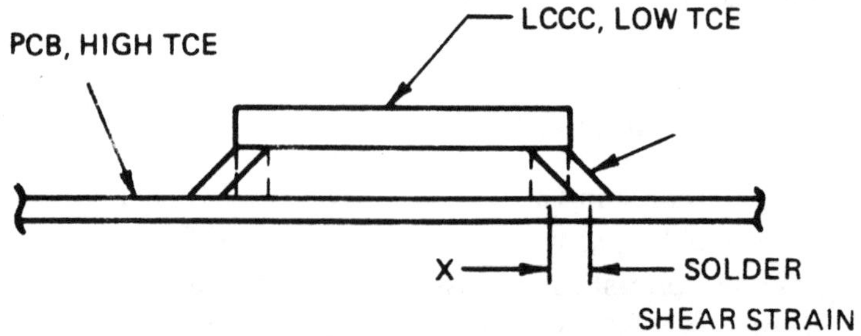

Figure 4.1 Differences in the TCE of the LCCC and the PCB force the solder to shear and creep.

through holes (PTHs). Therefore, in order to reduce the possibility of premature cracking of the PTHs, special precautions must be taken to increase the PTH copper strength when C-I-C planes are added to a PCB. This is usually accomplished by increasing the thickness of the copper in the PTH, which increases the strength and reliability.

Another method for avoiding solder joint failure is to add relatively long electrical lead wires to the LCCC, which will be compliant enough to allow the PCB to expand freely without inducing high shear stresses in the solder joints. This is shown in Fig. 4.3.

Surface mounted components for commercial applications often use plastic bodies to support the silicon chips. Plastics have a much higher TCE than the PCB along the x and y planes which again results in significant expansion differences in thermal cycling environments, as shown in Fig. 4.1. Again this can result in high solder joint shear stresses and early solder joint failures.

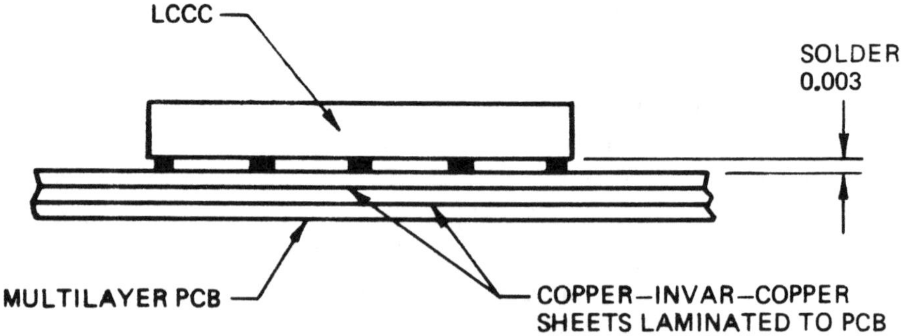

Figure 4.2 Copper-invar-copper sheets used to reduce the TCE of a multilayer PCB.

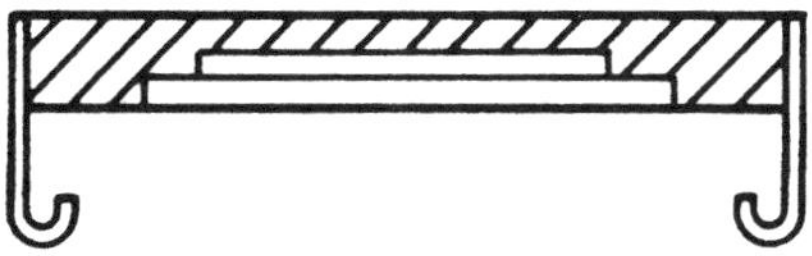

"J" WIRES

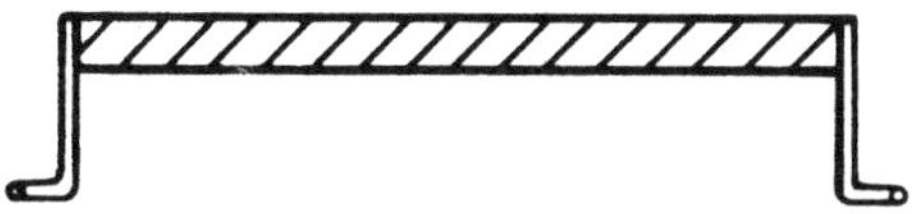

GULL WING WIRES

Figure 4.3 Electrical lead wires added to leadless ceramic chip carriers to reduce solder forces and stresses.

Many large component manufacturers are adding electrical lead wires to their plastic components, making them plastic leaded chip carriers (PLCCs). The addition of these lead wires reduces the forces and stresses in the solder joints and this increases the fatigue life of the electronic system.

Chip carrier components are normally associated with surface mounted PCBs. However, many plastic and ceramic chip carriers are being mounted on through-hole PCBs with the use of sockets. A large number of manufacturers have developed a wide variety of sockets that can accommodate almost every type of component made. Dimensional tolerances associated with the manufacturing of the components and the sockets can result in a very tight or a very loose fit so that every package might not be accommodated properly in every socket. Individual components and sockets must be tested to make sure they can work together properly.

Components mounted in sockets may not be suitable for use in severe military environments involving high levels of vibration and shock. These environments can result in large amounts of relative motion between the component and the socket, resulting in severe fretting corrosion at the socket electrical interface. This can result in rapid wear through the outer protective layers of gold and nickel on the electrical contacts. The result can then be a contact which is brass on brass. High friction and velocity at the contact interfaces then produces a black oxide that is not electrically conductive. This can produce an electrical open circuit for about one microsecond, and this can cause the loss of several bits of information in a fast switching circuit. In a slow analog circuit, such a time interval may not even be noticed.

This type of fretting wear on the contacts is hard to see by a visual inspection because polished brass looks very much like gold. However, if a five percent solution of sodium sulfide (in water) is brushed on the contacts, a black spot will appear in about four seconds if any copper base alloy is present. This is a non-destructive test because the solution is harmless and can be washed off with water. If a black spot does appear on any of the contacts, it shows that there is wear through the outer protective layers of gold and nickel, and that the component and the socket must be replaced [Steinberg (1989a)].

3 THE AIR FORCE AVIONICS INTEGRITY PROGRAM

The Air Force is attempting to achieve a dramatic increase in the integrity and reliability of its avionic equipment through a better understanding of the failure mechanisms that are often induced by careless designs and manufacturing processes and procedures. These requirements are outlined in an avionics integrity program (AVIP) specification (listed as MIL-A-87244) to ensure that the avionic systems will be able to achieve, over the life of the equipment, the desired performance and a minimum failure free operating period. The Air Force hopes to achieve this large improvement in reliability while holding down the acquisition and support costs.

An examination of a large number of avionics failures has shown that, except for voltage variations and out-of-tolerance conditions, most of these failures are mechanical in nature. The failures typically occur in component lead wires, solder joints, plated through holes, electrical connectors, harnesses, wires, adhesive bonded joints and hermetic seals. These failures, as pointed out by Steinberg (1988, 1989b, 1991), are often the result of poor design and manufacturing practices in a combination with severe thermal, vibration, shock, salt and humidity environments.

4 SLOW CYCLE AND RAPID CYCLE FATIGUE PROPERTIES OF SOLDER

Engelmaier (1982) has indicated that most of the electronic components used for surface mounted technology (SMT) applications rely upon the structural integrity of the solder to support these components in various thermal, vibration, and shock environments. Some exceptions to this are in the large components, such as transformers and hybrids, that may be bolted or cemented to the PCB. These components often provide heavy loads on their electrical lead wires, and solder joints are not capable of carrying high dynamic loads. Sometimes dummy lead wires can be added to large components to help carry high dynamic loads. Extra wires can reduce the load carried by each wire and this tends to reduce the probability of a failure.

Because solder is a strange material that can cause many different types of failures, the physical properties of solder must be investigated thoroughly to understand how some of these failures can be prevented. For example, experience has shown that solder can creep at a rate of about 0.002 to 0.003 in. per year when subjected to compression loads. A number of field failures have occurred due to solder creep where short circuits developed due to solder extruding across a 0.002 in. thick kapton insulating washer under a power transistor bolted to a PCB over a five year period. The PCB had a thin solder surface coating, and the kapton washer was used for electrical insulation protection. The thin kapton washer was also used to maintain the 0.002 in. gap between the PCB and transistor and to suppress radio frequency interference (RFI) and electromagnetic interference (EMI) locally. The action of the bolts clamping the transistor produced a compression force on the solder under the transistor, slowly extruding the solder until it bridged across the gap, producing a short circuit. The washer thickness was increased to 0.010 in. and the problem was solved, as the maximum solder extrusion was 0.005 in. and there were no adverse affects with RFI and EMI.

The solder that is most often used is eutectic solder, which is 63% tin and 37% lead. This solder has a melting temperature of only about 184° C, which is low enough to prevent thermal damage to sensitive electronic components when they are attached to the PCB. This solder has a tendency to cold flow and creep extensively when stress levels exceed values of about 800 pounds per square inch

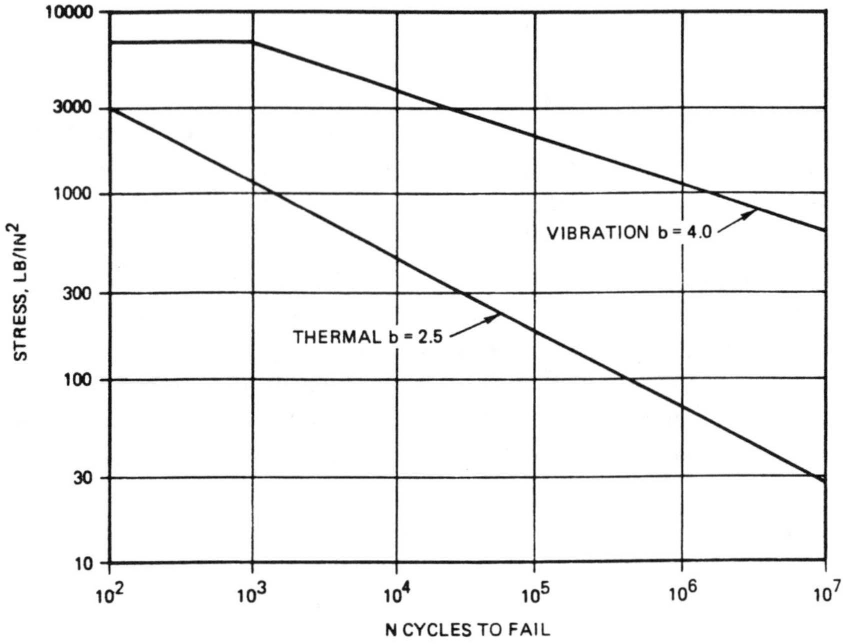

Figure 4.4 Vibration and thermal cycle fatigue, 63/37 solder. The data is for room temperature [Steinberg (1991)].

(psi) at room temperature conditions. These creep effects are temperature sensitive. They increase greatly at high temperatures above about 85° C. However, the creep effects are sharply reduced at low temperatures (below about −20° C). These creep characteristics can cause premature failures in the solder unless the stress levels are kept below values of about 400 psi in thermal cycling and vibration environments [Steinberg (1991)]. Although solder has an apparent static strength of about 6000 psi at room temperatures, the maximum allowable stress level should be limited to a value of about 1000 psi for vibration conditions where less than about one million stress cycles are involved. When more than 10 million vibration stress cycles are expected, the solder joint stress levels should be limited to about 500 psi [Steinberg (1989a)].

Thermal cycle fatigue in solder is often classified as slow cycle fatigue and vibration fatigue in solder is often classified as rapid cycle fatigue. These properties are shown in Fig. 4.4.

The strength of solder depends a great deal upon the cycle speed of the applied load, so the strength of solder appears to be time dependent. This is due to the creep properties of the material. A slow thermal cycle will allow more creep to occur, which is more damaging to the solder than a rapid thermal cycle where creep does not have a chance to occur. These characteristics are shown in Fig. 4.5.

Solder becomes very plastic when it is exposed to high temperatures and to high stresses. These conditions produce a high creep rate which permits the solder to relax and to relieve these high internal stresses. The approximate time it takes for the solder to relax and to relieve internal stresses depends upon the temperature, as shown in Fig. 4.6. High internal stresses can often be relieved in a matter of minutes at temperatures around 100° C but it may take a week or more to relieve high stresses when the temperatures are below −20° C. These considerations are important for determining the stress levels in solder joints when the fatigue life is calculated. For example, when the thermal stress in a solder joint is to be calculated for a specific temperature cycling environment, it is necessary to know how fast the temperature is changing. When the temperature change is rapid, the solder does not have a chance to creep and relieve the high stresses. Thus, only half of the full temperature cycling range may be used for the stress calculations. When

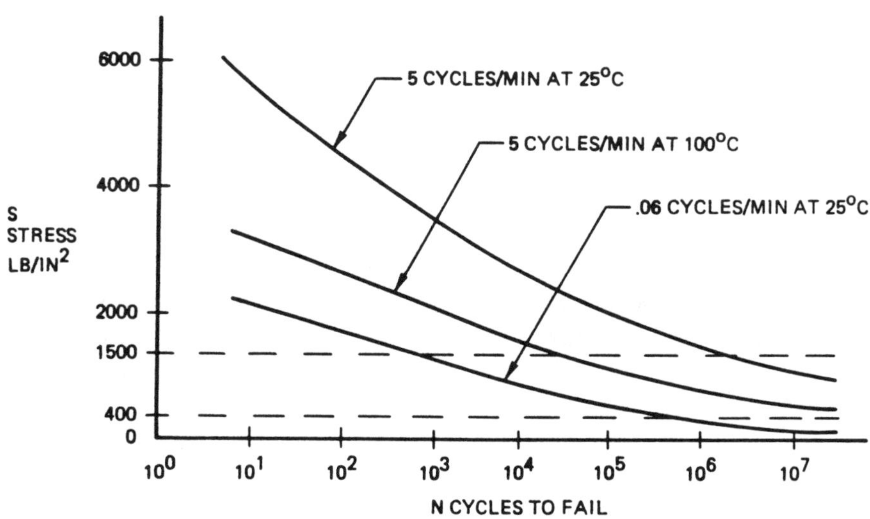

Figure 4.5 63/37 Sn/Pb Solder alternating lap shear stress versus temperature and frequency of applied load [Steinberg (1991)].

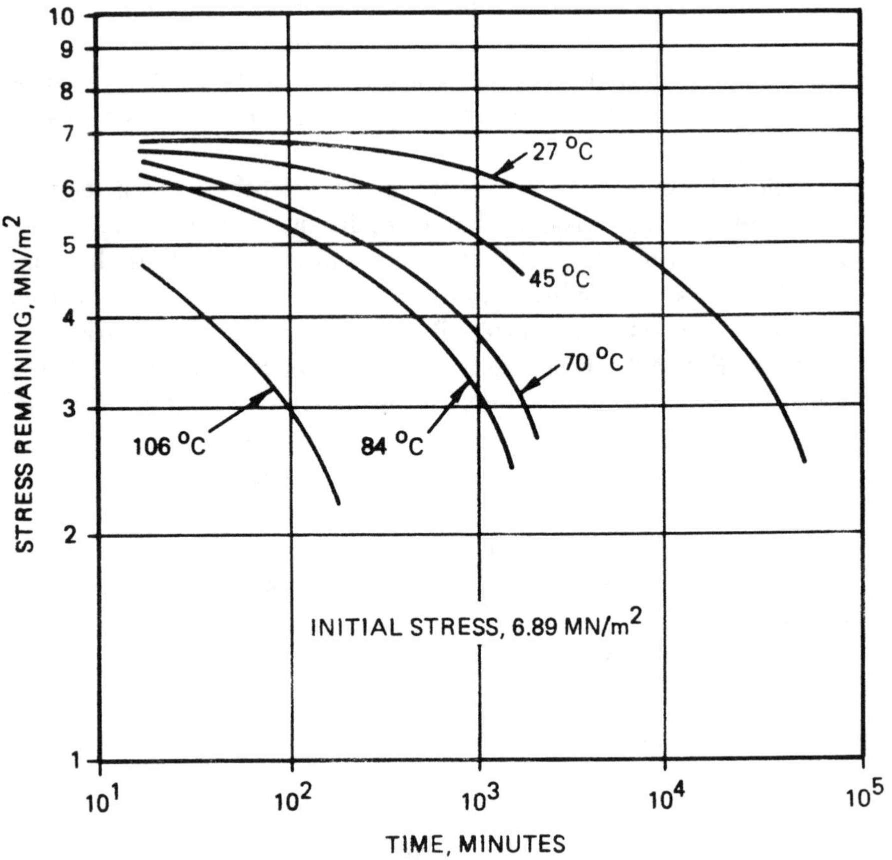

Figure 4.6 Relaxing of stresses due to creep in 60/40 tin/lead solder as a function of time and temperature [Massie (1987)].

the temperature change is rapid but followed by an extended period where the temperature is held constant, the creep properties will allow the solder to cold flow and relax the high stresses and strains to a near zero stress state [Evans (1989)]. The full temperature cycle range must then be used to determine the stress value in the solder because each cycle starts from a near zero stress condition.

This situation is similar to the stress state in a bar of steel that can be axially loaded to produce positive (tension) and negative (compression) stresses. Starting from a neutral point, a positive load on the bar produces a positive stress. Bringing the bar back to the neutral point brings the stress back to zero. A negative (compression) load on the bar produces a negative stress, and bringing the bar back to a neutral point also brings the stress back to zero.

The stress state in the steel bar can be compared to the stress state in a solder joint that will be rapidly temperature cycled from room temperature at 20° C to an upper level of 70° C and then to a lower level of −30° C. The same process is then repeated over the same maximum temperature change of 100° C for several hundred thermal cycles. The question here is what temperature cycle range should be used to determine the stress level in the solder joint; is it 50° C or is it 100° C?

In order to answer this question, it is necessary to go back to the steel bar and ask the same question: how is the stress determined? For the steel bar, the true stress level is determined from the neutral point to the maximum positive stress or from the neutral point to the maximum negative stress. The maximum positive stress and the maximum negative stress must not be added to obtain the total stress because this would result in a value that is twice the correct value. The same situation applies to the solder joint stress where the correct thermal cycling range that must be used is 50° C for the rapid temperature change condition.

The temperature cycling condition for determining the solder joint stress is exactly the same as the loading condition on the steel bar. In both cases the correct stress must be determined from the neutral point to the maximum positive stress, and from the neutral point to the maximum negative stress. This only applies to the solder joint stresses for the rapid temperature cycling condition, with no extended dwell periods, so that the solder does not have a chance to creep and to relax the high stresses and strains back to a near zero stress condition.

5 THERMAL EXPANSION FORCES IN SOLDER JOINTS AND LEAD WIRES

Electronic products are fabricated from a wide variety of plastics and metals which have significant differences in their physical properties. Metals (except for solder) are assumed to have linear properties, that is, the stress is proportional to the strain. Plastics, which are often used in printed circuit boards can exhibit non-linear characteristics when they are subjected to high temperatures or to high stresses. All of these materials have different expansion coefficients in different directions and at different temperatures. Thermal coefficients of expansion (TCEs) for the various materials must be known over the operating temperature range in order to determine the forces and stresses developed in these electrical structural elements.

Slight differences in the thermal expansions of electronic components and PCBs can produce high forces in the solder joints and electrical lead wires. These forces can lead to early fatigue failures if these elements are exposed to any type of thermal cycling environment. Thermal expansion forces can be estimated by assuming that the solder joints and lead wires act like linear structural members. Spring rates can then be determined as a function of the geometry and physical properties of the materials involved. Steinberg (1991) shows that under these conditions, the force developed in such a linear system can be expressed as

$$P = ky \tag{4.1}$$

where

P = Force developed in the structural members, lb
k = Spring rate of the structural members, lb/in
y = Displacement developed in the members, in

An examination of the structural system shows that the displacements developed in thermal cycling environments can be shown to be

$$y = \alpha L \Delta T \tag{4.2}$$

where

y = Thermal expansion displacement in the members, in
α = Thermal coefficient of expansion (TCE), in/in/°C
ΔT = Temperature change expected in the members, °C
L = Length of the structural members, in

When Eq. (4.2) is used in Eq. (4.1), it is observed that, because α and L are constant, the only possible way that, for a given ΔT, the force P can be reduced is to reduce the spring rate k. The spring rates of the various structural members must therefore be examined, to see how they affect the thermal forces and stresses in the system.

An examination of the solder joints forced to shear, as demonstrated in Fig. 4.1, shows that the relative displacement between the component and the PCB can be expressed in terms of the standard shear displacement relationship:

$$y = \frac{Ph}{AG} \tag{4.3}$$

where

y = Thermal expansion from Eq. (4.2), in
P = Force developed in solder, lbs
h = Height of solder joint, in
A = Cross section area of solder joint, in^2
G = Shear modulus of solder, lb/in^2 (psi)

The shear force developed in the solder joint can then be obtained:

$$P = \frac{AGy}{h} \tag{4.4}$$

This condition can be demonstrated with a sample problem.

6 SAMPLE PROBLEM — FORCE INDUCED BY A THERMAL EXPANSION MISMATCH

A hybrid component in a kovar case measuring 0.30 in. square is to be surface mounted on a polyimide PCB for operation in a rapid thermal cycling environment over a temperature range from $-30°$ C to $+70°$ C. Electrical interconnections are made with one solder pad at each of the four corners of the hybrid (a total of four solder pads). Each solder pad measures 0.030×0.080 in. with an expected height of 0.004 in. Determine if the proposed hybrid mounting is acceptable for the operating environment.

SOLUTION

Thermal expansion differences between the hybrid kovar case and the polyimide PCB will produce relative displacement differences that will produce shear forces in the corner solder joints. The greatest dimensional lengths will produce the greatest forces in the solder joints. Thus, as proposed by Engelmaier (1988), the diagonal dimension must be used to determine the maximum induced force. A worst case analysis is used here by assuming that the centroid of the solder pad is located at the corner of the hybrid and by assuming that all of the relative expansion differences will be reflected by shear displacements in the solder joints. Because all expansions take place with respect to the centroid of a structure, one half of the diagonal dimension of the hybrid must be used in determining the thermal forces developed in the solder joints.

The TCEs in the x-y plane for the polyimide PCB and the kovar hybrid are

$$\alpha_P = 13 \times 10^{-6} \text{ in/in/}^\circ \text{C}$$

and:

$$\alpha_K = 6 \times 10^{-6} \text{ in/in/}^\circ \text{C}$$

The solder shear modulus is

$$G = 1.2 \times 10^6 \text{ lb/in}^2$$

and the pertinent dimensions are the half diagonal:

$$L = \sqrt{2}(0.30)/2 = 0.2121$$

the expected solder joint height:

$$h = 0.004 \text{ in}$$

and the solder pad area:

$$A = (0.030)(0.080) = 0.0024 \text{ in}^2$$

The maximum temperature range is $70 - (-30) = 100^\circ$ C and this makes the neutral point to maximum temperature for a rapid cycle

$$\Delta T = \frac{100}{2} = 50^\circ \text{ C}$$

The thermal expansion difference is obtained from Eq. (4.2):

$$\begin{aligned} y &= (\alpha_P - \alpha_K)L\Delta T \\ &= [(13 - 6) \times 10^{-6}](0.2121)(50) \\ &= 7.425 \times 10^{-5} \text{ in} \end{aligned}$$

and with this in Eq. (4.4), the shear force is

$$\begin{aligned} P &= \frac{AGy}{h} = \frac{(0.0024)(1.2 \times 10^6)(7.425 \times 10^{-6})}{0.004} \\ &= 53.46 \text{ lb} \end{aligned}$$

In the worst case condition, only one solder joint will carry this shear load, which will result in the following shear stress.

$$S_s = \frac{P}{A} = \frac{53.46}{0.0024} = 22,274 \text{ lb/in}^2$$

This shear stress is well above the maximum desired solder joint shear stress of 400 psi. This indicates that rapid failures can be expected and that the proposed design is unacceptable.

Some changes in the proposed design can be investigated to see how they affect the forces and stresses. For example, what can be expected if constraining layers of copper-invar-copper are added to the polyimide PCB [Hastings (1988)]? The mismatch in the TCEs for the composite PCB and the kovar hybrid can probably be reduced to about two parts per million per °C (2 PPM/°C). This will reduce the expansion difference to about 21.2 millionths of an inch, resulting in a solder joint shear force of 15.3 lbs, and a solder joint shear stress of 6364 psi. This is still much too high and is still not acceptable.

Increasing the height of the solder joint will help to lower the solder joint shear stresses. If the solder joint height is increased to 0.010 in., with the C-I-C polyimide composite PCB the expansion difference will remain the same but the solder shear force will be reduced to 6.1 lbs. This will reduce the shear stress in the solder to 2545 psi which is still unacceptable.

Special "J" wires or "gull wing" wires with sufficient height may have to be added to the hybrid component to reduce the stress levels to 400 psi or less. This is described in the next section.

7 SAMPLE PROBLEM — REDUCING THE THERMAL EXPANSION FORCES IN SOLDER JOINTS WITH THE USE OF J-WIRES

Determine the shear forces developed in the solder joints of the previous problem for a rapid temperature cycle condition of 100° C when copper J-wires with a length of 0.060 in. and with a diameter of 0.014 in. are added to the hybrid mounted on a plain polyimide fiberglass PCB as shown in Fig. 4.7.

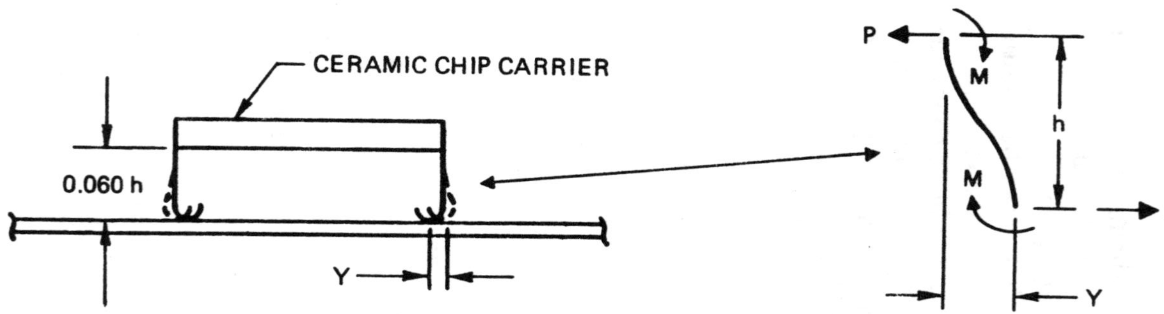

Figure 4.7 Differences in expansion between the PCB and the ceramic chip carrier force the lead wires to bend. The 0.060 represents the height in inches.

SOLUTION

Relative displacements will occur between the kovar hybrid and the plain polyimide glass PCB during thermal cycling. This relative motion difference will induce bending in the relatively long copper J-wire [Kotlowitz and Engelmaier (1989)]. The bending action of the wire will replace the shear action of the solder joint so Eq. (4.3) must be modified to reflect this bending. Because the wire will be rigidly clamped at the hybrid body and at the solder joint, the wire can be treated as a long beam clamped at both ends, and forced to displace laterally. The bending displacement for this condition is given by

$$y = \frac{Ph^3}{12EI} \tag{4.5}$$

where E is the modulus of elasticity and I is the moment of inertia. Solving for the bending force developed in the wire and the shearing force in the solder joint:

$$P = \frac{12EIy}{h^3} \tag{4.6}$$

Here, as before

$$\begin{aligned} \alpha_P &= 13 \times 10^{-6} \text{ in/in/°C} \\ \alpha_K &= 6 \times 10^{-6} \text{ in/in/°C} \\ L &= 0.212 \text{ in} \\ \Delta T &= [70 - (-30)]/2 = 50° \text{ C} \end{aligned}$$

and with these values, y was computed as

$$y = 7.425 \times 10^{-5} \text{ in}$$

The moment of inertia of the wire is

$$I = \pi d^4/64 = \pi(0.014)^4/64 = 1.886 \times 10^{-9} \text{ in}^4$$

and with the modulus of elasticity of copper taken as $E = 16 \times 10^6$ lb/in.^{2} and a height of J-wire, $h = 0.060$ in., the bending force in the copper wire which is the same as the shear force in the solder joint can be obtained from Eq. (4.6) as

$$P = \frac{12(16 \times 10^6)(1.886 \times 10^{-9})(7.425 \times 10^{-5})}{(0.060)^3} = 0.124 \text{ lb}$$

The shear stress in the solder joint becomes

$$S_s = \frac{P}{A} = \frac{0.124}{(0.018)(0.060)} = 114.8 \text{ lb/in}^2$$

Because this solder shear stress is well below 400 psi, the hybrid installation for the solder is acceptable [Steinberg (1991)].

Bending stresses in the copper J-leads must be determined to make sure that they are not excessive. The bending moment must be obtained first by considering the sum of the moments acting on the wire. This results in

$$M = \frac{Ph}{2} = \frac{(0.124)(0.060)}{2} = 0.00372 \text{ in-lb}$$

The J-wire bending stress can be obtained from the standard bending stress relation (the so-called flexure formula)

$$f_b = \frac{Mc}{I}$$

where

$$M = 0.00372 \text{ in-lb, the bending moment in J-wire}$$
$$c = 0.014/2 = 0.007 \text{ in, the wire radius}$$

and

$$I = 1.886 \times 10^{-9} \text{ in}^4\text{, the wire moment of inertia}$$

Thus, the bending stress is

$$S_b = \frac{(0.00372)(0.007)}{1.886 \times 10^{-9}} = 13,807 \text{ lb/in}^2$$

Because this is below the fatigue endurance limit of 15,000 psi for copper at about 100 million fatigue cycles, there should not be any problems with bending fatigue in the copper J-wire as long as there are no sharp and deep cuts or scratches in the copper wires [Steinberg (1989a)].

Extreme caution must be used when long J-wires are mounted on components that are cooled by conduction to the PCB. The long J-wires will increase the height of the air gap under the component and thus increase the temperature rise across the component interface for high power devices. Some type of interface pad may have to be used under the component to eliminate the air gap and to decrease the thermal resistance across the interface.

Any thermal pad under the component should have a low modulus of elasticity or a low z-axis TCE (or both) to reduce any high z-axis forces that may develop due to high thermal expansions.

8 PREDICTING THE FATIGUE LIFE IN ELECTRONIC ASSEMBLIES

Experience with military electronic equipment operating in severe thermal cycling and vibration environments has shown that the greatest failure rates occur in the electrical connectors and interconnects in the cables and harnesses. Some Air Force documents show failure rates as high as 30%. The next highest failure rates are often attributed to thermal cycling environments where temperature changes can induce dimensional changes in structural elements such as solder joints, lead wires, plated through holes and various types of electronic components. The simple act of turning the power on and off in a typical electronic system can introduce alternating stresses in the system. When the power is turned on, the system heats up and various structural elements within the system will expand. When the power is turned off, the system cools off and the same structural elements will shrink. The turn on and turn off cycle produces a stress cycle, and every stress cycle uses up a small part of the fatigue life of that system.

Metals and plastics can crack and fracture when they are subjected to repeated loads and stresses. These failures can occur at levels that are far below the static strength of the materials. There are many different theories as to how cracks first develop. However, it appears that submicroscopic cracks are initiated at some point after repeated alternating stresses, and these cracks grow, slowly at first, and then more rapidly until there is a sudden and complete fracture [Followell et al. (1991)].

When a small crack is first observed, it is usually a cause for concern because it often means that a failure will probably occur quite soon if the source of stress is not removed. Sometimes a small crack just stops growing, or the growth rate slows so much that the expected failures do not occur. The best policy to ensure the reliability of the electronic system is to assume that any visible crack will grow until a complete fracture occurs. A corrective action program must then be initiated to correct the deficiency and this can involve the design, the manufacturing process, or even the environment. Cracks in any element should be repaired, or reinforced if possible, or the element should be replaced to avoid future problems.

Alternating loads and stresses can be imposed on structural members in many different ways. Vibration and shock are usually considered to be the most damaging. However, Air Force reports show that thermal stresses are actually more damaging because they account for about 80% of Air Force avionics failures, while vibration and shock only account for about 20% of all failures [Steinberg (1991)]. An examination of a large number of avionics failures appears to show that thermal cycling environments often cause crack initiation. Because the process is very slow, often only a few cycles per day, this process may not result in a complete fracture. These fractures are usually the result of vibration or shock when they are also imposed on the structure. Vibration and shock usually involve very rapid stress reversals, often several hundred cycles per second, so the cracks propagate very rapidly to a complete fracture.

Storage environments are usually considered to be very benign for electronic equipment. However, long term storage for periods of five years or longer can involve several thousand diurnal thermal cycles that can extensively damage systems that are not designed properly or are not preprocessed properly for the long storage environments.

Lack of proper manufacturing processes and controls can be a significant contributor to early failures in electronic systems. Extra heavy cuts on a shaft being machined on a lathe have been known to introduce high alternating stresses in the shaft. High vibration and shock levels in the operating environment then add to the fatigue damage accumulated during the manufacturing phase until a premature fatigue failure occurs. These types of failures are very difficult to trace because the shafts may be fabricated by several different workers, on several different machines, which induce different amounts of damage to the shaft. Fatigue properties are generally available for many different metals and plastics from a variety of test laboratories and handbooks on metals and plastics. Most fatigue data are obtained from alternating stresses imposed on precision machined parts until failures occur. These data are plotted on log-log paper with stress levels (S) along the vertical axis and the number of cycles to fail (n) along the horizontal axis. The best average straight line is then drawn through the various scattered points, as shown in Fig. 4.8 [Steinberg (1991)].

The approximate fatigue life of a structural member may be determined if the magnitudes of the stress levels can be defined and if the general fatigue properties of the materials can be obtained. These are not easy tasks because the fatigue values for the various metals and plastics used in electronic assemblies are often non linear, and often vary greatly with temperature. Steinberg (1988) shows that when the materials are homogeneous, with linear properties, then the fatigue life can be estimated from

$$n_1 S_1^b = n_2 S_2^b \tag{4.7}$$

where

n_1	=	Number of stress cycles required for failure at stress condition 1
S_1	=	Stress magnitude at condition 1, lb/in^2
n_2	=	Number of stress cycles required for failure at stress condition 2
S_2	=	Stress magnitude at condition 2, lb/in^2
b	=	Slope of straight line used for fatigue curve

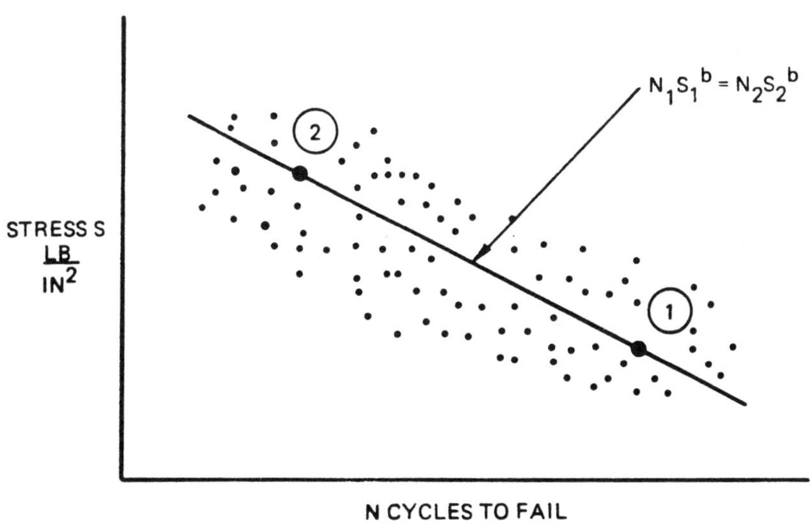

Figure 4.8 Typical *S* – *n* fatigue curves.

Different materials have different fatigue properties and these are reflected in the b exponent, which is very useful in predicting the approximate fatigue life of different systems. The slope of the fatigue curve must take into consideration the many different types of stress concentrations (such as holes and notches), and cuts and scratches that may be formed in the materials. Care must be used in the application of these stress concentrations, which are not usually considered in very ductile materials (such as solder), because they tend to cold flow and creep and relieve the high stresses and strains. In typical metals, more than five thousand stress reversals must be involved before stress concentrations are normally considered [Steinberg (1989a)]. Therefore, stress concentrations may not be considered in thermal cycling stress calculations, unless the electronic system is expected to experience a large number of stress cycles. Stress concentrations are more likely to be considered in vibration and shock environments. Because resonant frequencies of electronic assemblies are most often between 100 and 200 cycles per second, it does not take very long to accumulate several million stress cycles [Steinberg (1989a)].

In a linear system, the number of stress cycles is directly proportional to the time (t). If the time is doubled, the number of stress cycles is also doubled. This applies to thermal cycles as well as to vibration and shock cycles. Also, in a linear system, the stress, (S), will be directly proportional to the force P, as well as to the displacement (z). This application can also be utilized to examine vibration and shock accelerations in terms of gravity units (g) in the same equation. In a thermal cycle, the temperature change over the cycle is represented by the symbol (T).

Several relationships are apparent. For stress cycles (n) and temperature change, (T):

$$n_1 T_1^b = n_2 T_2^b \tag{4.8a}$$

For stress cycles (n) and displacement z:

$$n_1 z_1^b = n_2 z_2^b \tag{4.8b}$$

For stress cycles (n) and force (P):

$$n_1 P_1^b = n_2 P_2^b \tag{4.8c}$$

and for time to fail (t) and acceleration level (g):

$$t_1 g_1^b = t_2 g_2^b \tag{4.8d}$$

Some typical values of the fatigue exponent b, given by Steinberg (1991), are shown for several materials and conditions:

$$\begin{aligned} &\text{Solder joint vibration} && b = 4.0 \\ &\text{Solder thermal cycling} && b = 2.5 \qquad (4.9) \\ &\text{Component lead wire vibration} && b = 6.4 \end{aligned}$$

9 SAMPLE PROBLEM — ESTIMATING THE FATIGUE LIFE OF A SOLDER JOINT IN A THERMAL CYCLING ENVIRONMENT

The last part of the sample problem shown in Section 4.6 demonstrated that a solder joint shear stress of about 2545 psi can be expected for a C-I-C reinforced polyimide glass PCB when the solder joint height is 0.010 in. on a 0.30 in.^{2} hybrid component operating in a rapid cycling thermal environment over a temperature range from $-30°$ C to $+70°$ C. Estimate the number of fatigue cycles that the system can tolerate before solder joint cracks will occur. (Note that a solder height of 0.010 in. may be difficult to achieve. Typical solder joint heights for surface mounted devices are normally in the range of 0.003 to 0.004 in. [Engelmaier (1982)].)

SOLUTION

The approximate fatigue life of the hybrid solder joint can be estimated from the solder fatigue curve shown in Fig. 4.4. The curve can be used directly, or the fatigue life can be calculated by using some reference points from the curve. For example, the thermal cycle fatigue curve for solder shows that a fatigue life of about 80,000 cycles can be expected when the solder shear stress is about 200 psi. Using this as a reference point, the following information is available:

n_2 = 80,000 cycles, reference point for solder failure in thermal fatigue, Fig. 4.4

S_2 = 200 psi, stress, reference point for solder failure in thermal fatigue, Fig. 4.4

S_1 = 2545 psi, stress level calculated in solder joint

Substitute into Eq. (4.7) with exponent $b = 2.5$ from Eq. (4.9) to determine the approximate life of the solder joint:

$$n_1 = (80,000)\left(\frac{200}{2545}\right)^{2.5} = 139 \text{ cycles to fail}$$

10 ACCELERATED LIFE TESTING METHODS

Equation (4.7) can also be used for evaluating the effects of accelerated life testing when extended environments are involved. This condition occurs when there is the need to estimate the fatigue life of a system that will require extended testing periods in thermal or vibration environments. Tests are often run using higher stress levels, for shorter time periods, to try to generate the same damage as will occur in real environments where the stress levels are much lower but where the time periods are much longer. The cost of running a 15 year test to determine if a system will last for 15 years is extremely high. The desire is to run a compressed life test, where the same damage that is accumulated in the 15 year environment can be compressed to a 3 month test period. If this can be accomplished, a significant cost savings will result [Steinberg (1991)].

Accelerated testing techniques are based upon the concept of damage accumulation. If it is possible to define the amount of damage that is generated in different environments, then it may be possible to develop various tests that can closely duplicate the damage. This is the theory of accelerated testing where a high stress level is used for a short time period to accumulate the same damage as a lower stress level applied for a longer time period. Although there are no exact methods for duplicating the same amount of damage in two different environments, the method has been widely practiced by many different organizations for a large number of electronic systems with good success.

11 SAMPLE PROBLEM — RECOMMENDING THERMAL CYCLING TESTS

An electronic system must be capable of reliable operation in an airplane for a 15 year period. The airplane is expected to fly 4 times per week, for 52 weeks a year, for 15 years. The flights will be long enough for the equipment to reach a steady state temperature but not long enough for the solder joints to cold flow and creep to relax and strain relieve the solder stress to a near zero condition. The steady state temperature rise for the electronic equipment is expected to be 25° C. Recommend a short term test program that will duplicate the thermal cycling damage expected in the flight environment. Base the short term thermal cycling test program on a scatter factor (safety factor) of 2.0.

SOLUTION

Consider each flight to be the equivalent of one thermal cycle [Steinberg (1991)]. The number of thermal cycles that will be accumulated in a period of 15 years, with a scatter factor (safety factor) of 2.0 will be

$$n_2 = (4 \text{ flights/wk})\ (52 \text{ wk/yr})\ (15 \text{ yr})\ (2F) = 6240$$

and the expected steady state temperature rise will be

$$\Delta T_2 = 25^\circ \text{ C}$$

The accelerated thermal cycling test program must use a broader thermal cycling range to accumulate the same damage with fewer thermal cycles. Start the thermal cycling test program using a temperature test range from -44° C to $+100^\circ$ C, measured on one of the electronic components. The high and the low temperatures will not be maintained for periods long enough to allow the solder joints to cold flow and creep to a near zero stress condition. Therefore, only half of the full temperature cycling range will be used to represent the thermal stress from the neutral point to the maximum stress condition [Steinberg (1991)]:

$$\Delta T_1 = [100 - (-44)]/2 = 72^\circ \text{ C}$$

Substitute these values into Eq. (4.8a) to determine the number of thermal cycles required to generate the same damage as the actual flight environment with a scatter factor of 2.0:

$$n_1 = (6240)\left(\frac{25}{72}\right)^{2.5} = 443.3 \text{ or } 444 \text{ cycles}$$

The thermal cycling test program would probably recommend a total of 500 cycles using temperature controls obtained from thermocouples mounted directly on one or more components. Visual examinations of the solder joints should be made every 100 cycles using a 30 power microscope to inspect for visible cracks in the solder.

12 SURFACE MOUNTING AXIAL LEADED COMPONENTS

Axial leaded electronic components must be examined closely before they are surface mounted on a PCB. TCE differences between a transformer and a PCB have been known to lift solder pads off of the PCB in only 12 thermal cycles from $-55°$ C to $+95°$ C. Because the solder pads are only held to the PCB with adhesive, it does not take a large force or moment to fracture the adhesive bond with subsequent lift off of the solder pad from the PCB [Steinberg (1991)]. A lifted pad will result in a fractured copper circuit trace which will usually produce a hard failure in the electronic system.

Plated through holes are often used in the solder pads just to anchor the solder pads to the PCB and to prevent this lifting action. This will substantially increase the fatigue life of the assembly, typically to about 150 thermal cycles, where the solder joints may now start to fail in shear. Thermal stress calculations often show bending stresses in the electrical lead wires that are above the ultimate tensile strength of the wire. Test data usually shows that the lead wires very seldom fail unless there are sharp notches or deep cuts in the wires. It is the solder joints, not the wires, that typically fail in these components.

13 SAMPLE PROBLEM — FORCES AND STRESSES IN A SURFACE MOUNTED TRANSFORMER

A small transformer with axial leads must be surface mounted to a composite PCB with a 0.040 in. thick internal aluminum heat sink, which has two 0.025 in. thick PCBs rigidly bonded on each side of the sink. This results in an overall thickness of 0.090 in., as shown in Fig. 4.9. The composite PCB assembly must be capable of passing a proof test of 1000 rapid thermal cycles from $-54°$ C to $+100°$ C with no electrical operation. Is the proposed design acceptable?

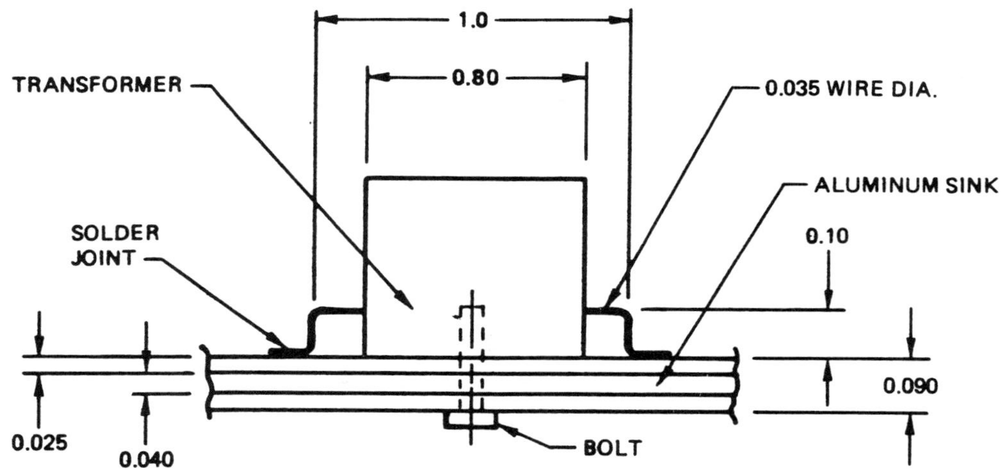

Figure 4.9 Dimensions of surface mounted transformer. Dimensions are in inches.

SOLUTION

Thermal expansion differences between the transformer, the aluminum heat sink and the epoxy fiberglass boards in the x and y-planes of the assembly will force the electrical lead wires to bend. This will produce a shearing force on the solder joints. Because all masses tend to expand and contract with respect to their center of mass, and because the transformer is symmetrical, only half of the transformer length need be used in the analysis.

The temperature cycling range from $-54°$ C to $+100°$ C covers a range of 154° C. Thermal stresses will be evaluated for a temperature range of half that value, or 77° C, in order to obtain the single amplitude stress from the neutral point to the maximum and the minimum temperature points. If the full temperature cycle range of 154° C is used for a rapid temperature cycle condition, and the solder does not have a chance to creep and relax internal stresses to a near zero condition, the resulting stresses will be double amplitude stresses, not single amplitude stresses.

A worst case condition in which all of the relative displacement between the transformer and the PCB is assumed to occur in the bending of the electrical lead wires will be examined. This condition is approximately true because the actual force in the wire is very small and the wire stiffness is very low. As a result, the wire force will have a neglegible effect on any expansion differences. Thermal expansions perpendicular to the plane of the PCB, along the z-axis, are ignored when transformers are bolted to the PCB. Mounting bolts can reduce the z-axis thermal expansions but they will not have much effect on the x and y-thermal expansions.

The difference in the thermal expansion, in inches, between the PCB and the transformer can be determined from

$$x = (\alpha_T - \alpha_P) b \Delta T \tag{4.10}$$

where α_T is the TCE of the transformer body and α_P is the effective TCE of the composite PCB determined from

$$\alpha_P = \frac{\sum_i A_i E_i \alpha_i}{\sum_i A_i E_i} \tag{4.11}$$

Here two materials make up the PCB, one material is aluminum with

$$\alpha_a = 23 \times 10^{-6} \text{ in/in/°C}$$

$$E_a = 10 \times 10^6 \text{ lb/in}^2$$

and

$$A_a = 0.040 \text{ in}^2$$

which is the area of a one inch wide strip. The other material is epoxy fiberglass with

$$\alpha_e = 15 \times 10^{-6} \text{ in/in/°C}$$
$$E_e = 2 \times 10^6 \text{ lb/in}^2$$

and

$$A_e = 0.050 \text{ in}^2$$

which is the area of a one inch wide strip of two layers of epoxy fiberglass.

The effective TCE for the PCB, α_P, may be determined by using these values in Eq. (4.11):

$$\alpha_P = \frac{(0.04)(10 \times 10^6)(23 \times 10^{-6}) + (0.05)(2 \times 10^6)(15 \times 10^{-6})}{(0.04)(10 \times 10^6) + (0.05)(2 \times 10^6)}$$
$$= 21.4 \times 10^{-6} \text{ in/in/°C}$$

With the TCE for the transformer taken as 35×10^6 in/in/°C, an effective transformer length of $0.80/2 = 0.40$ in. and a $\Delta T = [100 - (-54)]/2 = 77°$ C, Eq. (4.10) is used to determine the expansion difference between the PCB and the transformer body:

$$x_1 = (35 \times 10^{-6} - 21.4 \times 10^{-1})(0.40)(77) = 4.189 \times 10^{-4} \text{ in}$$

The expansion difference between the electrical lead wire and the PCB is obtained the same way. Observe, however that the TCE of the copper lead wire:

$$\alpha_w = 17 \times 10^{-6} \text{ in/in/°C}$$

is less than the TCE of the composite PCB, α_P. This will reduce the overall expansion difference. Thus with an effective length of lead wire of 0.10 in:

$$x_2 = (17 \times 10^{-6} - 21.4 \times 10^{-6})(0.10)(77) = -0.339 \times 10^{-4}$$

and the true expansion will be

$$x = 4.189 \times 10^{-4} - 0.339 \times 10^{-4} = 3.85 \times 10^{-4} \text{ in}$$

The electrical lead wires on the transformer are in the form of a square bent with clamped ends. The expansion will force the wires to bend as shown in Fig. 4.10.

Assuming that all of the relative expansion difference is reflected in the bending of the lead wire, the force developed in the wire will be, as shown by Steinberg (1991):

$$P = \frac{7.5 E_w I_w x}{L_w^3} \tag{4.12}$$

where

$$I_w = \frac{\pi d^4}{64} \text{ in}^4$$

is the moment of inertia of the copper wire of diameter, $d = 0.35$ in., and length, $L_w = 0.10$ in. Thus

$$I_w = \frac{\pi (0.035)^4}{64} = 7.366 \times 10^{-8} \text{ in}^4$$

and with the modulus of elasticity of copper taken as $E_w = 16 \times 10^6$ lb/in^2, Eq. (5.12) is used to determine

$$P = \frac{7.5 E_w I_w x}{L_w^3} = \frac{75(16 \times 10^6)(7.366 \times 10^{-3})(3.85 \times 10^{-4})}{(0.10)^3} = 3.40 \text{ lb}$$

The direct shear stress developed in the solder joint can be determined from the solder pad area estimated to be about 0.035 in. × 0.080 in.

$$f_s = \frac{KP}{A} \qquad \text{Direct solder shear stress}$$

where

$$\begin{aligned} P &= 3.40 \text{ lbs} \\ A &= (0.035)(0.080) = 0.0028 \text{ in}^2 \end{aligned}$$

and $K = 1.0$ is a stress concentration factor which can be ignored for very ductile materials with less than 5000 cycles. Thus, the solder shear stress will be

$$S_s = \frac{3.40}{0.0028} = 1214 \text{ lb/in}^2$$

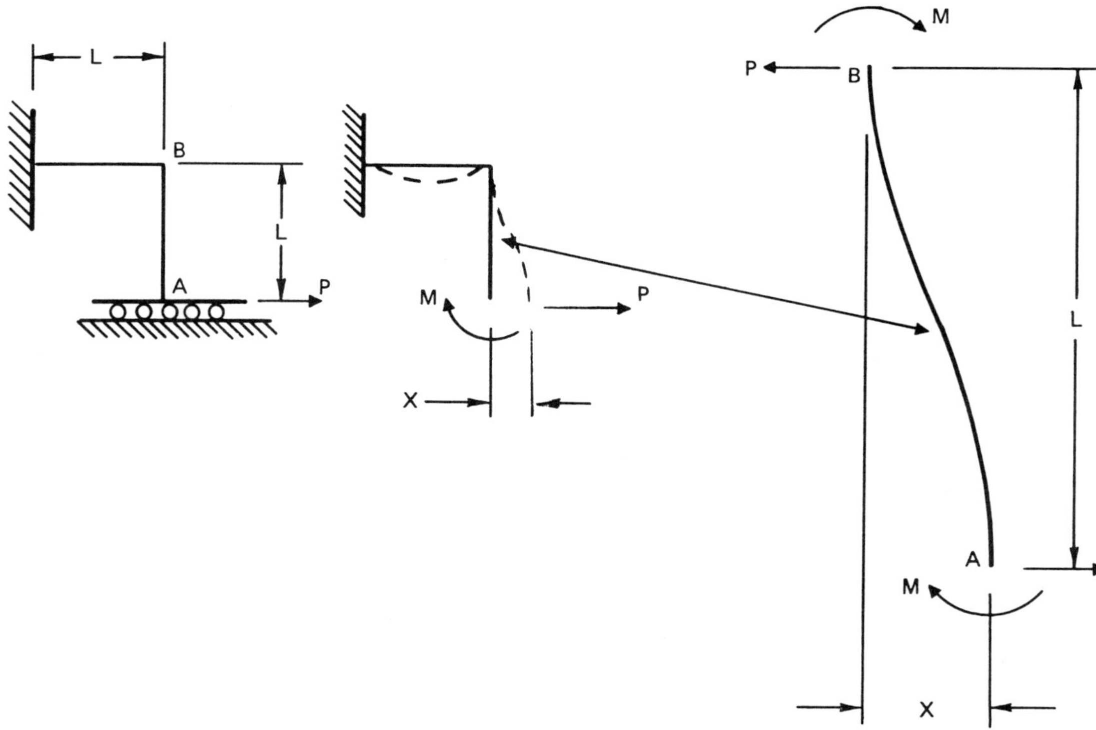

Figure 4.10 Differences in the horizontal thermal coefficients of expansion in the component and the PCB force the lead wires to bend.

The approximate fatigue life of the solder joint in the proposed thermal cycling environment can be estimated from Eqs. (4.8) and (4.9) where

$$\begin{aligned} n_2 &= 80,000 \\ S_2 &= 200 \text{ lb/in}^2 \\ S_1 &= 1214 \text{ lb/in}^2 \end{aligned}$$

Thus, the cycles to fail can be predicted to be

$$n_1 = (80,000)\left(\frac{200}{1214}\right)^{2.5} = 881$$

This is very close to the goal of 1000 thermal cycle life so only small changes will be required to increase the fatigue life to the desired value. For example, the electrical lead wire diameter can be reduced slightly to about 0.030 in. to reduce the wire spring rate and also the wire force. Of course, this change must also be acceptable as to electrical current carrying capacity.

The bending stress in the wire can be determined from the standard bending stress relation [Steinberg (1991)]:

$$S_b = \frac{KMc}{I}$$

where

$$\begin{aligned} M &= 1.2PL = 1.2(3.40)(0.10) = 0.408 \text{ lb-in} \\ c &= d/2 = 0.035/2 = 0.0175 \text{ in} \\ I &= 7.366 \times 10^{-8} \text{ in}^4 \\ K &= 1.0 \text{ Stress concentration factor} \end{aligned}$$

Thus

$$S_b = \frac{(0.408)(0.0175)}{7.366 \times 10^{-8}} = 96,932 \text{ lb/in}^2$$

This is well above the ultimate tensile strength of 45,000 psi which means that the wire will experience plastic bending. Testing experience has shown that the wire, in these circumstances, very seldom experiences a failure (when there are no sharp and deep cuts, scratches or nicks in the wire) due to the low number of stress cycles expected in thermal cycling environments. Moreover, because the solder is typically very plastic at high temperatures, extensive creep and relaxation will allow the wires to relax the high stresses to a near zero condition in a short time period.

14 CERAMIC CHIP CARRIERS AND PLASTIC CHIP CARRIERS

Leadless chip carriers are available in many different plastic cases and ceramic cases. Electrical lead wires are often added to both types of components to reduce thermally induced forces and stresses. When long lead wires are used to raise the component higher off of the PCB to reduce solder shear forces and stresses, thermal problems may develop in systems that use conduction cooled components. A longer lead wire increases the height of the air gap between the chip carrier and the PCB. If the heat flow path is across the air gap to the conduction cooled PCB, a large air gap results in a high thermal resistance, which can result in an excessive component junction temperature. Many applications may require the addition of a thermal pad with a high thermal conductivity to reduce the temperature rise across the component mounting interface. Care must be used in this type of application to provide a thermal pad with a low TCE and a low modulus of elasticity to reduce the thermal expansion forces along the z-axis, perpendicular to the PCB. If care is not used, high z-axis forces will tend to lift the component off of the PCB and this will load the solder joints in tension. Solder joints loaded simultaneously in tension and shear will fail more rapidly, which, of course, is undesirable.

Horton (1987) points out that electronic components that will be used in severe thermal, vibration and humidity environments must be packaged using materials that can survive these harsh conditions for long time periods without failing. One of the most severe environments is humidity, where high concentrations of water vapor are present in the form of a gas. Under these conditions, the water vapor obeys the standard gas laws where the size of the gas molecule is related to the molecular weight of the gas molecule [Steinberg (1991)]. Air, which is a mixture of oxygen and nitrogen, has a large molecule; the molecular weight of oxygen is 32 and the molecular weight of the nitrogen is 28. Water vapor has a molecular weight of only 18, indicating that the water vapor molecule is much smaller than the air molecule. This means that a system that is air tight may not be water vapor tight. Special materials must therefore be used in applications where high humidity conditions exist, to prevent or to reduce water related damage. This applies to the selection of plastic or ceramic components.

Electronic equipment that must operate with a high degree of reliability in high humidity environments should make use of ceramic materials such as aluminum oxide. These materials can provide a hermetic seal against the damaging effects of water vapor. Military electronic systems typically specify the use of hermetically sealed components to insure a long failure free operating life in severe military types of environments. Plastic materials are most often used for commercial electronic systems that operate in air conditioned and humidity controlled offices and factories. Prolonged operation of plastic chip carrier components in severe thermal cycling environments, with high temperature and humidity conditions, may lead to moisture damage in sensitive electronic systems.

15 SAMPLE PROBLEM — FORCES AND STRESSES IN THE LEAD WIRES AND SOLDER JOINTS OF A LARGE SURFACE MOUNTED COMPONENT

A large leadless ceramic chip carrier (LCCC) with 52 input/output (I/O) electrical contacts must be surface mounted on a conduction cooled composite PCB with an aluminum heat sink core 0.038 in. thick as shown in Fig. 4.11. Polyimide glass circuit boards, 0.028 in. thick, are bonded on either side of the aluminum sink to provide an overall PCB thickness of 0.094 in. The goal is to provide an installation which will result in a solder joint fatigue life of 1500 thermal cycles in a rapid temperature cycling "roof of design test" from $-54°$ C to $+100°$ C.

In order to ensure that the testing goal of 1500 thermal cycles can actually be achieved, the analysis will utilize a scatter factor of 2 to compensate for the large variations typically associated with fatigue life test data. This means that the theoretical calculations of the thermally induced stresses in the lead wires and solder joints must show minimum fatigue life values of 2×1500 or 3000 thermal cycles. Determine the type of mounting that should be provided for the LCCC component.

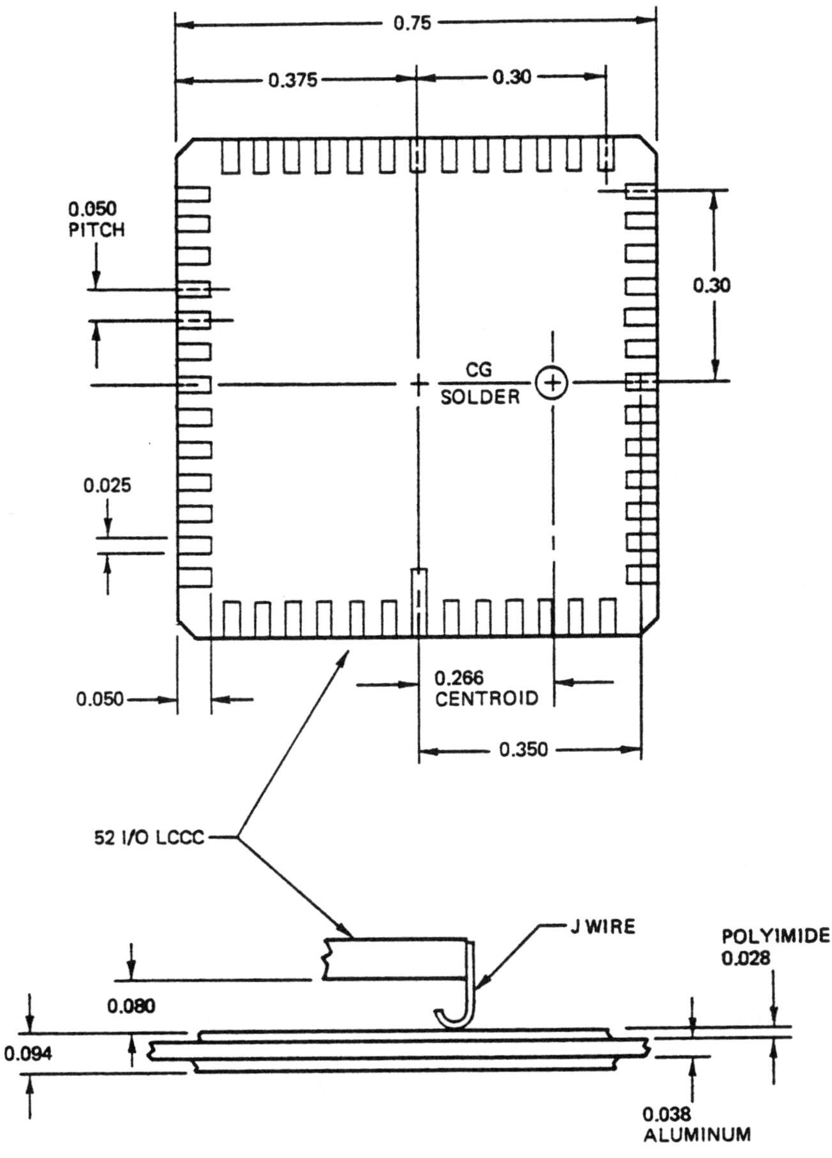

Figure 4.11 LCCC mounted on an aluminum composite PCB. Dimensions are in inches.

SOLUTION

Past experience with large ceramic chip carriers has shown that they will not survive many temperature cycles when they are mounted on plain polyimide glass circuit boards. The typical height of an average solder joint will usually be about 0.003 in. which is not adequate for extended operation over a wide temperature range. Wires will have to be added to the ceramic component to provide a compliant interface to the aluminum composite PCB. Because the component is quite large, a copper wire 0.080 in. long with a diameter of 0.014 in. is proposed in the form of a J-lead. This wire can be thermocompression bonded to the LCCC device or the wire can be attached using a high temperature solder that will not melt during the normal soldering operation with eutectic solder.

The maximum forces and stresses will develop in the corner lead wires and solder joints because they are the farthest away from the centroid (or the mass center) of the component body. All dimensions for thermal expansions can, therefore, be taken with respect to the center of the LCCC since it is symmetrical.

During the increasing temperature phase of the thermal cycle, expansion displacements will occur in proportion to individual TCE properties. Additional displacements and movements will occur due to the internal forces developed by the TCE expansion variations of the various materials used in the PCB. Expansion forces and displacements are examined in the x and y-axes only, and not in the z-axis, perpendicular to the PCB.

An examination of the mounting shows that the expansion of the PCB will be equal to the axial expansion of the ceramic body of the component plus the bending of the lead wire. The PCB has the highest TCE in the x-y plane so it will expand the most at the high temperature end. The expansion of the PCB will produce a lateral force in the lead wires which will force the wires to bend. This force will be carried to the LCCC and force the LCCC to stretch. There will be a reaction back to the PCB which will reduce the PCB expansion, which was treated as a contraction. This will result in the equilibrium equation shown as follows:

$$\alpha_C L_C \Delta t$$

and

$$\alpha_P L_P \Delta t$$

are the thermal expansions, in inches, of the LCCC and the PCB respectively,

$$\frac{P_C L_C}{A_C E_C}$$

is the forced stretch of the LCCC in inches,

$$\frac{P_P L_P}{A_P E_P}$$

is the forced contraction of the PCB in inches and

$$\frac{P_W L_W^3}{12 E_W I_W}$$

represents the bending of the wire in inches.

$$\alpha_C L_C \Delta t + \frac{P_C L_C}{A_C E_C} + \frac{P_W L_W^3}{12 E_W I_W} = \alpha_P L_P \Delta t - \frac{P_P L_P}{A_P E_P} \tag{4.13}$$

The length of the LCCC and the length of the PCB must be determined before Eq. (4.13) can be used to solve for the maximum force, P, produced in the wire and solder joint. The length at each solder joint must be used because each solder joint has a different location with respect to the center of the LCCC. However, only the maximum force in the wire and solder joint is desired. This maximum force must occur at the corner wires and solder pads because they are the farthest away from the center of the LCCC, so they will experience the maximum displacements.

The force at each lead wire and solder joint can be obtained by locating the centroid, or the center of gravity, for all of the solder joints with respect to the center of the LCCC, using the dimensions shown in Fig. 4.11 for a total of 25 solder pads. The centroid is located by:

13	pads	at	0.350	in	=	4.55
2	pads	at	0.300	in	=	0.60
2	pads	at	0.250	in	=	0.50
2	pads	at	0.200	in	=	0.40
2	pads	at	0.150	in	=	0.30
2	pads	at	0.100	in	=	0.20
2	pads	at	0.050	in	=	0.10
25	pads					6.65

The solder centroid is at

$$L = \frac{6.65}{25} = 0.266 \text{ in}$$

The first step is the determination of the TCE and modulus of elasticity of the composite aluminum PCB. With subscripts A and Po to designate aluminum and polyimide with two full internal copper plans:

$$\begin{aligned} \alpha_A &= 23 \times 10^{-6} \, \text{in/in/}^\circ\text{C} \\ \alpha_{Po} &= 13 \times 10^{-6} \, \text{in/in/}^\circ\text{C} \end{aligned}$$

$$\begin{aligned} E_A &= 10 \times 10^6 \, \text{lb/in}^2 \\ E_{Po} &= 3 \times 10^6 \, \text{lb/in}^2 \end{aligned}$$

$$\begin{aligned} A_A &= 0.038(1.0) = 0.038 \text{ in}^2 \\ A_{Po} &= 0.056(1.0) = 0.056 \text{ in}^2 \end{aligned}$$

Eq. (4.11) is employed to give

$$\begin{aligned} \alpha_P &= \frac{A_A E_A \alpha_A + A_{Po} E_{Po} \alpha_{Po}}{A_A E_A + A_{Po} E_{Po}} \\ &= \frac{0.038(10 \times 10^6)(23 \times 10^{-6}) + 0.056(3 \times 10^6)(13 \times 10^{-6})}{0.038(10 \times 10^6) + 0.056(3 \times 10^6)} \\ &= 1.993 \times 10^{-5} \, \text{in/in/}^\circ\text{C} \end{aligned}$$

as the effective TCE for the composite PCB.

The effective modulus of elasticity can be determined in simular fashion:

$$\begin{aligned} E_P &= \frac{A_A E_A + A_{Po} E_{Po}}{A_A + A_{Po}} \\ &= \frac{0.038(10 \times 10^6) + 0.056(3 \times 10^6)}{0.038 + 0.056} \\ &= 5.829 \times 10^6 \, \text{lb/in}^2 \end{aligned}$$

Now with the wire having a moment of inertia of $I_W = \pi d^4/64$, for 25 pads and a wire diameter of 0.014 in.,

$$I_W = 25 \left(\frac{\pi}{64} \right) (0.014)^4 = 4.714 \times 10^{-8} \, \text{in}^4$$

and with data for wire and ceramic:

$$\begin{aligned} E_W &= 17 \times 10^6 \, \mathrm{lb/in^2} \\ E_C &= 42 \times 10^6 \, \mathrm{lb/in^2} \\ L_W &= 0.080 \text{ in} \\ L_C &= 0.266 \text{ in} \\ L_P &= 0.266 \text{ in} \\ A_C &= (0.75)(0.040) = 0.030 \text{ in}^2 \\ \alpha_C &= 6 \times 10^{-6} \, \mathrm{in/in/^\circ C} \end{aligned}$$

use of Eq. (4.13) gives with $\Delta T = 77^\circ$ C and

$$A_P = 0.75(A_A + A_{P_o}) = 0.075(0.038 + 0.056) = 0.0705 \text{ in}^2$$

$$(6 \times 10^{-6})(0.266)(77) + \frac{0.266 P_C}{(0.030)(42 \times 10^6)} + \frac{(0.080)^3 P_W}{12(17 \times 10^6)(4.714 \times 10^{-8})}$$
$$= (1.993 \times 10^{-5})(0.266)(77) - \frac{0.266 P_P}{(0.0705)(5.829 \times 10^6)}$$

or

$$1.229 \times 10^{-4} + 2.111 \times 10^{-7} P_C + 5.324 \times 10^5 P_W$$
$$= 4.082 \times 10^{-4} - 6.473 \times 10^{-7} P_P$$

With $P_C = P_W = P_P = P$:

$$5.410 \times 10^{-5} P = 2.853 \times 10^{-4}$$

or

$$P = 5.274 \text{ lb}$$

on 25 wires and

$$P = 0.211 \text{ lb/wire}$$

The maximum wire and solder joint force will occur at the corner of the LCCC component. This requires the use of two correction factors.

1. Correction for the distance from the centroid of all of the solder joints to the solder joints at the edge of the LCCC:

$$C_1 = \frac{0.350}{0.266} = 1.316$$

2. Correction for the distance from the solder joints at the edge of the LCCC to the solder joint at the farthest corner of the LCCC:

$$C_2 = \sqrt{2} = 1.414$$

The maximum thermally induced force in the most critical lead wire and solder joint at the corner of the LCCC component can now be determined by multiplying the average force per wire by the correction factors:

$$P_{\mathbf{max}} = (0.211)(1.316)(1.414) = 0.393 \text{ lb}$$

The maximum shear stress expected in the solder joint can be determined from the approximate shear area of the solder joint:

$$S_s = \frac{KP}{A}$$

Here the estimated solder area is (0.018)(0.060) = 0.00108 in^2 and with the expected stress concentration factor $K = 1.0$, one obtains with $P = 0.393$ lb:

$$S_s = \frac{0.393}{0.00108} = 364 \text{ lb/in}^2$$

The approximate number of stress cycles expected to produce the first failures (usually as visible cracks in the solder) in the solder joints can be determined from Eq. (4.15) with $n_2 = 80,000$ cycles, $S_2 = 200$ psi, and $S_1 = 364$ psi:

$$n_1 = (80,000)\left(\frac{200}{364}\right)^{2.5} = 17,902 \text{ cycles to fail}$$

Because this value is well above the 3000 thermal cycle design goal for the calculated fatigue life, the design is acceptable for the solder joint shear stresses.

The bending stresses in the lead wires can be determined from the bending moment and bending stress relations with $P = 0.393$ lb and the height of the J-wire, $h = 0.080$ in. The maximum bending moment will be

$$M = \frac{Ph}{2} = \frac{0.393(0.080)}{2} = 0.0157 \text{ in-lb}$$

The flexure formula then gives with

$$I = \frac{\pi}{64} d^4 = \frac{\pi}{64}(0.014)^4 = 1.886 \times 10^{-9}\,\text{in}^4$$

the bending stress:

$$S_b = \frac{(0.0157)(0.014/2)}{1.886 \times 10^{-9}} = 58,272 \text{ psi}$$

The maximum bending stresses in the copper J-wires are well above the ultimate tensile strength of 45,000 psi. However, extensive testing experience with leaded components has shown that thermal cycling failures very seldom occur in the lead wires as long as there are no deep and sharp cuts, nicks or scratches in the wires. The wires will experience plastic bending which considering the low number of fatigue cycles expected, will not normally result in wire failures.

Virtually all of the thermal cycling failures are expected to occur in the solder joints of the surface mounted components. This assumes that the J-wires are securely attached to the LCCC edges with the mechanical strength necessary to survive the number of thermal cycles expected. Because there are many different methods of soldering and thermo-compression bonding of the wires to the component case, the method of attachment must be investigated to insure the integrity and reliability of the joint at the case.

16 CUMULATIVE DAMAGE CONCEPTS FOR ESTIMATING THE FATIGUE LIFE

Extensive testing experience and test data has shown that structures that are exposed to a large number of stress cycles will often fail at stress levels that are substantially below the ultimate strength of the material. This has resulted in a theory which states that every time a structural element is subjected to a stress reversal, a small part of the life is used. When all of the life is exhausted, the structure will fail. This simple theory is used extensively to determine the approximate fatigue life of many different types of structures exposed to alternating cycling stress environments such as vibration, shock, acoustics and thermal cycling. The damage that is accumulated in these environments is assumed to be linear. This means that the damage generated by some vibration exposure can be added to the damage generated by some thermal cycling exposure to obtain the total damage of the structure for estimating the fatigue life in the combined environments. This theory is known as "Miner's Rule," and sometimes as "Miner's Cumulative Damage Rule," to honor the one who was to develop this theory. This rule is usually expressed in the form of a damage ratio, R:

$$R = \frac{n_1}{N_1} + \frac{n_2}{N_2} + \frac{n_3}{N_3} + \cdots = 1.0 \qquad (4.14)$$

Here

n = Number of stress cycles actually accumulated at different stress levels 1, 2, 3, ···

N = Number of stress cycles required to produce a failure at stess levels 1, 2, 3, ···

R = Ratio of fatigue cycle damage accumulated for estimating the amount of life used.

The cumulative damage ratio can be used for finding the fatigue life of a structure operating in only one environment such as vibration. The different number of fatigue cycles for ratios 1, 2, and 3 might be based upon three different input g acceleration levels. The cumulative damage ratio can also be used for finding the fatigue life of a structure operating in several different environments, such as vibration, shock and thermal cycling. The ratio of the life expended in each environment would then be evaluated separately and then added together to obtain the total fatigue accumulation for all of the environments acting simultaneously.

Different fatigue cycle ratios are normally used depending upon the type of application and how the product will be used. Some type of safety factor is suggested since fatigue data has a wide scatter. For commercial applications when the public safety is not at risk, and where a television set or a radio is involved, an R value of 1.0 is suggested. Where the public safety may be involved, as in an airplane, automobile or train, an R value of 0.7 is suggested. Where critical life systems are involved, as in the space shuttle, a higher safety factor is recommended, so that an R value of 0.3 is suggested [Steinberg (1989a)].

Suggested fatigue cycle ratios R for some applications are

$$R = \begin{cases} 1.0 & \text{For commercial applications where the public safety is not involved} \\ 0.7 & \text{For applications where the public safety is involved} \\ 0.3 & \text{For applications where critical life systems are involved} \end{cases} \tag{4.15}$$

Steinberg (1989a, 1991) provides additional theory related to Miner's cumulative damage, with sample problems in the areas of combining vibration and thermal cycling damage and predicted fatigue life.

NOMENCLATURE

A	Area, in^2
α	Thermal coefficient of expansion, in/in/°C
AVIP	Avionics integrity program (US Air Force)
b	Fatigue exponent slope (slope of straight line on a log-log fatigue plot)
b	Effective length, in
C	Distance to neutral axis for bending, in
C-I-C	Copper-invar-copper
E	Young's modulus, lb/in^2
F	Scatter factor
G	Shear modulus, lb/in^2
g	Acceleration in gravity units, dimensionless
h	Height or thickness, in
I	Moment of inertia, in^4
k	Spring rate, lb/in
K	Kovar metal
K	Stress concentration factor
L	Length, in
LCCC	Leadless ceramic chip carrier
M	Bending moment, in-lb
N	Number of stress cycles to produce a failure
n	Number of stress cycles actually accumulated
P	Force, lbs
PCB	Printed circuit board

PLCC	Plastic leaded chip carrier
PTH	Plated through hole
R	Ratio of accumulated damage (Miner's ratio)
S	Stress, lb/in^2
S_b	Bending stress, lb/in^2
S_s	Shear stress, lb/in^2
T	Temperature, °F or °C
TCE	Thermal coefficient of expansion
x, y, z	Displacement, in

REFERENCES

Engelmaier, W (1988). Reliability of surface mounted assemblies: Controlling the thermal expansion mismatch problem, presented at the *SMART IV Conf*, Los Angeles, CA, Jan. 11–14.

Engelmaier, W (1982). Fatigue life of leadless chip carrier solder joints during power cycling, *Proc Technical Program 2nd Ann Intl Elec Packaging Soc Conf*, San Diego, CA, Nov 15–17.

Evans, J W (1989). An overview of thermally induced low cycle fatigue in surface mounted solder joints, *Surface Mount Tech Mag*, Feb.

Followell, D A, Liguore, S L, Perez, R, and Yates, W D (1991). Computer aided assessment of reliability using finite element methods, McDonnell Aircraft Co report sponsored by Rome Laboratories, Report MDC B2303, Feb.

Hastings, D W (1988). A military airborne application of LCCs on Cu/Invar/Cu and Kevlar based PWBs, Ford Aerospace and Communications Corp, Newport Beach, CA, Sep.

Horton, G (1987). Implementation of surface mount technology in high reliability products, presented at the *NEPCON West Conf* Anaheim, CA, Feb.

Kotlowitz, R W and Engelmaier, W (1989). Impact of lead compliance on the solder attachment reliability of leaded surface mounted devices, AT&T Bell Laboratory Report, Whippany, NJ, Oct.

Massie, S (1987). Lead stress analysis of a surface mounted relay, *Elec Packaging and Production Mag*, May.

Steinberg, D S (1991). *Cooling Techniques for Electronic Equipment*, 2nd Ed, John Wiley & Sons Inc, New York.

Steinberg, D S (1989a). *Vibration Analysis for Electronic Equipment*, 2nd Ed, John Wiley & Sons Inc, New York.

Steinberg, D S (1989b). Tougher tests for military equipment, *Machine Design Mag*, May 25.

Steinberg, D S (1988). Tools for implementing AVIP, presented at *9th Ann IEEE/AESS Dayton Chap Symp*, Nov 30.

SOURCES OF ADDITIONAL INFORMATION

Agrawala, B N (1985). Thermal fatigue damage in Pb-In solder interconnections, *23rd Reliability Physics Symp*, 198–205.

Baker, E (1975). Stress relaxation in tin-lead solders, *Mat Science & Eng*, **38**, 241–247.

Berkehile, M H (1967). Investigation of solder cracking problems on printed circuit boards, Marshall Space Flight Center, Huntsville, AL, N67-36783.

DeVore, J A (1982). Fatigue resistance of solders, *Proc NEPCON*, Cahners Co, Des Plaines, IL.

Hall, P M (1983). Solder attachment of ceramic chip carriers, *Solid State Tech*, 103–107.

Harris, C M and Crede, C E (1976). *Shock and Vibration Handbook*, McGraw-Hill Book Co, New York, NY.

Howes, M A H (1976). A study of thermal fatigue mechanisms, thermal fatigue of materials and components, *ASTM STP 612*, 86–105.

Kubik, E C and Li, T P L (1982). Thermal shock and temperature cycling effects on solder joints of hermetic chip carriers mounted on copper thick films, *Intl J for Hybrid Microelect*, **5** (2), 314–321.

Lawson, L (1986). Thermal fatigue testing of 97Pb-3Sn solder, MS thesis, Northwestern Univ, Evanston, IL.

Levine, E and Ordonez, J (1981). Analysis of thermal cycle fatigue damage in microsocket solder joints, *IEEE Trans Comp Hybrids & Manuf Tech* **CHMT-4**, 515–519.

Manko, H H (1967). *Solders and Soldering*, McGraw-Hill Book Co, New York, NY.

Suhir, E (1988). On a paradoxical phenomenom related to beams on elastic foundation, *ASME J Appl Mech*, **55**, 818–821.

Timoshenko, S P and Goodier, J N (1970). *Theory of Elasticity*, McGraw-Hill Book Co, New York, NY.

Chapter 5

BIBLIOGRAPHY OF HEAT TRANSFER IN ELECTRONIC EQUIPMENT — 1989

R. E. Simons
IBM Corporation
Poughkeepsie, New York

1 INTRODUCTION

For more than 50 years, the thermal control of electronic components and equipment has been an important area for the application of advanced heat transfer techniques. During this period improved methods of heat removal and thermal analysis played a pivotal role in achieving the high reliability, increased power density, and physical miniaturization which characterize modern day electronics. The continued development of electronic devices and packages may be expected to place even greater demands on the art and science of thermal management.

The importance of thermal management today is attested to by the number of papers and articles published each year on various aspects of heat transfer in electronic equipment. A bibliography prepared by Antonetti and Simons (1985) covering the period 1970-1984, listed 237 publications related to various aspects of heat transfer in electronic equipment. A more recent annotated bibliography by Simons (1988) reported on 93 publications in this area in 1986. In a similar bibliography covering 1987 and 1988 Simons (1990) reported on 296 papers, articles, and patents. An annotated bibliography discussing nearly 200 papers, articles, and patents published in 1989 is provided here. For the reader who is also interested in the broad area of heat transfer and its more fundamental aspects, excellent reviews and discussions of the 1988 and 1989 literature were provided by Eckert et al. (1989, 1990).

The material presented here was obtained from technical journals, trade magazines, conference proceedings, and U.S. patents. The journals and magazines surveyed included: *Applied Energy (UK)*, *ASME Journal of Electronic Packaging*, *ASME Journal of Heat Transfer*, *Bulletin of the International Centre for Heat and Mass Transfer*, *Computers & Structures (UK)*, *Electronics and Communications in Japan*, *Electronic Engineering (UK)*, *Electronic Packaging and Production*, *Energy and Automation*, *Heat Transfer Engineering*, *IEEE Journal of Quantum Electronics*, *IEEE Transactions on Components, Hybrids, and Manufacturing Technology*, *IEEE Transactions on Electron Devices*, *International Journal of Heat and Mass Transfer*, *International Journal of Heat and Technology (UK)*, *Japanese Journal of Applied Physics*, *Journal of Electronic Materials*, *Journal of the Institute of Energy*, *Mechanical Engineering*, *Numerical Heat Transfer*, *Printed Circuit Design*, *Soldering & Surface Mount Technology (UK)*, and *Test & Measurement World.*

Some of the papers discussed here were presented at conferences held in 1989. Conferences which provided publications pertaining to the cooling of electronic equipment included: the Annual Reliability and Maintainability Symposium (Atlanta — January 24–26), the IEEE 5th Semiconductor Thermal and Temperature Measurement Symposium (San Diego — February 7–9), the National Electronic Packaging and Production Conference — NEPCON WEST (Anaheim — March 7–9), the IEEE 11th Custom Integrated Circuits Conference (San Diego — May 15–18), the IEEE 39th Electronic Components Conference (Houston — May 22–24), the National Electronic Packaging and Production Conference — NEPCON EAST (Boston — June 13–15), the National Heat Transfer Conference (Philadelphia — August 6–9), the 9th International Electronics Packaging Conference (San Diego — September 11–13), the 11th International Telecommunications Energy Conference (Firenze, Italy — October 15–18), the ISHM International Symposium on Microelectronics (Baltimore — October 24–26), and the ASME Winter Annual Meeting (San Francisco — December 10–15).

The sections that follow provide groupings of the papers, articles, and patents in 11 topic areas covering the state of the art, materials, thermal contact resistance, air cooling, liquid cooling, thermal measurements and sensing, thermal analysis techniques, package cooling techniques, system cooling techniques, cooling devices/hardware, and thermal assembly processes. The content or contribution of each paper is briefly highlighted.

2 STATE OF THE ART

Three papers are included here which addressed the state of the art of heat transfer in electronic equipment, providing an overview of past and recent developments and applications. A paper by Hannemann (1989) sought to "provide an answer to an important question for the practitioners of electronics thermal control technology: what are the practical limits to the use of forced air cooling, and will a departure from this method of thermal control be required in the foreseeable future for mini and microcomputers?" In this paper he noted that "current generation high-end mainframe computers have chip powers approaching 20 W and array heat fluxes of perhaps 3.0 W/cm^2" as shown in Fig. 5.1, and that "advanced cooling techniques including liquid cooling and phase change cooling will become more important for a specialized segment of the computer market." Hannemann then assessed the limitations of air cooling in comparison with other limiting physical factors (i.e., "tiling limit" and "wiring limit") in the design of electronic package assemblies. He concluded that "there is no fundamental reason for departing from forced air cooling for mini and microcomputers in the foreseeable future."

A paper by Tousignant and Maddox (1989) provided a review of different cooling methods covering a range of applications. Comparisons were made in terms of heat transfer coefficient for applications utilizing air cooling, indirect liquid cooling, and direct immersion cooling. They concluded that "although direct immersion systems must be sealed to prevent loss of the dielectric fluid, the systems can be relatively simple and the packaging density can be immensely improved over other cooling systems."

A paper by Incropera (1989) reviewed the available data, correlations, and predictions pertaining to liquid immersion cooling of single or multichip configurations, with or without extended surfaces. The heat transfer options considered included natural convection, mixed convection, forced convection, jet impingement, pool boiling, and forced convection boiling. Consideration was also given to two-

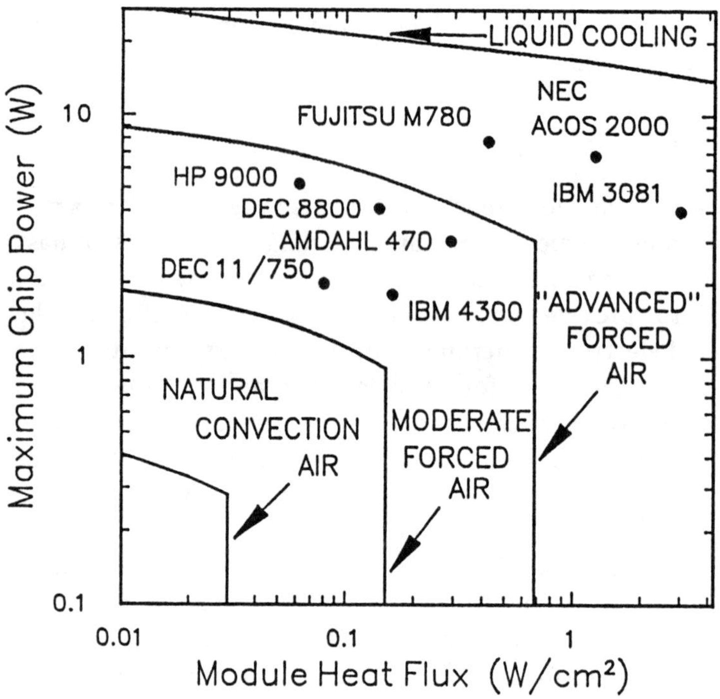

Figure 5.1 Cooling technology map [Hannemann (1989)].

phase thermosyphons and submerged and vapor space condensers in relation to electronic cooling applications. It was concluded that although the existing convection heat transfer knowledge base is by no means complete, "it allows the packaging engineer to assess, at least in an approximate sense, the relative units of various cooling options, several of which permit the dissipation of large heat fluxes ($> 100\text{W/cm}^2$) in a dielectric coolant."

3 MATERIALS

The choice of materials is an important consideration in the thermal design or evaluation of an electronics package. Before heat can be removed from any external surface it must be transported to the surface by the process of thermal conduction. In many electronic modules containing integrated circuit chips, thermal conduction through the substrate is the principal path for heat flow from the chip to the external module surface. Consequently, the thermal conductivity of the substrate material is a property of principal interest in thermal design. Of course, other material characteristics such as thermal coefficient of expansion, electrical insulation capacity, low dielectric constant, and mechanical strength are equally important in terms of other design considerations.

An article by Spitz (1989) provided a survey of the various ceramics currently used, and being developed for use as substrates for hybrid and integrated circuit packages. Although alumina (Al_2O_3) is still the most widely used substrate material, efforts to develop improved substrate designs using materials such as aluminum nitride (AlN), silicon carbide (SiC), and beryllia (BeO) continue. As shown in Fig. 5.2, these materials offer significantly higher thermal conductivity than does alumina, and in the case of aluminum nitride and silicon carbide, a thermal coefficient of expansion much closer to that of silicon. The range of thermal conductivities shown for the various ceramics represents the range of values derived from the papers discussed in this section. The thermal coefficients of expansion (TCE) were taken from the paper by Rabinovich (1989) which also discussed glass-ceramic substrate materials. Although glass-ceramic materials exhibit very low thermal conductivities (e.g., 2–5 W/mK), as a result of lower dielectric permittivity they offer lower electrical propagation delay, and are of interest as substrate materials for high performance packages. A paper by Suryanarayana (1989) presented and discussed the results of a survey conducted on the physical properties of several of the new ceramic systems including carbides, borides, and nitrides, as they appeared in recent literature. Dettmer et al. (1989) cited large variations in thermal conductivity values claimed by different vendors of AlN and SiC materials as the motivation to independently measure the thermal conductivity of actual substrate materials. The paper describes the technique and apparatus used to measure thermal conductivity, and the values obtained for samples of Al_2O_3, AlN, BeO, and SiC.

Two papers by Johnson (1989) and Jensen et al. (1989) assessed the performance of substrate materials using thermal test packages. Johnson used resistor test modules (made of AlN and BeO) mounted on a controlled temperature chuck to obtain comparative measurements of thermal resistivity. The study conducted by Jensen et al.

utilized test packages consisting of a 6.1 × 6.1 mm thermal chip attached to a 5 × 5 cm substrate. Substrate materials evaluated were aluminum nitride, alumina, epoxy-glass, refractory-glass, and a composite consisting of alumina fibers in a polyimide matrix. The cooling test conditions included natural and forced convection with air, and a liquid cooled heat sink arrangement providing a forced flow of a glycol solution over the back of the substrate. The experimental results obtained were compared with the results of a previously reported finite element numerical analysis.

A paper by Beasley (1989) discussed a study of the use of metal matrix composite materials for standard electronic module (SEM) frames used in avionic and shipboard electronics systems. In these systems a printed circuit board (PCB) carrying integrated circuit components is attached to each module frame. In addition to providing structural support, the module frame provides a thermal conduction path to cooling ducts in the walls of the card cage housing. The objective of the study was to optimize the thermal performance of the module frame by maximizing the thermal conductivity of the

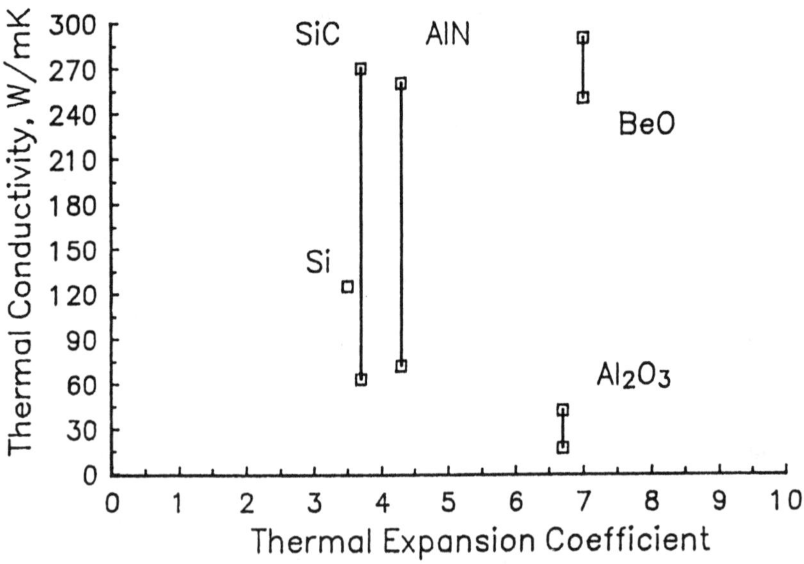

Figure 5.2 Thermal characteristics of selected ceramics [Adapted from Spitz (1989) and Rabinovich (1989)].

material while tailoring the TCE to match that of a ceramic substrate.

4 THERMAL CONTACT RESISTANCE

In virtually all electronic packages and equipment, there are physical, mechanical, or metallurgical joints between different materials or parts of the structure across which heat is conducted. The thermal resistance across such interfaces is often referred to as a thermal contact resistance, and is generally a complex function of the geometric, physical, and thermal properties of the contacting solids and the interstitial substance.

As noted in the previous section, in a majority of integrated circuit packages today, the principal heat flow path is from the chip into the substrate to which it is bonded. It is therefore, important that the thermal contact or "die bond" resistance between the chip and its substrate carrier be as small in magnitude as possible. A variety of methods have been used to bond the chip to the substrate including organic glues, and hard or soft solders. A paper by Estes and Pernice (1989) reported on a study conducted to investigate and compare the electrical and thermal stability of third generation organic die attach adhesives when used to attach the collector of a typical medium power bipolar transistor (i.e., NPN 2N3055). Based upon the results of this study it was concluded that "a solder attach produced the most reliable electrical and thermal bond, both initially and after thermal aging at 150° C. However, some of the silver filled epoxies examined performed nearly as well as solder and might be considered as its substitute in this and other medium power devices."

Within the widely used plastic dual-in-line packages (PDIPs), heat from an integrated circuit chip must cross a number of thermal interfaces before exiting the package. A study reported by Childres and Peterson (1989) employed a numerical finite element model of an 18 pin PDIP (see Fig. 5.3) to predict the interface pressures within the package. The analysis performed assumed that the model cooled from a uniform curing temperature to an imposed operating temperature distribution. The predicted pressures were combined with a previously developed correlation to determine the thermal contact conductance at each interface within the package. The authors stated

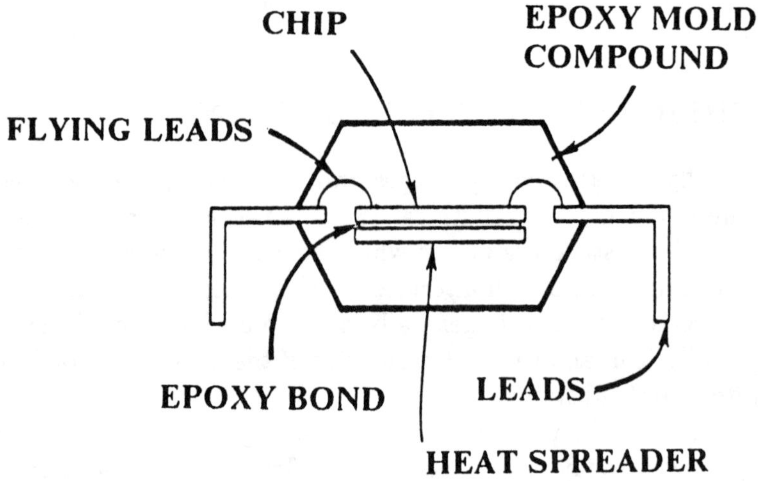

Figure 5.3 Cross-section of 18 PDIP and finite element model [Adapted from Childres and Peterson (1989)].

that "the results indicate that the summation of the contact resistances may comprise as much as 43% of the total package thermal resistance for the packages investigated."

For some large multi-chip modules with many chips, the substrate can no longer be effectively used as the principal path for thermal conduction to the external package surface. In such cases, a method is needed to remove heat from the back of the chip. One method of heat removal from the back of the chip is that provided in the thermal conduction modules (TCMs) used in the IBM 308X and 3090 series of processors. These TCMs are filled with helium gas, and heat is removed from the chips by spring-loaded metal pistons with a crowned tip (as shown in Fig. 5.4). Heat transfer from the chip occurs both through the solid contact and through the surrounding helium filled gap, and the thermal resistance, R_{c-p}, between a chip and its piston is a matter of some importance. Considering this cooling approach, Fisher and Yovanovich (1989) studied the chip to piston contact and developed a theoretical model to predict the effect of a surface layer on the thermal constriction resistance of the chip to piston contact. Comparisons of model results with experimental results showed good agreement for light loads in the elastic range. To further reduce the piston-to-chip thermal contact resistance a larger crown radius may be used at the piston tip, resulting in smaller interface gaps. As these gaps become smaller thermal conduction through the intervening gap (i.e., helium) may be affected by "rarefied gas" phenomena effectively reducing the thermal conductance across the gap below the usual Fourier Law based conduction equations. An important parameter in estimating this effect is the thermal accommodation coefficient (TAC) which represents the extent to which the exchange of energy takes place between the gas and the solid. Song and Yovanovich (1989) proposed an experimental method to estimate TAC values for given gas-solid combinations. The results of experiments performed to obtain TAC values for helium, argon, and nitrogen in interfaces formed by stainless steel or nickel surfaces were reported.

A paper by Majumdar and Tien (1989) discussed the analysis of "heat flow through imperfect contact of rough surfaces and interstitial gases" by means of a network of thermal resistances. It was noted that the analysis "accounts for the effects of load, surface hardness, surface roughness and thermal conductivities and proves that the exponent of load is related to the spectrum of surface roughness."

A paper by Naik et al. (1989) described experimental measurements of the thermal interface resistance for transistors bolted to

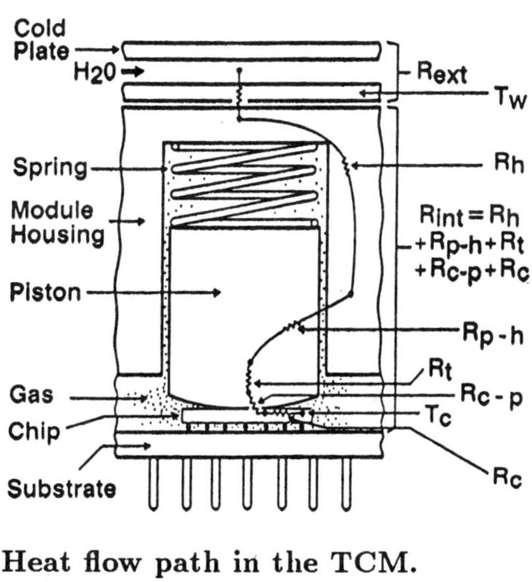

Figure 5.4 Heat flow path in the TCM.

air-cooled heat sinks. Results were presented for heat sinks made of copper and duralumin.

5 AIR COOLING

Although more cooling designs are appearing that utilize some form of liquid cooling, air continues to be the most widely used cooling medium in electronic equipment. The principal advantages of cooling with air are its ready availability and ease of application. For applications with low power and low packaging densities, natural convection cooling with air is often used. The advantage offered by natural convection cooling is that no fan or blower is required to move the air. Instead, air movement is induced by density differences resulting from the heat given off by the electronic components. Two papers reported in this section are in the area of natural convection.

In the first paper, Witzman et al. (1989) reported the results of an experimental study "to determine the temperature field created by by uniform arrays of electronic devices mounted on a printed circuit board." The printed circuit boards were mounted within a natural convection cooled cabinet which was modified to make it possible to take infrared pictures of the board. Based upon their results the authors concluded that "buoyancy induced forced convection corrected by the conduction through the printed circuit board (when heat is spread from the device into the board) represents the most adequate physical model to be used in problems involving "free convection" heat dissipation inside electronic cabinets."

In the second paper, Leung and Probert (1989) addressed the problem of enhancing natural convection heat transfer through the use of a finned heat sink, or as they referred to it a "heat exchanger." Specifically, the paper reports on experiments to investigate the effects of geometric parameters (i.e., fin height, thickness, and spacing) on the rate of heat transfer from vertical fins protruding upward from a horizontal base. The authors concluded that the separation between fins is the "prime geometric variable," and recommended non-dimensional heat transfer correlations to be used for the fully developed region (GrPr $\leq$ 500) or for the single-plate region (GrPr $>$ 500).

For packaging situations which require more cooling capability or a greater degree of control, forced convection heat transfer with air is often chosen. Forced air cooling offers the ability to achieve higher heat transfer coefficients than natural convection, is relatively insensitive to packaging orientation, is less affected by heating of the air by upstream components, and in general is more easily controlled. The remaining papers in this section address forced convection cooling with air.

Forced air cooled packages or systems are often made up of arrays of vertically (or horizontally) stacked printed circuit cards as shown in Fig. 5.5. Mounted on these cards are arrays of electronic circuit modules. Efforts continue to be directed at further understanding and characterizing heat transfer for such configurations. Lehmann and Wirtz (1989) conducted experiments to investigate "the effect of variations in stream-wise spacing and component length on convection from rectangular, surface mounted components in a channel flow." Heat transfer rates and pressure losses were measured in a two-dimensional channel with periodically spaced rectangular elements on one wall as shown in Fig. 5.6, indicating the parameters. The results were correlated (as shown in Fig. 5.7.) in terms of an average dimensionless convective resistance defined as

$$R_{c-a} = \frac{k}{H_1 h_{c-a}}$$

The correlations suggested by Lehmann and Wirtz are based on a Reynolds number formulated using H_1. Over the Reynolds number range of 1000 to 3000, for $0 \leq H_2/H_1 \leq 0.25$, $b/H_1 = 0.5$:

$$R_{c-a} = 5.5\mathrm{Re}^{-0.65}$$

and for $0.25 \leq H_2/H_1 \leq 1.0$, $0.25 \leq b/H_1 \leq 0.75$:

$$R_{c-a} = 4.84 \left(\frac{b}{H_1}\right)^{0.11} \left(\frac{H_2}{H_1}\right)^{-0.14} \times \mathrm{Re}^{-0.65}$$

In another paper Anderson (1989) described an experimental study involving regular arrays (i.e., in-line arrays with uniform row and column spacing) of non-uniformly heated modules in forced convection channel flow. The superposition method described by Moffat

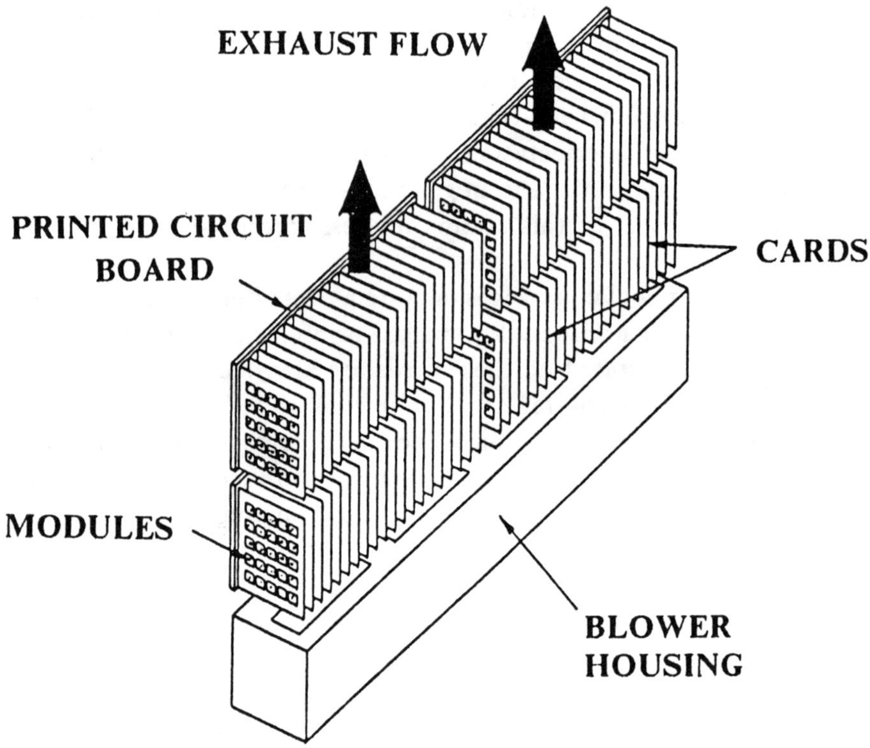

Figure 5.5 Typical forced air cooling packaging configuration.

and Anderson (1988) was used to develop a model to predict temperatures for the arrays of components studied. It was noted that the correlations developed are specific to the type of geometry tested, but, the method used may applied to other geometries. In many actual air cooling applications, a variety of modules having different shapes and dimensions may be packaged on a printed card or board resulting in both non-uniform module spacing and height. Wirtz and McAuliffe (1989) addressed the problem of "convective heat transfer from small surface-mount devices which are located in the flow recirculation zone downstream from a row of larger pin-mount modules which span the coolant passage." Based upon experiments a correlation was developed to estimate the local heat transfer coefficient in the recirculation region downstream from a package row.

A paper by Hall (1989) considered the effect of altitude (i.e., air pressure and density) on forced convection cooling performance. Hall described identical heat transfer experiments conducted in Austin, Texas (altitude — 198 m) and Boulder, Colorado (altitude — 1646 m). Altitude correction factors for ambient air preheat and module thermal resistances measured at 198 meters were determined for altitudes of 915, 1646, and 2135 meters. An example was provided illustrating the use of the correction factors to apply data obtained at 198 meters to predict a component operating temperature at 2135 meters.

To provide adequate cooling, it is often necessary to achieve heat transfer rates higher than those normally attainable by forced convection with air. In these situations, heat transfer may be enhanced by increasing the surface area available for heat transfer, or by increasing the heat transfer coefficient, or both. The remaining papers

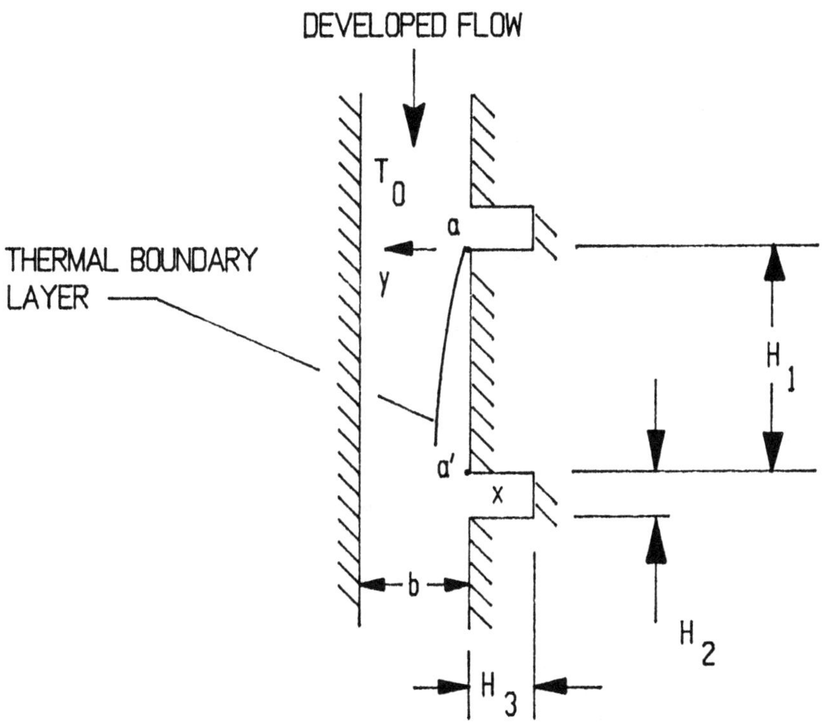

Figure 5.6 Two-dimensional channel flow geometry [Lehmann and Wirtz (1989)].

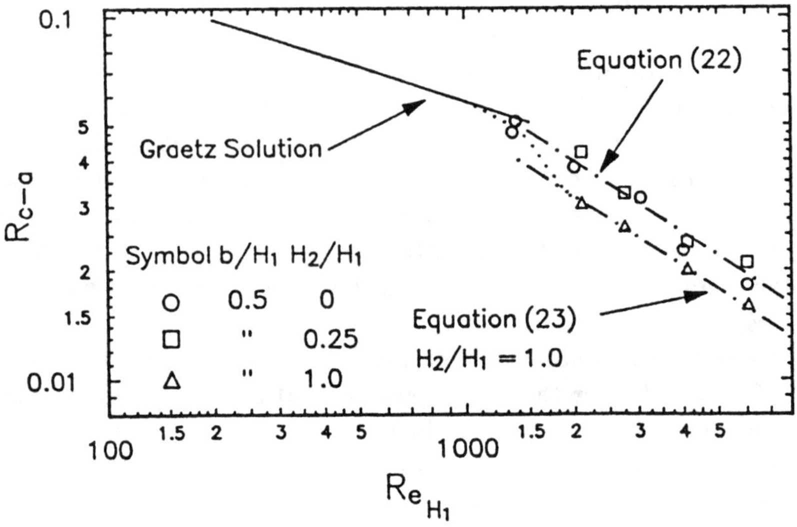

Figure 5.7 Average convective resistance variation with component Reynolds number [Lehmann and Wirtz (1989)].

in this section address some of the techniques for increasing forced convection air cooling capability.

Finned heat sink structures are often used to increase the convective heat transfer area of electronic circuit modules. Lau and Mahajan (1989a) reported heat transfer data (including the effects of tip clearance) for fin arrays with two different fin heights and with "cut fins." A design equation to predict the effect of the tip clearance on heat transfer coefficients for fin arrays with different heights, spacings, and thicknesses was also derived. Lau and Mahajan (1989b) reported further research investigating the heat transfer performance characteristics of convoluted fin designs in terms of pressure drop, thermal resistance, and Nusselt number. The authors concluded that "significant gains in thermal performance can be obtained with high fin density heat sinks for a moderate increase in pressure drop," and "for higher thermal performance, tip clearance should be kept to a minimum." Schulenberg et al. (1989) conducted an evaluation of air cooled microchannel heat sinks. A total of twenty, 2 in. wide heat sinks, with 1 in. tall copper and aluminum fins were tested. The flow channel widths varied from 0.032 to 0.050 in., and the lengths varied from 0.50 to 4.0 in. Experimental data obtained were compared with an analytical model developed to predict the thermal performance of microchannel heat sinks. You (1989) also reported the results of an experimental study to determine the heat transfer coefficients for air flowing through a microchannel heat sink. Heat transfer data were obtained over a range of Reynolds numbers ($8000 < Re < 30000$) for a heat sink geometry consisting of twenty channels, each with a depth of 15 mm and a width 3 mm. The heat transfer results were shown to be in good agreement with analytical heat transfer solutions for a circular tube with a constant heat flux at the wall.

One technique that may be used to obtain higher heat transfer coefficients with air is jet impingement cooling. Hollworth and Durbin (1989) reported the results of experiments conducted to determine the performance of arrays of low velocity air jets used to cool a simulated electronics package. The test system shown in Fig. 5.8 included a plenum with 40 clusters of 2 x 2 arrays of circular jet orifices opposite a simulated PCB. Mounted on the simulated PCB were 40 simulated electronic modules (25 mm x 25 mm) in 8 rows of 5 elements each, which was a geometric configuration Wirtz et al. (1985) studied under channel flow cooling conditions. A comparison of the array-averaged jet cooling data shown in Fig. 5.9, with the channel flow correlation obtained by Wirtz et al. indicates that the impingement cooling system clearly provides substantially higher heat transfer coefficient at the same mass flow rates. In another paper Hollworth et al. (1989) discussed the results of another jet impingement cooling study. In this study, spent cooling fluid (i.e., air) was vented through holes drilled through the PCB midway between adjacent modules, instead of being vented at one end of the channel between the PCB and plenum plate as in the previous study. Venting through the board resulted in 10 to 20% higher system pressure loss and pumping requirements. Two papers by Hamadah (1989a, 1989b) addressed jet impingement cooling with air. The first paper reported the results of an experimental study of the heat transfer characteristics between a square array of round air jets and a square block ($L = 50.8$ mm) simulating an electronics module. The second paper reported the results of experiments to determine the heat transfer characteristics arrays of circular air jets

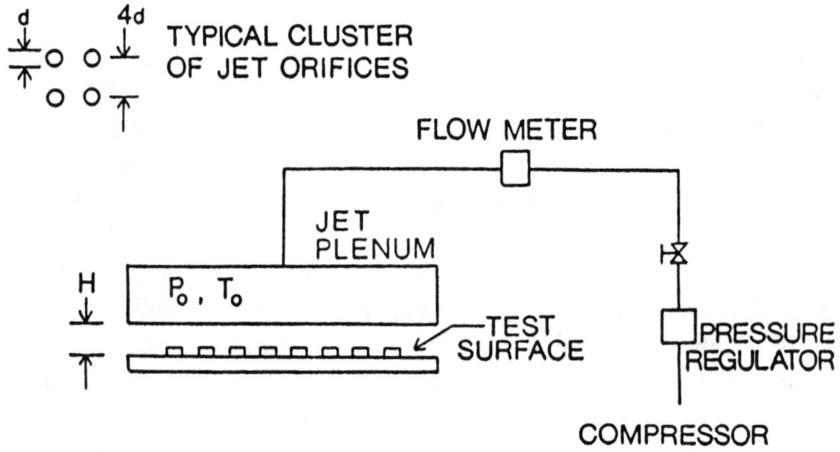

Figure 5.8 Air jet impingement test cooling system [Adapted from Hollworth and Durbin (1989)].

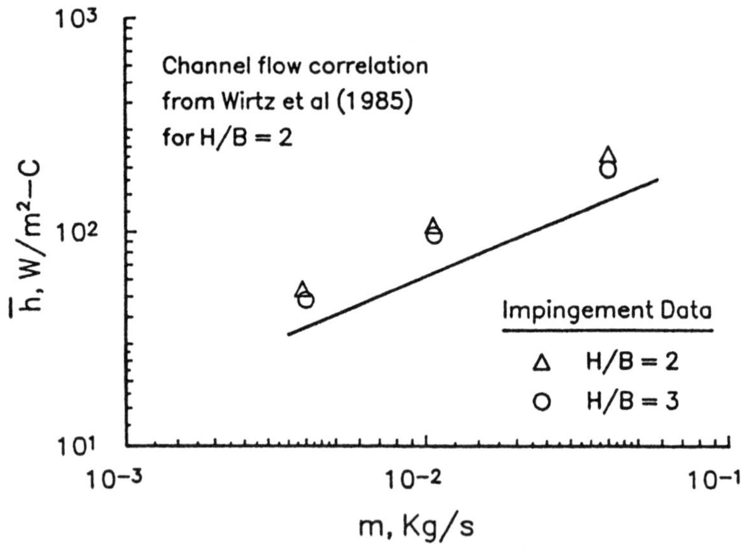

Figure 5.9 Comparison of jet heat transfer data with channel flow correlation [Hollworth and Durbin (1989)].

impinging on a 7×9 array of square blocks ($L = 25.4$ mm) simulating an array of electronic modules. In the conclusions, it was noted that "at the same mass flow rate, impingement heat transfer coefficients are 60–80% higher than those of the channel flow. However, at the same pumping power, channel and impingement heat transfer coefficients are virtually identical."

Another technique which has been proposed to obtain higher heat transfer coefficients with air, is the use of fluidized beds. Brown and Jasper (1989) performed experiments to evaluate the cooling of electronic circuit boards by directly immersing them in fluidized beds of dielectric particles as shown in Fig. 5.10. Figure 5.11 indicates that the heat transfer coefficients obtained in the fluidized bed were six to seven times greater than those obtained for conventional forced convection cooling with air alone. The authors noted that "although impressive, this enhancement is much smaller than the factor of 50 predicted by flat plate theory and correlations available for fluidized beds." This difference was attributed to the very important part that conduction into the PCB plays in increasing the effective heat transfer area for the case of forced convection without particles. Mazenq et al. (1989) also conducted experiments to investigate the cooling of electronic devices in a simple fluidized bed. Their results indicated approximately a five times increase in heat transfer coefficient for a metal-clad resistor cooled in the fluidized bed, in comparison to cooling by forced convection with air alone.

6 LIQUID COOLING

Package cooling capability can be significantly improved when liquid cooling is used. Liquid cooling applications can generally be categorized as indirect or direct. An indirect scheme is usually considered to be one in which the liquid does not contact the electronics or the packages housing the electronics. Since there is no electrical contact, water can be used as the coolant, taking advantage of its superior thermophysical properties. Computers such as the IBM 308X/3090 series, the NEC SX-2000, and the FUJITSU FACOM M-780 utilize indirect liquid cooling schemes. In each of these machines, water is circulated through cold-plate structures in contact with the modules to be cooled. For direct liquid cooling the electronics or packages within which they housed are in direct and intimate contact with the cooling liquid. Direct liquid cooling offers the opportunity of removing heat directly from the chip, or surface of the heat generating device, and transferring it into the liquid with virtually no intervening thermal conduction resistance paths. Although the only computers in which direct liquid cooling has been used are the CRAY-2 and ETA-10 supercomputers, interest in direct liquid cooling remains high. The current level of interest in direct immersion cooling as a means to support high heat fluxes is attested to by the 17 papers reported in this section.

A paper by Wadsworth and Mudawar (1989) described experiments performed to investigate single-phase heat transfer with FC-72 liquid slot jets impinging on simulated electronic chips. The test vehicle consisted of a self-contained cooling module housing a 3 × 3

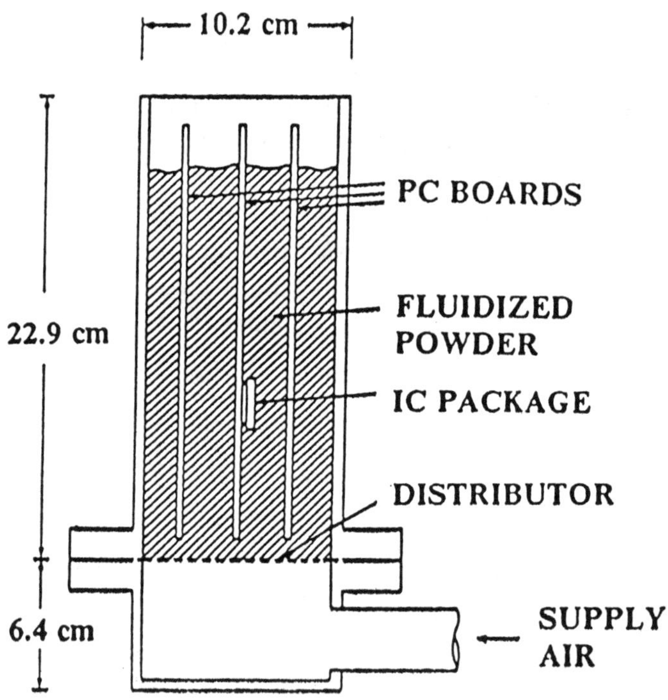

Figure 5.10 Enclosure for fluidized powder immersion cooling of electronic components [Adapted from Brown and Jasper (1989)].

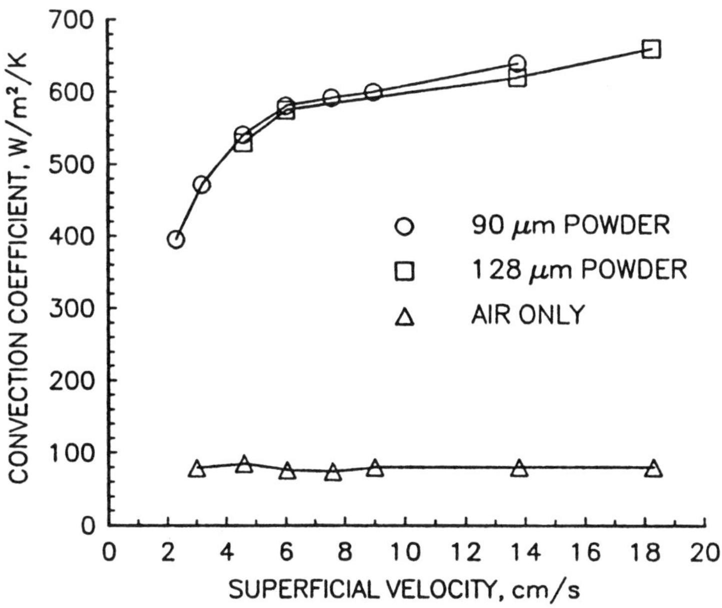

Figure 5.11 Heat transfer coefficient at surface of IC package versus superficial velocity through enclosure [Adapted from Brown and Jasper (1989)].

array of simulated electronic chips. A correlation for the convective heat transfer coefficient as a function of jet width, heater length, flow velocity, and fluid properties was obtained.

Bergles (1989) provided an overview of the challenge posed by enhanced heat transfer with phase change (i.e., pool boiling, forced convection boiling, and condensation) for a variety of applications including cooling of microelectronic devices. It was noted that for microelectronic cooling there is a particular need to elevate the peak nucleate or burnout heat flux. The performance of some of the techniques reported to accomplish this, and also reduce the wall superheat, is shown in Fig. 5.12. With regard to the enhancement of pool boiling heat transfer, it was concluded that structured surfaces have great promise; however, boiling curve hysteresis must be reduced, peak heat fluxes must be increased, and predictive methods must be developed.

Several papers reported on investigations of pool boiling utilizing fluorocarbon coolants (e.g., FC-72 and FC-87). In terms of electrical and chemical characteristics, the fluorocarbon family of coolants are generally considered to be the most suitable for direct immersion cooling of electronic devices and packages. These Fluorinert liquids developed by 3M Company, do not contain any chlorine atoms in their molecular structure, and should not be confused with the chloroflurocarbon compounds (i.e., CFCs) associated with the problem of atmospheric ozone depletion. Two papers by Mudawar and Anderson (1989a, 1989b) discussed studies of high heat flux cooling of a 12.7 mm × 12.7 mm simulated electronic chip by means of pool boiling. The first paper reported the effects of coolant variation (i.e., FC-72 or FC-87), pressurization, subcooling, and surface augmentation on boiling performance. The heat flux variation with superheat reported for one of the augmented surfaces is shown in Fig. 5.13.

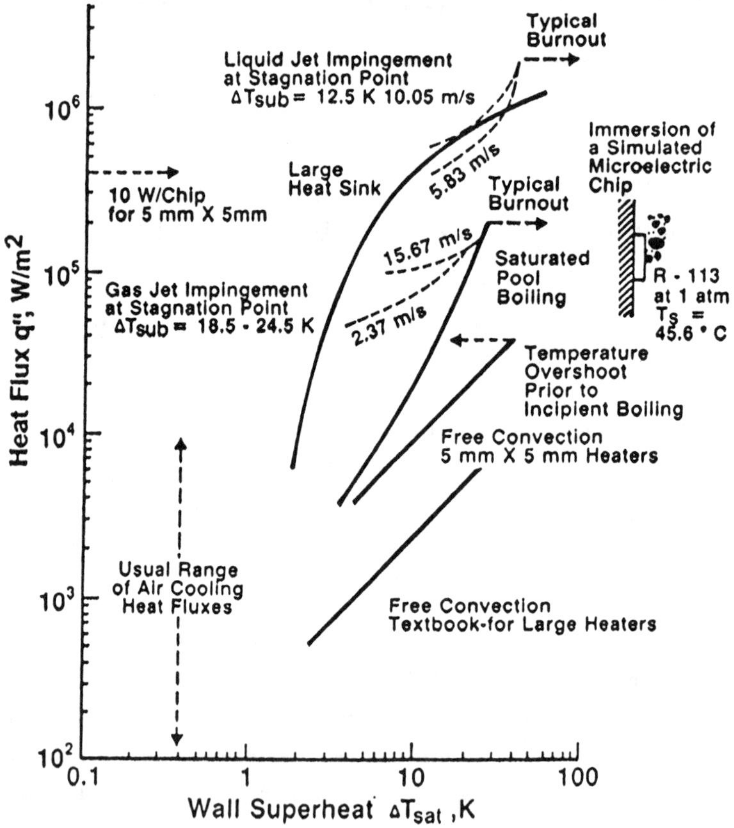

Figure 5.12 Heat flux ranges for immersion cooling of microelectronic chips [Adapted from Bergles (1989)].

In this case a boiling heat sink consisting of four cylindrical pins attached to the simulated chip surface was tested in a saturated pool of FC-72 (T_{SAT} = 56° C) and FC-87 (T_{SAT} = 30° C), attaining peak heat fluxes of 102 and 89.8 W/cm^2 respectively. Without enhancement, peak saturated pool boiling heat fluxes in fluorocarbon liquids are typically in the range of 15–20 W/cm^2. Other surface enhancement geometries studied included a pyramid shaped finned stud and a square finned stud. The second paper reported the results of further work to examine enhancement of pool boiling heat transfer with other extended surface geometries. The geometries investigated included single studs, 2 × 2 and 3 × 3 arrays of pin fins, and enhanced stud surfaces in the form of axial microgrooves, radial microgrooves, and microstuds. Another paper by Anderson and Mudawar (1989) provided further results of experimental studies of boiling heat transfer from a simulated microelectronic component immersed in FC-72.

A paper by Nakayama et al. (1989) presented the results of experiments and numerical modeling to study the temperature transient for a microporous stud cooled by boiling in FC-72. The stud was made up of several copper plates bonded together to form a porous structure with microchannels in three mutually orthogonal directions as shown in Fig. 5.14. The stud had a 10 mm × 10 mm square base for attachment to a microelectronic component. For heat loads in the range of 10 to 80 W, the steady-state heat transfer performance of the stud leveled off for lengths greater than 5 mm. The thermal resistance appeared to range from about 0.55° C/W at 10 W to 0.26° C/W at 80 W.

To accommodate higher heat fluxes than are attainable in pool boiling, flow boiling may be used. Maddox and Mudawar (1989a) reported on both single-phase forced convection and flow boiling heat transfer enhancement using three low profile, microstud surfaces of different heights. A paper by Mudawar and Maddox (1989) discussed experiments performed to determine the parametric influences of velocity and subcooling on critical heat flux (CHF). For these experiments FC-72 was used to cool a single 12.7 mm × 12.7 mm simulated electronic chip in the wall of a vertical rectangular channel. In another paper Mudawar and Maddox (1989b) reported further studies to investigate techniques for enhancement of CHF on a simulated

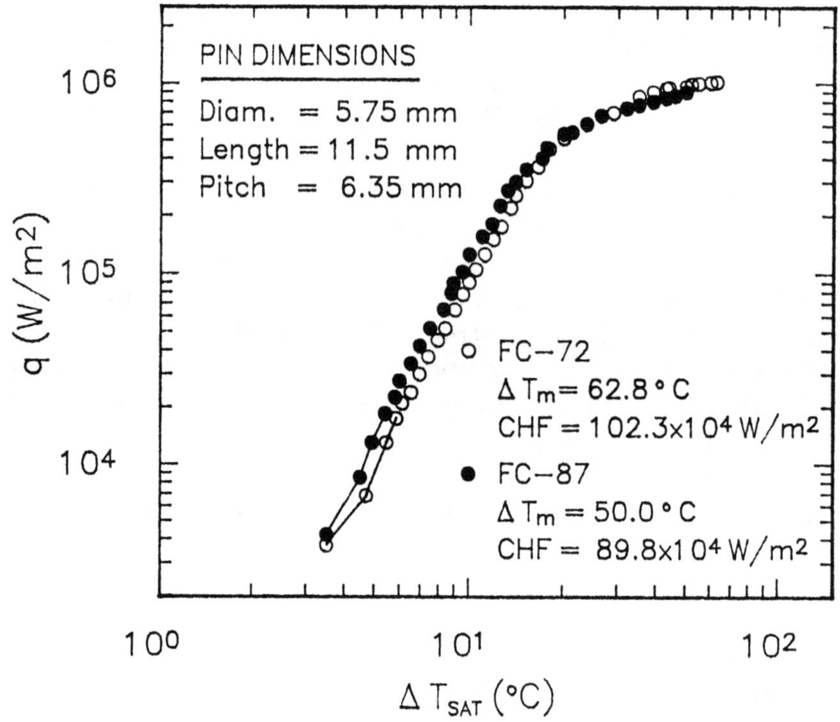

Figure 5.13 Saturated pool boiling performance of a 4 pin heat sink surface [Mudawar and Anderson (1989a)].

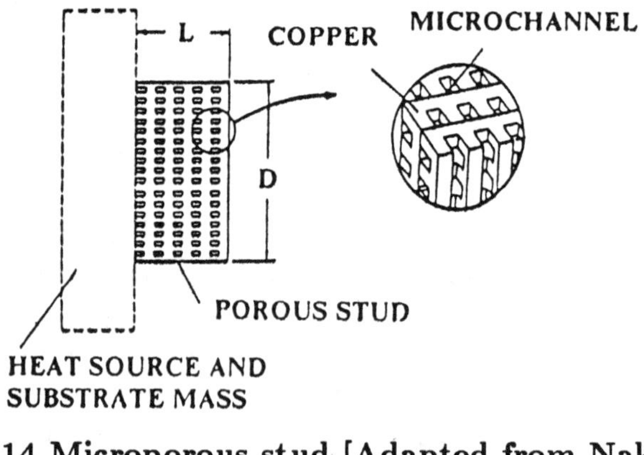

Figure 5.14 Microporous stud [Adapted from Nakayama et al. (1989)].

electronic chip. Results were reported for a smooth surface, and for surfaces with low-profile microgrooves, low-profile microstuds, and high-profile pin fins. Strom et al. (1989) conducted a study to determine the CHF condition for subcooled flow boiling using R-113 coolant. Ten heated "fingers" were mounted in the wall of a flow channel to simulate an in-line array of electronic chips. Boiling curves and CHF data were reported for elements flush with the wall, and elements slightly protuding (i.e., 0.8 mm) above the wall. Lee and Simon (1989) investigated flow boiling heat transfer from small heat sources (3 mm × 1 mm and 3 mm × 3 mm) to subcooled FC-72. It was reported that the "influences of velocity and subcooling on the nucleate boiling curve for these small heater sizes are essentially the same as in the literature for larger heated areas."

Galloway and Mudawar (1989) conducted experiments to investigate a novel flow boiling cooling concept. As shown in Fig. 5.15, the proposed concept employs a rotational stirrer to circulate cooling fluid around the interior of a vessel with electronic chips mounted on the walls. As explained by the authors, "in the two-phase regime, bubbles nucleating from the surface cavities are removed by the high velocity stream-wise liquid flow past the surface, and by the curvature of the test vessel creating a radial buoyancy force which pulls the bubbles toward the liquid/vapor interface." It was further explained that "as the rotational speed increases, higher stream-wise and radial forces promote more efficient removal of vapor bubbles from the surface, suppressing bubble accumulation and delaying CHF." An example of the effect of rotational speed, ω, on heat transfer rates observed for FC-86 coolant is shown in Fig. 5.16.

Another form of flow boiling that has attracted the interest of heat transfer researchers is the use of impinging liquid jets. Yao et al. (1989) conducted experiments to investigate cooling with sprays of liquid FC-72 droplets impinging on a high heat flux surface. The peak heat fluxes reported were in the range of 15 to 30 W/cm^2. A paper by Nonn et al. (1989) reported the results of a study examining the jet impingement cooling performance of a 50% volumetric mixture of FC-72 and FC-87. Cooling tests were carried out for jet diameters of 0.5 and 1.0 mm, and for 1, 4, and 9 jets impinging on a single 12.7 mm × 12.7 mm simulated electron chip heat source. It was

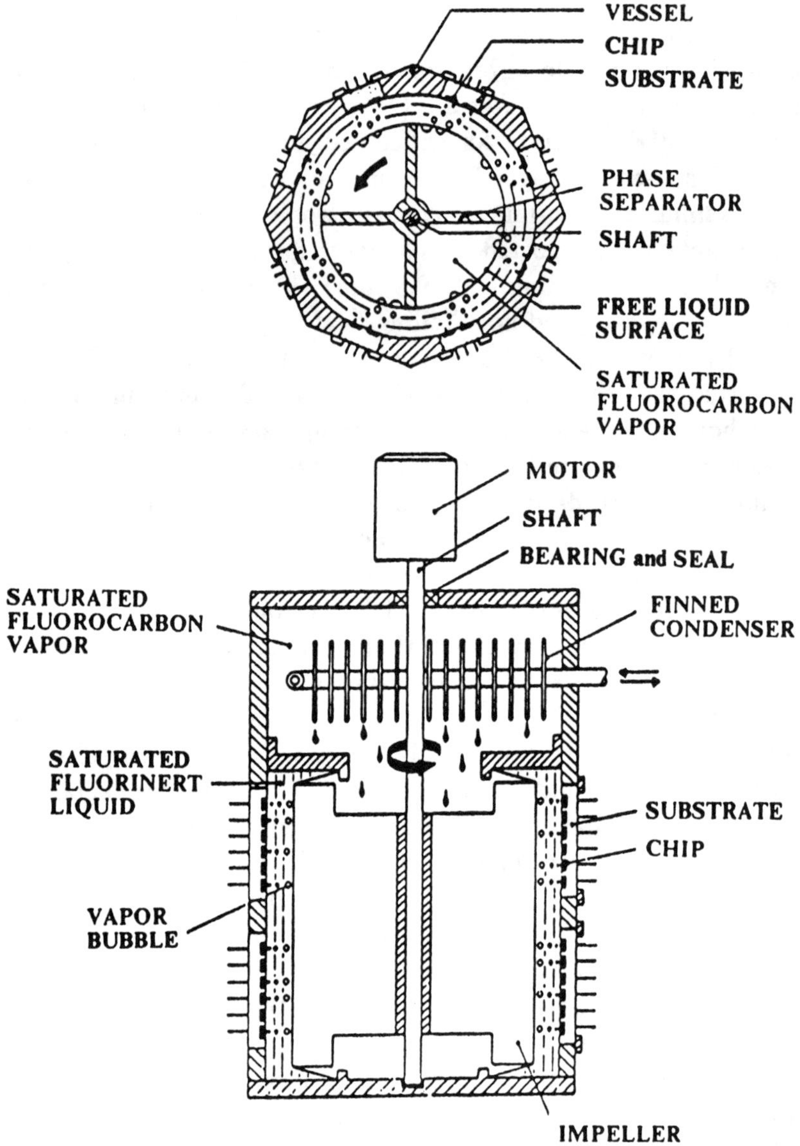

Figure 5.15 Schematic diagram of the rotating stirrer cooling concept [Galloway and Mudawar (1989)].

concluded that, "by comparing the present data for the mixture with that for FC-72, it is clear that the cooling of the heat source improves by about 10 to 15° C throughout the boiling regime, although CHF is lower due to the lower saturation temperature of the mixture." Jaeger et al. (1989) reported the results of liquid jet impingement cooling experiments utilizing a 75 mm diameter silicon wafer with a centrally located thin film nichrome heater element. The test chamber was part of a closed loop refrigeration system using Freon-12 as a coolant.

Two papers addressed direct immersion cooling with liquid nitrogen (LN_2). Cokmez-Tuzla et al. (1989) investigated LN_2 boiling characteristics with buoyancy induced flow over two 10 mm × 10 mm silicon wafers, mounted facing each other 10 mm from the bottom of a 40 mm tall flow channel. Measured CHF values were reported to be 12–15 W/cm^2. Yokouchi et al. (1989) reported the enhancement

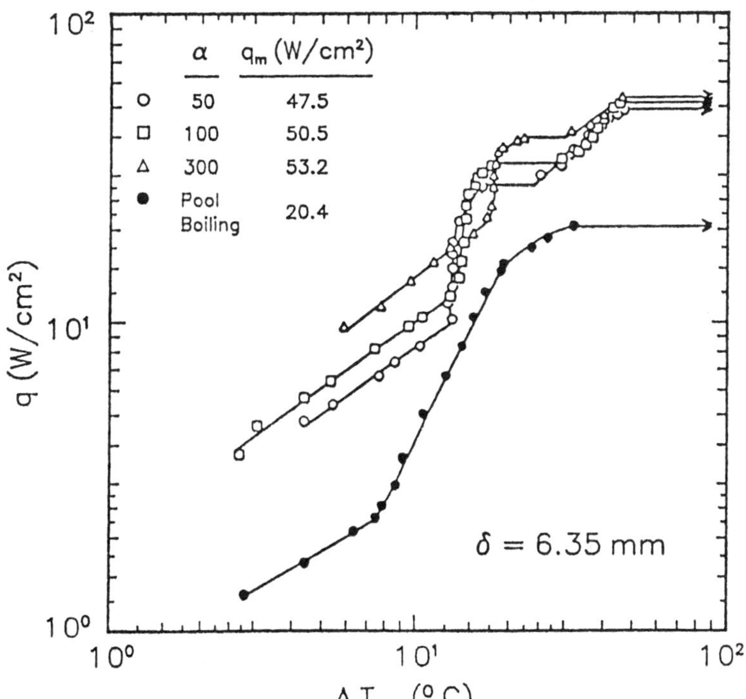

Figure 5.16 Effect of rotational speed on heat transfer with FC-86 [Adapted from Galloway and Mudawar (1989)].

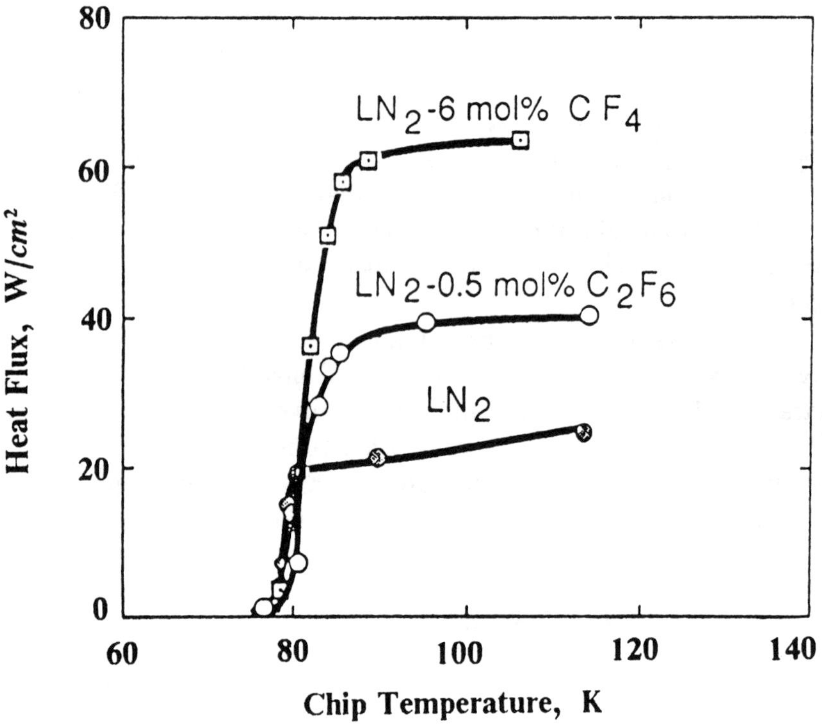

Figure 5.17 Effect of adding fluorocarbons to LN_2 on pool boiling performance [Adapted from Yokouchi et al. (1989)].

of LN_2 pool boiling characteristics by adding fluorocarbons. Boiling experiments were conducted using gallium arsenide (GaAs) chips immersed in pure LN_2, LN_2 mixed with C_2F_6 (0.5% mole fraction), and LN_2 mixed with CF_4 (6% mole fraction). As shown in Fig. 5.17, the peak heat fluxes were increased to 40 and 65 W/cm^2 respectively by the addition of the fluorocarbons.

7 THERMAL MEASUREMENTS AND SENSING

Thermal measurements play an important part in the successful design of cooling schemes for electronic components and systems. Accurate measurements of temperature and power are necessary to characterize the thermal resistance of components from the chip heat source to the cooling sink. This also includes accurate measurement of coolant flow and temperature conditions. In addition, temperature measurements may be helpful in determining the individual thermal resistances in one or more heat flow paths within a component assembly. Temperature measurement of prototype hardware has been a long-standing method to confirm the adequacy of a thermal design. The continuing trend towards higher power and packaging densities only serves to increase the need for accurate thermal measurements in verifying the adequacy of a cooling design. Thermal measurements can also play an important role in assessing the quality of incoming vendor components and outgoing manufactured electronic equipment.

For many years, thermocouples have provided a very useful device for measuring temperature, and they are still commonly used to measure fluid and surface temperatures. Resistance temperature detectors (RTDs) and thermistors may also be used to obtain the same temperature measurements as thermocouples. An article by Bahniuk (1989) described the characteristics of both RTDs and thermistors. In some cases, it may be desirable to measure temperatures at many points on an electronic package, or even obtain a complete temperature "map." Infrared (IR) surface temperature measurements may be used to obtain such temperature maps. Sihlbom and Svärd (1989) discussed the use of an IR scanner to map PCB temperatures. The discussion included a method to compensate for measurement errors due to variations of emissivity over the object surface, and an IR method to determine PCB temperatures based upon a thermal decay process. A paper by Briles and Bennett (1989) discussed a calibration procedure developed to improve the accuracy and precision of a commercial IR microscope.

Temperature sensitive electrical parameters (TSEP) of a device or circuit on a chip are often used to obtain a temperature measurement, or even a thermal resistance measurement. Three papers in this section addressed such techniques. Tarter (1989) described the use of an MOS thermal test circuit to obtain average values of thermal impedance for MOS as well as bipolar and other families of active devices. Estreich (1989) described a DC electrical method (DCEM) for measuring the thermal resistance of GaAs MESFETs using the temperature dependence of the gate metal resistivity. Pshaenich (1989) described the design and development of a high current, ΔV_F, transient thermal resistance tester.

Negus et al. (1989) described both experimental and modeling techniques used to characterize the thermal resistance of microwave bipolar devices. The temperature measurement techniques employed included infrared, liquid crystal transition, and TSEP methods. The liquid crystal transition technique was "found to be much more accurate than IR microscopy for microwave applications because of its increased resolution," and the modeling technique compared extremely well with the liquid crystal measurements of temperature.

A paper by Sammakia et al. (1989) described a transient technique to measure the temperature coefficient of resistance of thin film resistors on ceramic substrates. The method employed offered a faster alternative to the conventional method which required heating the resistor to a known higher temperature in an oven.

For many years the thermal resistance of air-cooled packages has been broken down into an internal resistance from junction to case (R_{jc} or Θ_{jc}) and an external thermal resistance from case to air (R_{ca} or Θ_{ca}). Internal thermal resistance is often treated as constant and independent of external cooling conditions (e.g., air flow, packing density on PCB, etc.). Unfortunately, this is often not the case, and internal thermal resistance measured under one cooling or packaging condition will exhibit a different value under other conditions. A paper by Bar-Cohen et al. (1989) discussed this problem and proposed a new approach providing "a universal, though modified, Θ_{jc} for use with a particular package in every possible packaging configuration." In order to reflect multiple internal thermal paths and the flow of heat according to boundary conditions imposed on the various surfaces, and still be free of system specific heat transfer coefficients (or contact resistance at the points of package attachment), it was proposed to represent chip temperature by

$$T_j = m \sum_{k=1}^{m} A_k T_k + A_{m+1} Q$$

It was explained that "re-expressing the chip temperature relation in a thermal resistance form results in

$$A_{m+1} = \frac{T_j - \sum A_k T_k}{Q}$$

$$\Theta_{jc} \geq A_{m+1}$$

where Θ_{jc} can be defined as a modified junction-to-case thermal resistance. Since the sum of the A coefficients is unity when the "case" temperature is uniform, i.e., $T_1 = T_2 \cdots$, the modified junction-to-case thermal resistance is found to reduce to the common definition of Θ_{jc}, which is thus found to equal A_{m+1}." The necessary A coefficients for a specific package could be determined and reported by vendors marketing the package, to allow users to better predict T_j for a specific application environment. In another paper Biskeborn (1989) discussed the need and a method to extract "true" module and "localized" external thermal resistances from test vehicle measurements.

In addition to the need for accurate measurements of temperature and power, it is also important that the cooling condition under which the characterization of an electronic component is performed be controlled and repeatable. In the case of forced air-cooled components this may be accomplished using a low-speed wind tunnel in which the electronic component is mounted. Keil and Berger (1989) described the use of a low-velocity (0–3 m/s) automated wind tunnel to obtain total thermal resistance measurements (i.e., R_{ja}) for a variety of integrated circuit packages. Results were presented comparing the performance of standard packages and packages with heat spreaders and heat sinks.

Thermal sensing can also play an important role in protecting electronic components and packages from excessive temperatures during operation in product environments. As a result of the high power densities used in many components and packages today, a cooling failure (e.g., loss of airflow) could result in physical destruction of the affected component, making thermal protection an essential element of cooling system design. An article by Swager (1989) discussed commercially available thermostats which can be used to implement temperature sensing for control/warning directly on a PCB.

8 THERMAL ANALYSIS TECHNIQUES

As the interval between successive generations of electronic equipment decreases, the time available for performing prototype tests on next generation equipment has decreased. This has led to an increased emphasis on the use of analytic and numerical models for the analysis and prediction of package temperatures and thermal performance. Sophisticated thermal analysis of proposed electronic packages is rapidly becoming the rule, rather than the exception. In many cases automated thermal modeling tools are being integrated into the computer aided engineering (CAE) and computer aided design (CAD) systems used for electrical design and physical layout of electronic packages. Articles by Comerford (1989), Wellins and Sullivan (1989), and Yao (1989), discussed the trend towards greater automation of thermal analysis and design for printed circuit boards. Papers by Weiss (1989), Weiss et al. (1989), and LaBonte (1989a) also discussed the use of thermal analysis tools to model PCB thermal performance. Another article by LaBonte (1989b) discussed the PCB thermal analysis problem in terms of the various sources of error and their relationship to failure rate prediction. A paper by Jacobson (1989) further emphasized the importance of accurate determination of chip junction temperatures in order to obtain accurate chip reliability estimates. Closely associated with the thermal analysis of PCBs is the optimum placement of components on a PCB. A paper by Osterman and Pecht (1989) discussed the mathematical theory for component placement to maximize reliability based upon the thermal response of conductively cooled PCBs. Another paper by Gore et al. (1989) described an approximate numerical solution approach to modeling thermal conduction in a PCB. The paper also describes experiments and measurements made to validate the model for incorporation in a CAD program for PCB design.

Relatively simple thermal networks are sometimes used to model the heat flow paths in an electronic package. Fukuoka and Ishizuka (1989) described an application of the thermal network method, and proposed a simple thermal model as a preliminary thermal design analysis tool for multichip packages. The model proposed is a non-steady-state model which includes thermal conduction, thermal radiation, and natural convection heat transfer. A paper by Vogelsong and Brzezinski (1989) discussed modifications to the SPICE (simulation program with integrated circuit emphasis) program to perform a simultaneous simulation of an electrical system and a thermal network. The method employed implicitly defined each electrical device's parameters in terms of instantaneous junction temperatures to avoid the inaccuracies caused by the constant temperature assumption normally used by SPICE. Somos et al. (1989) discussed thermal modeling of the two-dimensional heat flow in a GTO (gate turn-off) with a highly interdigitated emitter, using a thermal analog which had the form of an electrical transmission line.

A number of papers discussed the use of finite element methods (FEM) to model various aspects of component cooling and thermal behavior. A paper by Subramanyam and Rolph (1989) described the use of the ADINA-T thermal analysis program to study the thermal behavior of a direct mount chip attach system which utilized a beryllia heat spreader. The importance of (1) boundary conditions for heat sinking, (2) active regions in the semiconductor, (3) thermal behavior of the attachment layers, and (4) thermal properties of materials, in the analysis are discussed in the paper. A paper by Cooke and Lee (1989) presented the results of a numerical thermal analysis of a 144 lead plastic flat-pack (PFP) VLSI chip package using a FEM model. Tseng (1989) discussed the development and application of a FEM code to the thermal conduction analysis of a multilayered IMPATT (impact avalanche transit-time) device. The code was based on a FEM program previously developed for a large mainframe. The original program was broken into several segments in such a way that each segment was small enough to run in an IBM PC/AT's main memory, while the other segments were idling in the secondary memory (i.e., floppy-disk drives). Di Barba and Savini (1989) developed a FEM model of a freely suspended (i.e., without card or board) plastic leaded chip carrier (PLCC) cooled by natural convection and radiation. The modeling results reproduced the decrease in junction-to-ambient thermal resistance with increased chip power, which has been observed in experimental data. A paper by Lau (1989) described an analysis using a 3-D nonlinear finite element method to determine thermal stresses and strains in a surface mount assembly with a plastic-quad-flat-pack (PQFP) chip carrier. Aakalu (1989) described "the use of finite element models to aid in solving a thermally induced high stress problem inside of a cold plate-cooled step-down transformer for a switcher type power regulator." Papers by Schnipke et al. (1989) and Hayward (1989) described the use of a computational fluid dynamics (CFD) code called FLOTRAN to model the cooling performance of a pin grid array package with a stacked-fin heat sink. The model provided both a description of the flow field and a conjugate thermal analysis of the entire package. Rajakumar and Johnson (1989) used the ANSYS program to create a simulation of the buoyancy-induced flow around microelectronic components mounted on a circuit board. Results were presented in terms of plots of both velocity vectors and temperature contours in the flow channel, as well as averaged component Nusselt numbers. A paper by Hadim et al. (1989) described a software package written to interface finite element codes with finite difference thermal network analyzers for detailed thermal analysis of microelectronic equipment.

Several other papers discussed the use of finite difference models to study various aspects of component cooling and thermal behavior. Two papers by Darveaux et al. (1989a, b) discussed the thermal analysis of an advanced multichip package design using a two-dimensional finite difference model. The multi-chip package modeled consisted of a 5 × 5 array of 1 × 1 cm chips on a 102 × 102 mm AlN substrate, with an indium die bond between each chip and a SiC cap containing water-cooling channels. A paper by Khalilollahi and Sammakia (1989) described a two-dimensional numerical analysis of a predominately natural convection channel flow, with an imposed weak forced convection field. The channel walls modeled consisted of four uniform heat flux segments separated by adiabatic zones. A paper by Mahaney et al. (1989) presented the results of a numerical study of mixed convection flow in a horizontal rectangular duct with a 3 × 4 array of heat sources flush mounted in the bottom wall. The results indicated that "the variation of the row-average Nusselt number with Reynolds number exhibits a minimum, suggesting that heat transfer may be enhanced due to buoyancy-induced secondary flow by reducing the flow rate and hence the pumping requirements."

Two papers discussed the use of a commercial finite control volume code (PHOENICS) to model air cooling of electronics packages. Agonafer and Moffatt (1989) used the code to model the flow over 2-D blocks simulating electronic modules in a channel flow. The numerical results were compared with experimental results reported by other researchers. Linton and Agonafer (1989) described the use of the code to model flow distribution, convection coefficients, turbulence effects, and temperatures in a personal computer. Cadre and Viault (1989) discussed the use of another finite control volume code called THEBES to model air flows in electronic frames cooled by natural or forced convection. Examples of thermal optimization studies using THEBES were given for two types of racks for packaging telecommunications electronics equipment.

Asako and Faghri (1989) employed a numerical model to study the heat transfer characteristics associated with fully developed laminar flow through a channel with square heat dissipating blocks mounted on one wall of the channel. Calculations were performed for Reynolds Numbers from 100 to 1000. Results were reported in terms of temperature and heat flux distributions on the block(s), average Nusselt Numbers, and thermal resistances. Rizk and Kleinstrauer (1989) used the SIMPLE[1] algorithm developed by Patankar and Spalding (1972, 1980) to analyze steady laminar forced convection cooling of discrete heated blocks in open and porous matrix channels. It was concluded that for a highly porous matrix (e.g., volumetric porosity = 0.3), an acceptable $\Delta p/L$ is obtained at Re = 2000, along with increases in heat transfer over an open channel of 46% for water and 58% for air.

Four papers discussed conjugate heat transfer models. Zebib and Wo (1989) analyzed heat transfer from a silicon chip embedded in a ceramic chip carrier mounted on a printed circuit board. The printed circuit board formed the lower wall of a laminar flow duct, and the model assumed a two-dimensional flow field. Krause et al. (1989) analyzed forced convection heat transfer from small isothermal sources embedded in a large substrate using the finite element method. Comparisons were made with a previous study of the same problem by Ramadhyani et al. (1985) using a finite difference method. Lee and Yovanovich (1989) described a quasi-analytical conjugate model developed to predict the thermal characteristics of a vertical plate with discrete heat sources cooled by natural convection and radiation.

The last two papers in this section describe analyses of thermal conduction in chip-substrate structures and heat spreaders respectively. Lee et al. (1989) described an algorithm for the accurate integration of the temperature solution of a four-layer plate structure with infinite lateral boundaries. Huang and Lee (1989) described a 3-D analytical solution used to solve for the temperature distribution

---

[1]SIMPLE is an acronym for semi implicit method for pressure-linked equations.

in a generalized heat spreader (i.e., substrate and heat sink) for an electronic module.

9 PACKAGE COOLING TECHNIQUES

Specific cooling techniques in electronic equipment applications can yield equipment sizes that vary from the very small to the very large. But, whether the application involves a single natural convection-cooled heat producing component, or a high power, liquid cooled, multi-chip module, the objective remains the same: to allow thermal energy to flow from source to sink within the constraints of the specified temperature, space, and often cost. This section lists papers which describe a variety of package cooling evaluations and applications.

At the component level, a number of papers discussed the thermal performance evaluation of air-cooled components and packages. A paper by Estes (1989) presented the results of a combined experimental and analytical investigation of the thermal resistance of a chip-on-board package in which semiconductor chips are mounted directly to a PCB. The design factors considered included the effects of thermal planes, thermal vias, chip encapsulation, and single-sided versus double-sided mounting of chips. Blanco and Mansingh (1989) reported the results of a study of the thermal characteristics of buried resistors laminated within the layers of printed circuit boards cooled by natural convection and radiation. The study included both finite element modeling and experiments using an infrared camera to determine the temperature distribution of the resistors. A paper by Mahalingham (1989) provided an assessment of the thermal characteristics of surface mount plastic packages under both natural and forced convection cooling conditions. Results were presented for several package configurations including small outline integrated circuit (SOIC) packages, plastic leaded chip carriers (PLCCs), and plastic quad flat packs (PQFPs). Yokono and Ishizuka (1989) described numerical modeling and experimental studies of the thermal performance of a quad flat package (QFP) with a forced convection air-cooled omnidirectional heat sink. A conjugate approach was employed to simultaneously model flow and heat transfer in the cooling air stream and heat conduction within the package and heat sink. A paper by Lee (1989) discussed experimental thermal evaluation procedures and results for forced convection cooled 144 pin plastic flat packs soldered on a PCB. Bosler and Buechler (1989) reported the results of experimental studies evaluating direct air cooling of standard electronic modules (SEM) in a format B configuration used in military electronics applications. These modules (or cards) consist of leadless chip carriers (LCCs) mounted on double sided ceramic substrates on 0.4 in. centers in a test card cage accommodating 10 modules. The effects of component size, air mass flow rate, and heat flux density were shown. McShane et al. (1989) discussed package design considerations related to pin grid array packages and the use of a heat spreader to improve cooling performance. Use of a heat spreader reduced "worst case" (i.e., mean + 3 standard deviations) thermal resistance from 53° C/W to 32° C/W, a 40% improvement. Azar et al. (1989) reported the results of an investigation to assess the thermal sensitivity of simulated circuit packages to cooling condition (i.e., natural or forced), board orientation (i.e., vertical or horizontal), and board spacing. The results of this study indicated that "overall thermal performance can be affected by as much as 25% due to non-convection mode related variations."

A number of inventions pertaining to improved control or enhancement of air cooling for electronic packages were patented. Bonnefoy (1989) described an air-ducting scheme for providing "fresh" cooling air across columns of VLSI packages with heat sinks. As shown in Fig. 5.18, air is supplied via a duct at the top of the enclosure in parallel across modules arranged in columns. The heated air is then exhausted via the two return ducts above the columns of modules. Hamburgen (1989) disclosed an air impingement cooling scheme utilizing arrays of nozzles fabricated in a printed circuit board underneath each component to be cooled. In the embodiment shown in Fig. 5.19, two printed circuit boards are arranged back-to-back, with the enclosed space in between serving as an air distribution plenum for the arrays of nozzles. Phelps and Ward (1989) disclosed a molded plastic package utilizing a clip or through bolt to provide a thermal path from an internal heat spreader to an external heat sink. Ocken et al. (1989) disclosed a heat sink assembly for a plurality of transistors, which includes a specially formed heat

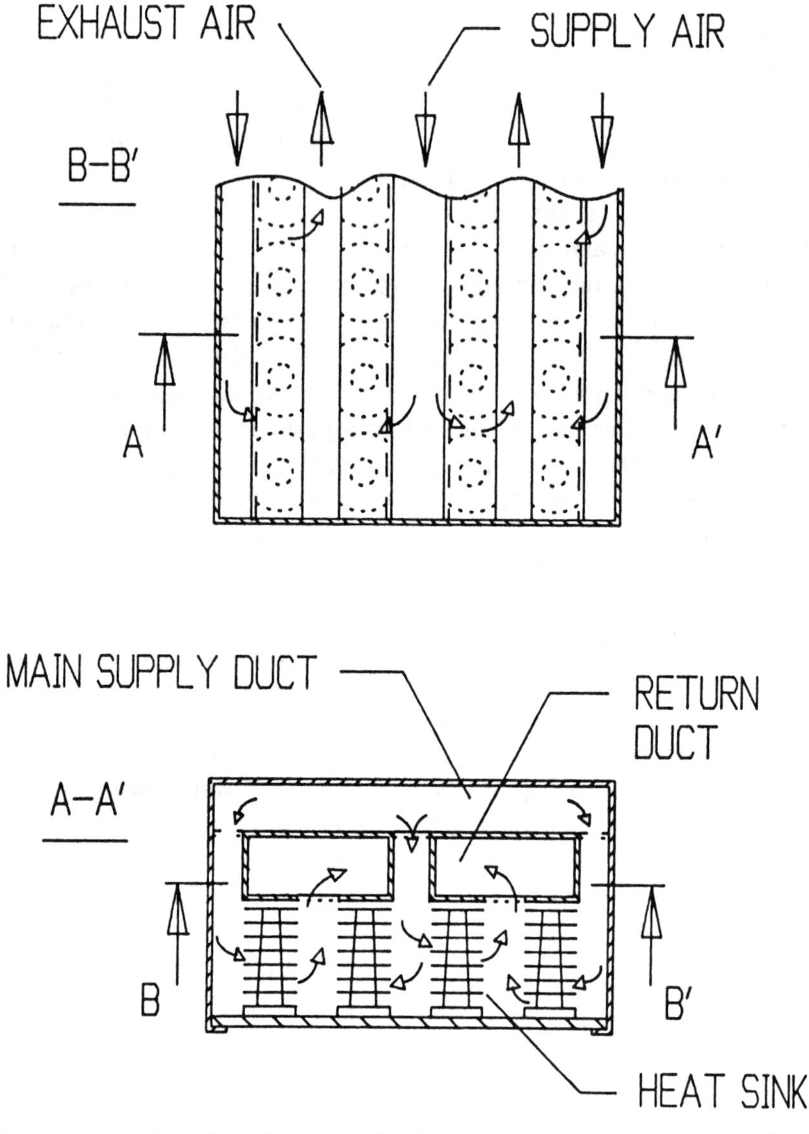

Figure 5.18 Device for ventilating components arranged in rows on a substrate [Adapted from Bonnefoy et al. (1989)].

sink and specially formed spring clips to hold the transistors on the heat sink. Friedman (1989) described a packaging concept utilizing a heat spreader attached to each chip on a printed circuit substrate, to provide a good thermal path to an air cooled heat sink.

Two articles by Wall (1989a, b) described the use of a "liquid heat sink" to provide a thermal path, from modules on a printed circuit board within a cabinet, to a wall of the cabinet. As shown in Fig. 5.20 the "liquid heat sink" consists of a flexible plastic pouch containing a fluorocarbon coolant. Heat is transferred from one surface to another by means of natural convection. As noted by the author, temperature differences on a typical board were reduced from 25–30° C under forced air cooling, to about 8° C with the "liquid heat sink."

Two papers addressed thermal management of surface mounted components on conduction cooled SEM printed circuit boards. In this application two printed circuit boards are bonded to opposite sides of a thin metal core. The metal core is mechanically clamped at each end to the card rack walls which may contain convective cooling channels, and is intended to provide a low thermal resistance path for heat flow from from the surface mounted components. Lycans et al. (1989) investigated the use of a "thermal bridge" in the form of a plug of material inserted between the base of an electronic component and the surface of the PCB, and the use of "thermal vias" in the PCB beneath the components. It was found "that components with thermal bridges operated significantly cooler than those without." Nagurny (1989) discussed thermal modeling and testing of surface mounted components on conduction cooled SEMs. This paper also noted the importance of providing an effective thermal path from the bottom of the component to the surface of the PCB.

For applications with a higher packaging density and a higher heat flux, it is sometimes necessary to resort to water-cooling to provide the required control of component operating temperature. Several papers discussed applications using water-cooling. Suzuki et al. (1989) described the scheme used to cool super integration modules (SIMs) in the Fujitsu FACOM VP2000 supercomputer. Each 300 × 300 mm SIM holds up to 144 single-chip micropackages (SMPs) with a heat dissipation of 30 watts and a surface heat flux of 10 W/cm^2. Within a SIM each SMP is cooled by a water-jet contained

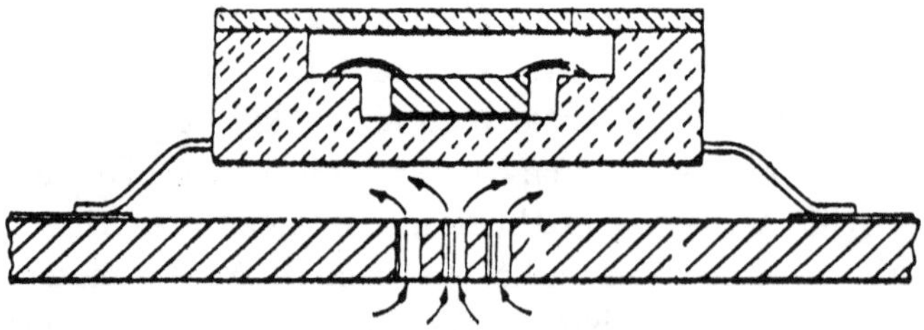

Figure 5.19 Air impingement cooling scheme using arrays of nozzles fabricated in printed circuit boards [Adapted from Hamburgen (1989)].

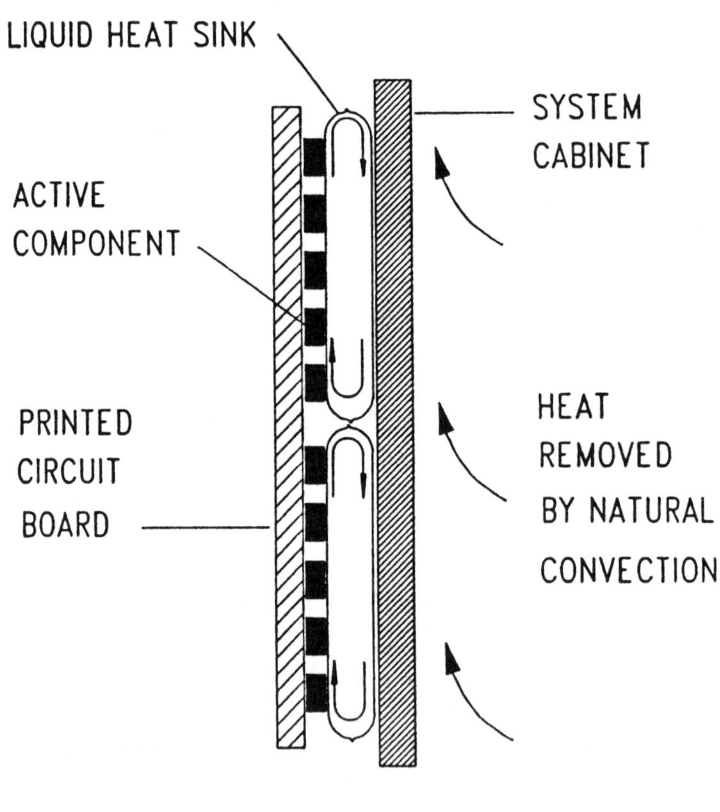

Figure 5.20 "Liquid heat sink" used to provide thermal path from circuit modules to cabinet wall [Adapted from Wall (1989)].

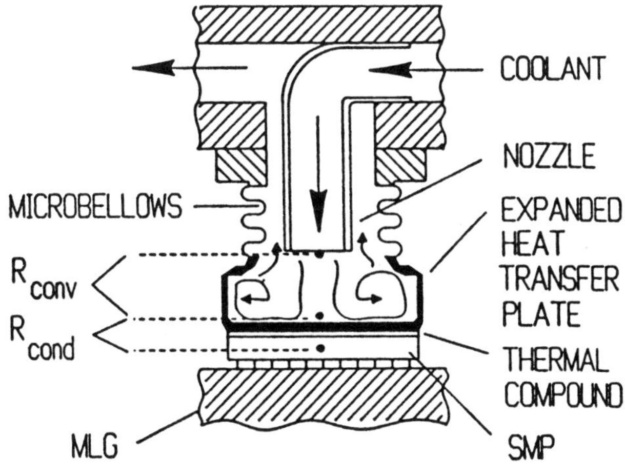

Figure 5.21 Water-jet cooling scheme used to cool SMP chip packages within super-integration modules in Fujitsu FACOM VP-2000 [Adapted from Suzuki et al. (1989)].

within a microbellows assembly in contact with the SMP as shown in Fig. 5.21. The reported thermal resistance from chip-to-water per SMP is 0.56° C/W. A paper by Hiraki et al. (1989) described a new water cooled switching power supply which utilizes both air and water-cooling and is one-third the size of a conventional air-cooled power supply. The water-cooled section of the power supply handles approximately 70% of the total heat load. Missaggia et al. (1989) discussed the cooling of a two-dimensional GaInAsP/InP diode laser array with a power dissipation of 500 W/cm^2. The approximately 1 × 4 mm laser array was soldered to a silicon microchannel heat sink containing 40 channels (100μm wide and 400μm deep) covering an area 8 mm wide × 10 mm long. A value of 0.04° C cm^2/W for the overall thermal resistance per unit area of the silicon heat sink was deduced from experimental data.

A number of inventions were patented which describe techniques to further extend chip cooling using indirect liquid cooling. Herrell and Gibson (1989) described a liquid cooled integrated circuit package. As shown in Fig. 5.22. a heat sink with liquid-cooling microchannels is positioned opposite heat dissipating chips. Manifolds are provided to direct coolant to and from the microchannels. Microcapillary slots may be formed on the heat sink surface opposite the chip to receive a liquid (e.g., oil) to generate capillary forces between the heat sinks and chips to facilitate heat transfer. Komoto et al. (1989) described the liquid cooling system for integrated circuit chip packages shown in Fig. 5.23. In this package chips are mounted on one side of a substrate and a thermally conductive plate is joined to the opposite side of the substrate by a thermally conductive bonding material such as solder or a silver epoxy resin. The thermally conductive plate provides a flat surface for attaching the removable cold plate to the module package. Three patents by Nakanishi et al. (1989), Yamamoto et al. (1989), and Tustaniwskyj and Halkola (1989) described bellows or "bellows-like" structures used to pro-

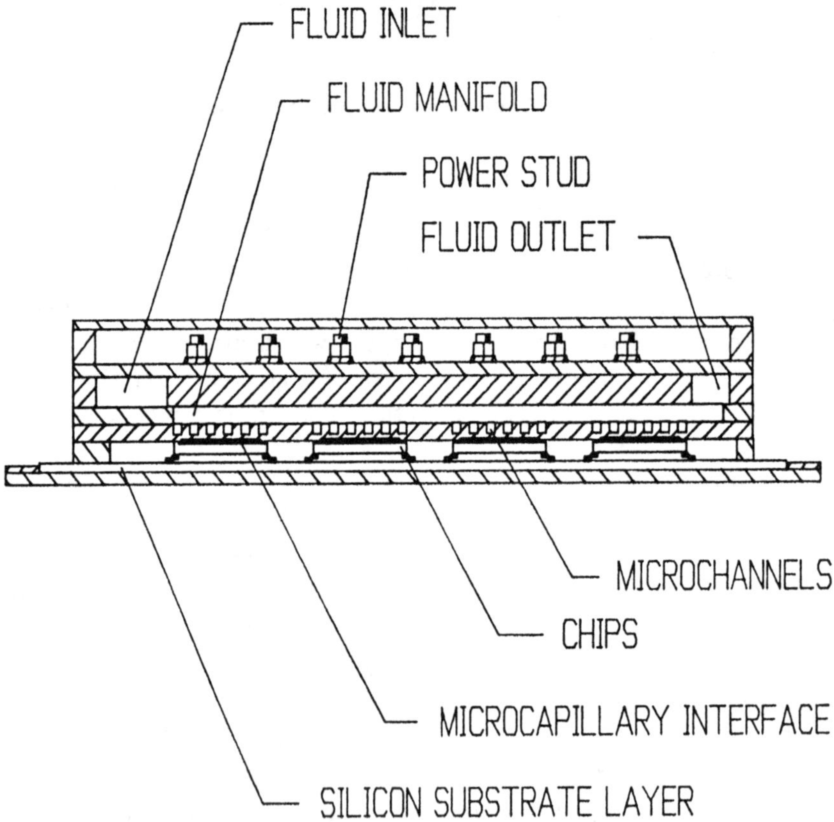

Figure 5.22 Cross-section of a liquid-cooled integrated circuit package with microchannels and microcapillary slots [Adapted from Herrell and Gibson (1989)].

vide water-cooling for arrays of chips on a substrate. Parks (1989) disclosed a three-dimensional microelectronic package structure consisting of a stack of "wafer packs" and "cavity wafers." As shown in Fig. 5.24, chip carriers are mounted on the "wafer packs," and the "cavity wafers" form a "plurality of cavities and tunnels for providing coolant to the cavities" to cool each chip carrier. As stated in the patent "the cooling capability of the three-dimensional package of the present invention is 5000 watts at a flow rate of 1.5 gallons per minute for the system or 0.01 gallons per minute per chip." Another patent by Nelson and Gupta (1989) described an "end-fed" liquid-cooled heat sink for attachment to an electronic chip.

Saaski et al. (1989) disclosed a heat pipe integrated within a module to transfer heat away from chips. One embodiment of the invention is shown in Fig. 5.25. The chip is cooled by evaporation of the working fluid and the resulting vapor is condensed at the fluted condenser surface in the cap. A porous wick is provided between the cap and the chip to facilitate return of coolant to the chip by capillary action. The top surface of the chip is grooved to form an inverted liquid meniscus causing the liquid adjacent to the top of the component to readily evaporate. In another patent Smolley (1989) described a high density package utilizing chips within a planar enclosure containing a fluorochemical coolant. Each chip is directly bonded to a stainless steel pad which accommodates the differential coefficient of expansion between the chips and the housing, and allows for wire bonding interconnections from each chip to a PCB. The pads also act as wicks drawing the surrounding coolant to the bottom side of the chips, in addition to the coolant in contact with the sides and top of the chips. Heat is removed from the chips by boiling, and from the package enclosure by an air-cooled finned compact heat exchanger attached to the housing wall. Package cooling techniques encompass not only chip, module, and printed circuit board applications, but also applications providing temperature control of other electrical devices used in peripheral equipment. A paper by Eibeck et al. (1989) discussed a thermal model of a disk drive which was developed to evaluate the feasibility of using a thermal prediction code in the design process. Experimental measurements of the temperatures within an actual disk

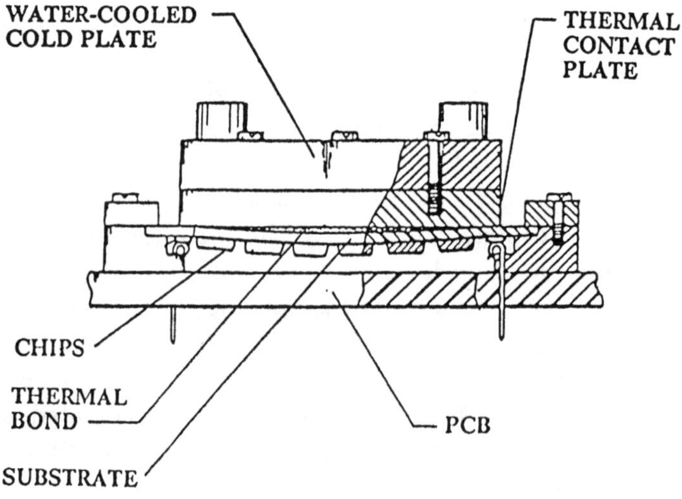

Figure 5.23 Cross-section of liquid-cooling system for integrated circuit chip packages [Adapated from Komoto et al. (1989)].

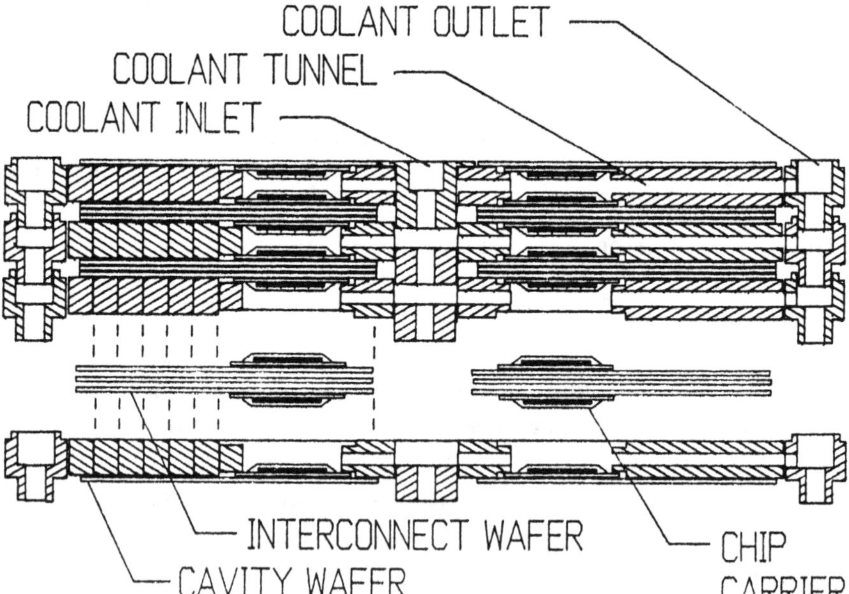

Figure 5.24 Cross-section of three-dimensional microelectronic package structure for semiconductor chips [Adapted from Parks (1989)].

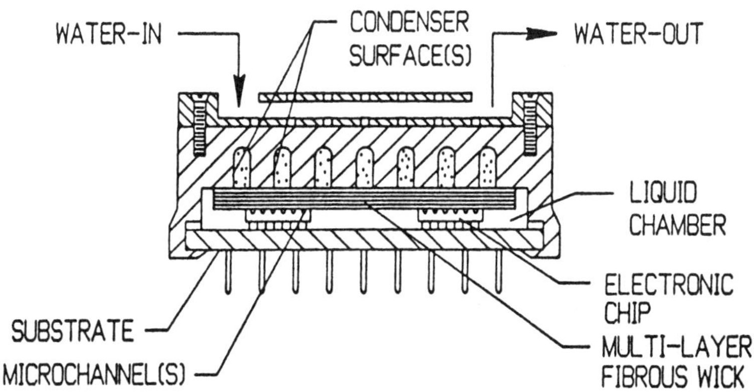

Figure 5.25 Cross-section view of integral heat pipe module [Adapted from Saaski et al. (1989)].

drive were made to compare with the model's predictions. It was concluded that the thermal model provided an excellent design tool to evaluate the impact of different designs on the resulting temperature distributions, time constants, and thermal expansions within a disk drive. A paper by Moore (1989) discussed the problems associated with the incorporation of mass storage devices (e.g., Winchester disk drives, floppy disk drives, and cartridge tape backup devices) designed for environmentally controlled computer rooms and in telecommunications equipment designed to operate in more severe thermal environments. In the last paper of this section, Tanabe et al. (1989) discussed a simulation model used to study the heating-melting phenomena occurring during a high-speed thermal transfer printing process used in some compact printers.

10 SYSTEM COOLING TECHNIQUES

As challenging as the package cooling problem may be, an equally challenging problem is that of the cooling system design. Usually, an electronic system will consist of many components or packages. The electronic system could be a small personal computer with a few natural convection cooled printed circuit boards, or a large mainframe computer containing both air-cooled and water-cooled package assemblies. In either case, a system cooling design is required to assure that the cooling fluid can be supplied to each component or package at the proper flow rate and temperature to provide adequate package cooling. The papers covered in this section discuss the various techniques used in air and liquid cooling systems.

Two publications addressed the problem of air cooling computers. An article by Wilson (1989) described the thermal management and design process used for the Honeywell Bull DPS 8000 mainframe. A paper by Chao (1989) provided an overview of thermal management for Data General ECLIPSE MV/40000 Series superminicomputers. This overview addressed thermal design at the component, PCB, and system levels. The paper also considered the introduction of an evaporator coil of an air-conditioner to pre-cool the air entering the card cage. A paper by Chida et al. (1989) described the heat dissipation techniques used in a distributed power supply system for communications equipment. In this application, heat pipes are used to transport heat from power voltage converters mounted on shelves within a rack, to natural convection-cooled heat sinks mounted on the back panel of each shelf. In addition three patents described inventions relating to air-cooled electronic packages. Foley et al. (1989) disclosed an environmentally sealed enclosure containing PCBs and a fan. As shown in Fig. 5.26 the side walls and rear walls of the enclosure are formed of thin (0.020 in.) corrugated sheets of aluminum providing both internal and external surface enhancement for the enclosure. The fan provides forced convection heat transfer between the PCBs and the air, and between the air and the internal walls of the cabinet. The outer walls of the cabinet are cooled by natural convection of air. Zushi et al. (1989) described the use of a honeycomb-like flow partition placed between vertically stacked mother boards with PCBs. The air entering the flow partition from the lower mother board may exhibit a non-uniform temperature distribution as a result of non-uniform component powers on the PCBs. The walls forming the honeycomb-like flow partition are alternately sloped to the left and to the right to cause changes in the direction of the emerging air streams, and induce flow mixing to promote a more uniform cooling air temperature distribution before flowing over the upper mother board. Gabuzda and Terrell (1989) disclosed a directed air management system (DAMS) that provides parallel cooling air flow to pin grid array (PGA) modules on printed circuit cards mounted vertically on a backplane. Air is distributed via a miniplenum facing each printed circuit card as shown in Fig. 5.27. A jet of cooling air is directed at a radial fin heat sink mounted on each module. The air-jet holes in each miniplenum are appropriately sized to provide the desired cooling flow to each module.

A patent publication by Kawashima et al. (1989) described an apparatus for circulating cooling fluid through cooling modules (i.e., cold plates) for cooling electronic components. As shown in Fig. 5.28, the apparatus consists of multiple parallel coolant loops each containing two pumps and a heat exchanger, with a common expansion tank for all the loops. The inventors explained that "the cooling fluid is supplied independently to the respective cooling systems, in such a manner that the pressure and flow of the cooling fluid in each cooling system are not affected by the other cooling systems, but in

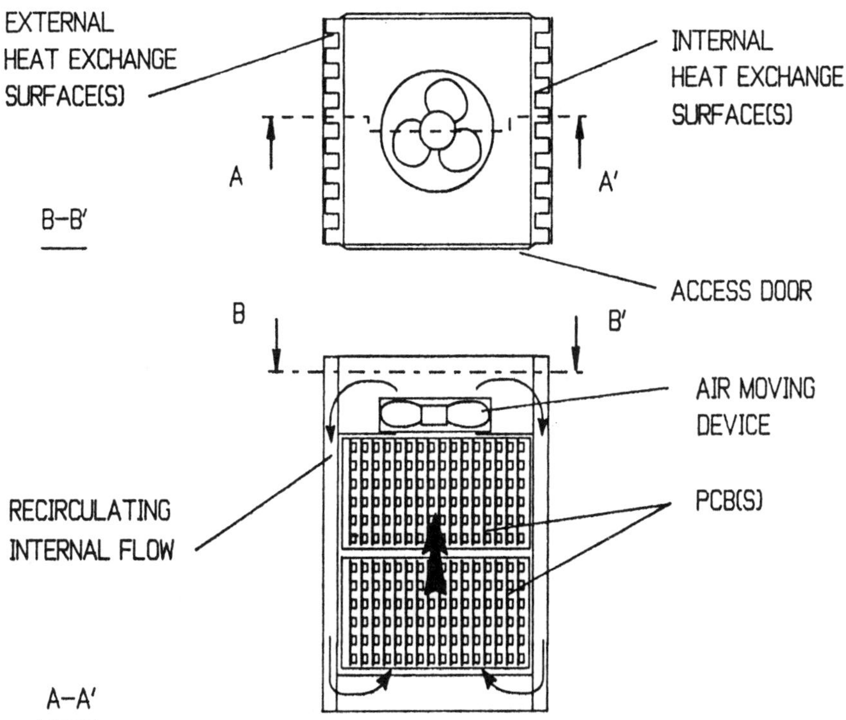

Figure 5.26 Environmentally sealed electronics enclosure with high surface area walls for heat exchange [Adapted from Foley et al. (1989)].

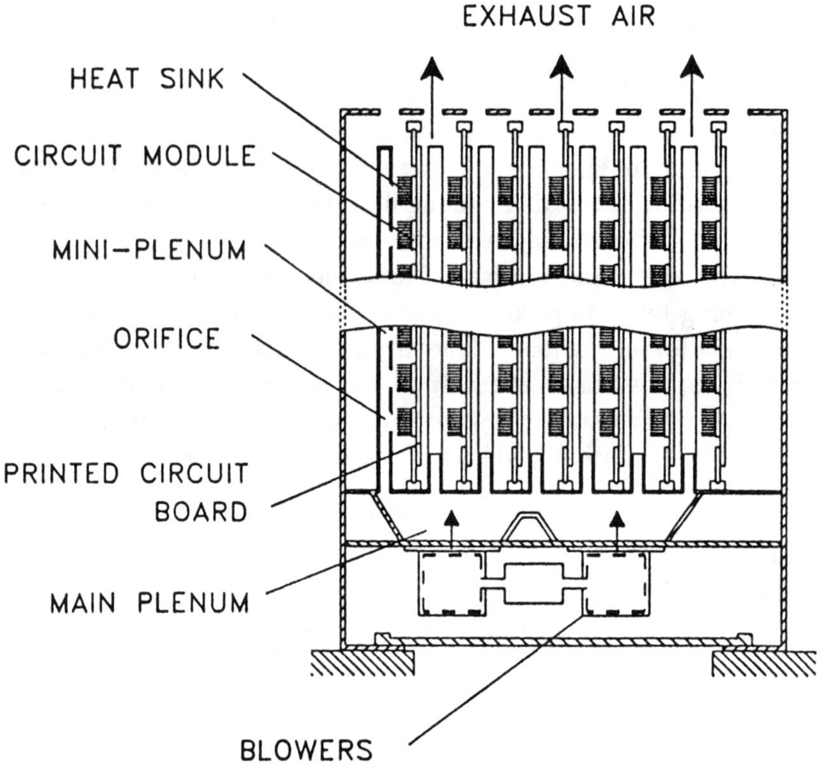

Figure 5.27 Directed air management system (DAMS) for parallel impingement cooling of electronic modules [Adapted from Gabuzda and Terrell (1989)].

accordance with the needs of the cooling module in question in the system."

One paper and a patent were related to applications of cryogenic cooling for electronic equipment. A paper by Carlson et al. (1989) discussed the ETA10 liquid-nitrogen-cooled supercomputer system. The supercomputer was developed using CMOS circuit technology, with as many as 2000 circuit chips of up to 20,000 gates, each operating at a temperature below 90 K (approximately doubling the speed of the circuit chips). It was further explained that "by choosing liquid-nitrogen operation, the performance of CMOS is increased to levels similar to other high-speed circuit technologies, and reliability of the circuits and hardware is enhanced at the low temperatures

(77 K)." Several topics related to thermal design were addressed including: (1) device packaging and interconnection in a cryogenic environment, (2) device cooling in liquid nitrogen, (3) design of a cryogenic enclosure for a 700 W central processor, and (4) the system level cryogenic refrigeration system. A patent by Hohenwarter and Grange (1989) disclosed a system for delivering a cryogenic fluid to cool electronic devices in an open-cycle environment. The system was described as consisting of a dewar for storing the cryogenic fluid, a cryogenic module to hold the device to be cooled, a fluid transfer line to deliver the cryogen from the dewar to the device (within the cryogenic module), locking and coupling mechanisms, and a subsystem to control flow of the cryogen from the dewar.

The remaining three papers in this section each addressed installation level aspects of electronic equipment cooling. A paper by Kolte and Zirath (1989) outlined a comparison between three cooling system approaches for application to telephone exchanges. Two of the approaches involved heat rejection from the cooling air at a point some distance removed from the electronic racks being cooled. The third approach utilized natural convection cooling of the electronic racks and removal of the heat from the exhaust air by means of water-cooled coils directly above the aisles of each electronic rack. The latter approach was determined to be the most advantageous with respect to: (1) power consumption, (2) emergency cooling (in event of power failure), (3) switching room working environment, (4) space, and (5) extendibility. A paper by Allen (1989) described the highly standardized, modular cooling system approach adopted by British Telecom to provide reliable cooling for digital telephone switching installations. These units were designed primarily to provide cooling by drawing in cold outside air and passing it through a filter into the telephone exchange room. The unit incorporates a refrigeration loop for use when the outside air is either too warm or too humid to use directly for cooling the electronics (less than 30% of the time). Under this circumstance, a damper is closed to allow only a small fresh air make-up to pressurize the room, the refrigeration compressor is turned on, and the outside air is then used only to cool the condenser. Prudhoe and Doukas (1989) discussed the use of phase change chemicals (PCCs) as a means to provide a temporary cooling temperature limiting system in the event of an instal-

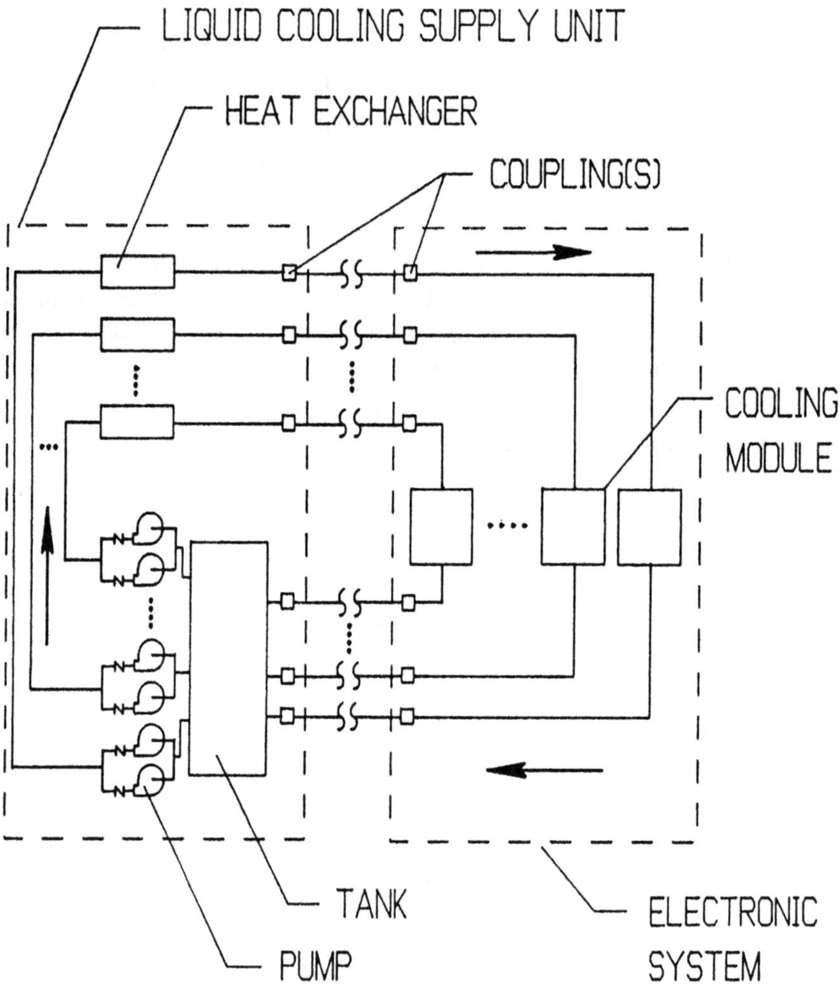

Figure 5.28 Liquid cooling supply unit with multiple, independent liquid cooling loops [Adapted from Kawashima et al. (1989)].

lation air-conditioning failure. Under normal operating conditions, the PCCs contained in banks of tubes, are "pre-frozen" by the air conditioning system cooling the electronic telephone equipment. If the air-conditioning fails, much of the heat generated from the electronic equipment and dissipated to the circulating air is absorbed by the PCCs, slowing down the rate of temperature increase in the equipment room. The PCCs appear to be particularly attractive in this application due to their high volumetric heat storage capacity and their latent heat of fusion associated with the phase change.

11 COOLING DEVICES/HARDWARE

A variety of devices and hardware may be used to control temperatures and improve cooling capability in electronic equipment. Fans and blowers are used to provide cooling air flow over power dissipating components and packages. Heat sinks and heat pipes are used to transport heat more effectively from source to sink. Mechanical or thermoelectric refrigeration devices are used to pump heat or provide sink temperatures below ambient. The papers in this section discuss the design, evaluation, and application of some of these devices and hardware.

Two patents described air moving devices for electronic cooling applications. Ohbayashi et al. (1989) disclosed a new cross-flow blower design which incorporates the elements of the drive motor in such a way that the overall size of the blower is considerably reduced. Cox (1989) disclosed a miniaturized fan for mounting directly on the surface of a printed circuit board to provide cooling air flow for components mounted on the same board.

Two patents described air-cooled heat sinks for application to electronic components. Funari et al. (1989) disclosed an electronic package assembly consisting of a circuit chip, a flexible means of interconnecting the chip to a printed circuit board, and a heat sink for direct attachment to the chip. The heat sink design incorporates pliant "legs" to permit the heat sink to be depressed downward to meet and form an adhesive joint with the chip. Jones and Bulmahn (1989) disclosed an air-cooled heat sink consisting of two U-shaped members extending from a common mounting base. Each U-shaped member is adapted to retain an electronic component between the inner leg and outer leg.

The application of heat pipes to cooling electronics was discussed in two papers. A paper by Thomson et al. (1989) described the characterization and modeling of a flat plate heat pipe (FPHP) as a thermal design option for the high power amplifier (HPA) of the Mobile Communication Satellite (MSAT). A thermal model was created incorporating the electronic box, the FPHP, and the radiator panel. The model incorporated FPHP evaporator heat transfer coefficients which were determined experimentally over a range of power inputs and temperatures. Results obtained using the model demonstrated that significant improvements in weight and reliability could be obtained in this application using a FPHP instead of a conventional solid aluminum spreader plate. Tanzer and Numa (1989) reported on a comprehensive test program to demonstrate the thermal performance of heat pipes for a particular avionics application. During this program the SEM format E printed wiring board heat pipe substrate was tested while being subjected to dynamic environments (e.g., random vibration, shock, and sustained acceleration) typical of advanced military aircraft, such as the Advanced Tactical Fighter. The PWB heat pipe substrates were shown to operate successfully during and after exposure to these stringent military aircraft environments.

Trade press articles by Buchmann (1989) and Levine (1989) discussed applications of solid state thermoelectric or Peltier cooling. As noted by Levine, some of the advantages offered by thermoelectric cooling are temperature control below ambient, precise and stable temperature control, ability to heat or cool, and reduced space, size, and weight. Levine also provided an example of thermoelectric module selection using a typical vendor chart, and a brief discussion of installation methods.

12 THERMAL ASSEMBLY PROCESSES

Although not directly related to electronic equipment cooling, the area of thermal assembly processes warrants inclusion here. Many of the steps and processes in the manufacture and assembly of semiconductor chips and packages require controlled application of heat and temperature. Failure to achieve the requisite degree of control may be reflected in poor quality and degraded reliability.

Two of the papers included here discussed thermal and mechanical models used to study aspects of the assembly process. Simon et al. (1989) presented the results of "a parametric study of thermal stresses in plastic packages that are induced during the die attach and encapsulation fabrication steps. Finite element models of plastic VLSI packages were used to vary package design parameters (i.e., die thickness, die bond materials and thickness, lead frame thickness, and package thickness and width/length) over typical ranges." Glaser and Juaire (1989) also used a FEM model to study the temperature and stress distribution in a typical PLCC package subjected to surface mount soldering conditions. Simulation results were obtained for both a vapor phase and an infrared soldering process.

A paper by Lea (1989) provided a comparative appraisal of three heat transfer fluids specifically marketed in the UK for vapor phase soldering. The fluids considered were two Fluorinerts (FC-70 and FC-3511) marketed by the 3M Company, and LS230 marketed by Montedison Ltd. The appraisal included considerations of heat transfer properties, solder reflow properties, operating power requirements, rosin flux solubility, thermal degradation (in terms of boiling point stability, toxicity, and corrosivity), and fluid consumption. It was noted that each of the three fluids studied can be successfully used to solder electronic assemblies using vapor condensation heat transfer in either batch loading or on-line facilities, and that the choice of fluid will usually depend mainly on the soldering temperature and the cost of fluid replenishment.

A paper by Spigarelli (1989) addressed the problem of achieving thermal separation during soldering attachment or rework of surface mount components. Thermal separation implies reducing or minimizing the thermal (i.e., temperature) exposure of adjacent components and their leads. The use of specifically directed heating, heat shields, and active cooling to increase thermal separation between a component undergoing a soldering operation and components adjacent to that site was discussed.

Another paper by Spigarelli (1989) focused on soldering techniques for attachment of tape automated bonding (TAB) components. This paper discussed an experimental study to characterize a new contact heating technology referred to as "convector" heating. The factors considered in the study included convector material and design, application temperature, temperature uniformity, thermal separation, and thermal response of the solder joint. Evans et al. (1989) reported the results of a study of the application of the hot bar, or thermode, solder reflow method for attachment of TAB components. The hot bar method is used to form an interconnecting solder joint between the component and PCB by contacting the lead/solder/pad structure of the component with a heated bar, or thermode. Mathematical modeling and empirical measurements were combined in the paper to develop recommendations for thermode configuration and desirable material properties, and to provide an understanding of the factors affecting thermode performance.

The remaining two papers discussed infrared soldering techniques. Chan (1989) reported the results of an evaluation of several infrared reflow machines for use with molded circuit boards. It was concluded that "with care in setting up the temperature zones, molded circuit boards made from engineered thermoplastic polyethersuphone can be successfully reflow soldered." An article by Avramescu (1989) provided a brief description of the methods for controlling and improving temperature uniformity during infrared reflow soldering of printed circuit boards.

REFERENCES

Aakalu, N G (1989). Thermal induced stress modeling in a switcher transformer, *Proc 9th Ann Intl Elec Packaging Soc (IEPS) Conf*, **1**, San Diego, CA, Sep, 334–341.

Agonafer, D and Moffatt, D F (1989). Numerical modeling of forced convection heat transfer for modules mounted on circuit boards, *ASME*, **HTD-121**, Dec, 1–5.

Allen, P A (1989). The development of high reliability modular cooling systems in British Telecom, *Proc 11th Intl Telecom Energy Conf*, Firenze, Italy, Oct.

Anderson, A (1989). The prediction of temperatures in non-uniformly heated arrays of uniform geometry but variable spacing, *Proc 9th Ann Intl Elec Packaging Soc (IEPS) Conf*, **2**, Sep, 899–912.

Anderson, T M and Mudawar, I (1989). Microelectronic cooling by enhanced pool boiling of a dielectric fluorocarbon liquid, *ASME J Heat Trans*, **111**, Aug, 752–759.

Antonetti, V W and Simons, R E (1985). Bibliography of heat transfer in electronic equipment, *IEEE Trans Components, Hybrids, Manuf Tech*, **CHMT-8** (2), 289–295.

Asako, Y and Faghri, M (1989). Three-dimensional heat transfer analysis of arrays of heated square blocks, *Intl J Heat and Mass Trans*, **32** (2), Feb, 395–405.

Avramescu, S (1989). Improving temperature uniformity in infrared reflow soldering of PCBs, *Electron Manuf*, Feb, 24–26.

Azar, K, Develle, S E, and Manno, V P (1989). Sensitivity of simulated circuit pack thermal performance to geometric arrangement variation, *Proc 5th Semiconductor Thermal and Temp Symp*, San Diego, CA, Feb, 112–119.

Bahniuk, D E (1989). The changing face of temperature measurement, *Machine Design*, Feb, 111–114.

Bar-Cohen, A, Elperin, T, and Eliasi, T (1989). θ_{JC} characterization of chip packages — justification, limitations, and future, *Proc 5th Semiconductor Thermal and Temp Symp*, San Diego, CA, Feb, 1–4.

Beasley, K G (1989). Thermal management of U.S. Navy's standard electronic module using metal matrix composites, *Proc 9th Ann Intl Elec Packaging Soc (IEPS) Conf*, Vol. 2, 953-960.

Bergles, A E (1989). The challenge of enhanced heat transfer with phase change, *Intl J Heat and Technology (UK)*, **7** (3–4), 1–12.

Biskeborn, R G (1989). Thermal resistance extraction method for electronic packages, *Proc Natl Elec Packaging and Production Conf — NEPCON East*, Boston, MA, Jun, 168–176.

Blanco, R and Mansingh, V (1989). Thermal characteristics of buried ressistors, *Proc 5th Semiconductor Thermal and Temp Symp*, San Diego, CA, Feb, 25–29.

Bonnefoy, J (1989). Device for ventilating components arranged in rows on a substrate, U.S. Patent 4,835,658, May.

Bosler, A and Buechler, T (1989). Navy format B, standard electronic module level thermal testing and analysis, *Proc 9th Ann Intl Elec Packaging Soc (IEPS) Conf*, **1**, San Diego, CA, Sep, 418–429.

Briles, S D and Bennett, G A (1989). Calibration procedure for IR surface-temperature measurements, *Proc 5th Semiconductor Thermal and Temp Symp*, San Diego, CA, Feb, 147–151.

Brown, R C and Jasper, S S (1989). Immersion cooling of electronics in fluidized beds of dielectric particles, *Heat Transfer Eng*, **10** (3), 36–42.

Buchmann, M (1989). Space-saving Peltier cooling elements, *Energy and Automation*, **21** (2), 20–21.

Cadre, M and Viault, A (1989). Thermal simulations for electronic equipment using the software package THEBES, *11th Intl Telecom Energy Conf*, Firenze, Italy, Oct, 1–8.

Carlson, D M, Sullivan, D C, Bach, R E, and Resnick, D R (1989). The ETA10 liquid-nitrogen-cooled supercomputer system, *IEEE Trans Electron Devices*, **ED-36** (8), Aug, 1404–1413.

Chan, B (1989). The evaluation of various infrared reflow machines for use with molded circuit boards, *Proc 9th Ann Intl Elec Packaging Soc (IEPS) Conf*, **1**, San Diego, CA, Sep, 188–198.

Chao, S-L (1989). Cooling the MV/40000 series superminicomputer systems, *Proc 9th Ann Intl Elec Packaging Soc (IEPS) Conf*, **1**, San Diego, CA, Sep, 69–79.

Chida, J, Ohara, M, Satoh, M, and Usui, Y (1989). A distributed power supply system using new heat dissipation techniques, *Proc 11th Intl Telecom Energy Conf*, Firenze, Italy, Oct.

Childres, W S, and Peterson, G P (1989). Quantification of thermal contact conductance in electronic packages, *Proc 5th Semiconductor Thermal and Temp Symp*, San Diego, CA, Feb, 30–36.

Cokmez-Tuzla, A F, Tuzla, K, and Chen, J C (1989). Natural circulation boiling for cooling of chips in cryogenic nitrogen, *ASME*, **HTD-111**, Aug, 1–5.

Comerford, R (1989). Circuit simulation opens a new frontier: Thermal effects, *Test and Measurement World*, Jan, 61–69.

Cooke, J A and Lee, S W (1989). Finite element thermal analysis of 144 pin plastic flat packs, *Proc 5th Semiconductor Thermal and Temp Symp*, San Diego, CA, Feb, 59–62.

Cox, N (1989). Miniaturized fan for printed circuit boards, U.S. Patent 4,885,488, Dec.

Darveaux, R, Hwang, L-T, Reisman, A, and Turlik, I (1989). Thermal analysis of a multi-chip package design, *J Electronic Materials*, **18** (2), 267–274.

Darveaux, R, Hwang, L-T, Reisman, A, and Turlik, I (1989). Thermal/stress analysis of a multi-chip package design, *Proc 39th Elec Comp Conf*, Houston, TX, May, 668–671.

Dettmer, E S, Romenesko, B H, Charles, H K, Carkhuff, B G, and Merrill, D J (1989). Steady state thermal conductivity measurements of AlN and SiC substrate materials, *Proc 39th Elec Comp Conf*, Houston, TX, May, 551–556.

Di Barba, P and Savini, A (1989). Nonlinear thermal behavior of a plastic leaded chip carrier package, *IEEE Trans Comp, Hybrids, Manuf Tech*, **CHMT-12** (4), Dec, 741–744.

Eckert, E G, Goldstein, R J, Pfender, E, Ibele, W E, Patankar, S V, Simon, T W, Decker, N A, Lee, H, Girshick, S L, Strykowski, P J, Tamma, K K, Bar-Cohen, A, and Heberlein, J V (1990). Heat transfer — A review of 1989 literature, *Intl J Heat and Mass Transfer*, **33** (11), Nov, 2349–2437.

Eckert, E R G, Goldstein, R J, Pfender, E, Ibele, W E, Patankar, S V, Simon, T W, Decker, N A, Lee, H, Girshick, S L, Scott, C J, Strykowski, P J, and Tamma, K K (1989). Heat transfer — A review of 1988 literature, *Intl J Heat and Mass Transfer*, **32** (12), Dec, 2211–2280.

Eibeck, P A, Clauss, N S, and Cohen, D J (1989). Prediction of temperature distributions and thermal expansions within a fixed-disk drive storage system, *ASME J Elec Packaging*, **111**, Sep, 220–227.

Estes, R C (1989). Thermal characterization of chip-on-board packaging mechanisms, *ASME*, **HTD-111**, Aug, 183–190.

Estes, R C and Pernice, R F (1989). Die attach adhesives — Evaluation of VceSat and θ_{JC} performance in power devices, *Proc 1989 Intl Symp on Microelec*, Baltimore, MD, Oct, 664–669.

Estreich, D B (1989). A DC technique for determining GaAs Mesfet thermal resistance, *Proc 5th Semiconductor Thermal and Temp Symp*, San Diego, CA, Feb, 136–139.

Evans, H E, O'Hara, J P, and Viswanadham, P (1989). Hot bar attachment: Tab applications, *Proc 9th Ann Intl Elec Packaging Soc (IEPS) Conf*, **2**, Sep, 1108–1127.

Fisher, N J and Yovanovich, M M (1989). Thermal constriction resistance of sphere/layered flat contacts: Theory and experiment, *ASME J Elec Packaging*, **111**, May, 249–256.

Foley, J M, Pluze, R L, Larson, R L, and Doumani, G (1989). Heat exchange element, U.S. Patent 4,860,824, Aug.

Friedman, H S (1989). Thermal package for electronic components, U.S. Patent 4,887,147, Dec.

Fukuoka, Y and Ishizuka, M (1989). An application of the thermal network method to the thermal analysis of multichip packages (proposal of a simple thermal analysis model), *Japanese J Appl Physics*, **28** (9), Sep, 1578–1585.

Funari, J, Green, M C, Reynolds, S D, and Sammakia, B G (1989). Electronic package with improved heat sink, U.S. Patent 4,849,856, Jul.

Gabuzda, P G and Terrell, S V (1989). Directed air management system for cooling multiple heat sinks, U.S. Patent 4,851,965, Jul.

Galloway, J E and Mudawar, I (1989). Boiling heat transfer from a simulated microelectronic heat source to a dielectric liquid film driven by a rotating stirrer, *ASME*, **HTD-111**, Aug, 67-78.

Glaser, J C and Julaire, M P (1989). Thermal and structural analysis of a PLCC device for surface mount processes, *ASME J Elec Packaging*, **111**, Sep, 172–178.

Gore, J P, DiMarzo, M, Resch, C, and Pecht, M (1989). Verification of thermal analysis of PWBs for RAMCAD, *Proc 1989 Ann Rel and Maint Symp*, Atlanta, GA, Jan, 474–478.

Hadim, A, Chang, A T, Chu, A, and Yskamp, A (1989). An interfacing software package for thermal analysis: Application to microelectronics, *ASME J Elec Packaging*, **111**, Mar, 54–60.

Hall, D A (1989). Prediction of electronic component temperatures at high altitudes using low altitude measurements, *Proc 5th Semiconductor Thermal and Temp Symp*, San Diego, CA, Feb, 121–125.

Hamadah, T T (1989a). Air jet impingement cooling of an array of simulated electronics packages, *ASME*, **HTD-111**, Aug, 97–105.

Hamadah, T T (1989b). Impingement cooling of a simulated electronics package with a square array of round jets, *ASME*, **HTD-111**, Aug, 107–112.

Hamburgen, W R (1989). Apparatus and method for cooling electronic component packages using an array of directed nozzles fabricated in the circuit board, U.S. Patent 4, 839,774, Jun.

Hannemann, R (1989). Thermal control for mini- and microcomputers: The limits of air cooling, *Bulletin Intl Center for Heat and Mass Transfer*, **3**, 65–83.

Hayward, J D (1989). Design of optimization of a VLSI heat sink using computational fluid dynamics, *Proc 9th Ann Intl Elec Packaging Soc (IEPS) Conf*, **1**, San Diego, CA, Sep, 328–333.

Herrell, D J and Gibson, D A (1989). Method of assembling a fluid-cooled integrated circuit package, U.S. Patent 4,860,444, Aug.

Hiraki, M, Kano, T, Kizu, H, Satoh, H, Yamashita, S, Kobayashi, K, Tanaka, H, Arai, Y, Satoh, K, and Fuchigami, K (1989). Water-cooled switching supply, *Proc 11th Intl Telecom Energy Conf*, Firenze, Italy, Oct.

Hohenwarter, G K and Grange, J A (1989). Cryogenic fluid delivery system, U.S. Patent 4,870,830, Oct.

Hollworth, B R and Durbin, M (1989). Impingement cooling of electronics, *ASME*, **HTD-111**, Aug, 89–96.

Hollworth, B R, Maxwell, K A, and Ukeiley, L S (1989). Impingement cooling of electronics: Effects of venting through circuit boards, *Proc 9th Ann Intl Elec Packaging Soc (IEPS) Conf*, **1**, San Diego, CA, Sep, 80–96.

Huang, H-C and Lee, R-S (1989). Thermal analysis and optimal design of general heat spreader, *Proc 9th Ann Intl Elec Packaging Soc (IEPS) Conf*, **1**, San Diego, CA, Sep, 317–327.

Incropera, F P (1989). Liquid immersion cooling of electronic components, *Bulletin Intl Center for Heat and Mass Transfer*, **3**, 29–64.

Jacobson, D W (1989). Importance of accurate thermal analyses, *Proc 1989 Ann Rel and Maint Symp*, Atlanta, GA, Jan, 465–469.

Jaeger, R C, Goodling, J S, Baginski, M E, Ellis, C D, Williamson, N V, and O'Barr, R M (1989). High heat flux cooling for silicon-on-silicon packaging, *Proc 5th Semiconductor Thermal and Temp Symp*, San Diego, CA, Feb, 104–111.

Jensen, R H, Andrejack, G A, Button, D P, and Bydal, B A (1989). Comparative thermal performance of various substrate materials in a simple packaging application — Actual versus predicted, *Proc 39th Elec Comp Conf*, Houston, TX, May, 572–576.

Johnson, M A (1989). The performance of electronic devices mounted on aluminum nitride, *Proc 39th Elec Comp Conf*, Houston, TX, May, 562–565.

Jones, J E and Bulmahn, D J (1989). Heat sink for an electronic device, U.S. Patent 4,872,505, Oct.

Kawashima, H, Okada, T, and Yamamoto, H (1989). Apparatus for supplying cooling fluid, U.S. Patent 4,865,123, Sep.

Keil, R W and Berger, D W (1989). Thermal characterization of integrated circuit packages in an automated wind tunnel, *Proc 9th Ann Intl Elec Packaging Soc (IEPS) Conf*, **2**, 937–952.

Khalilollahi, A and Sammakia, B G (1989). Transient mixed convection in air adjacent to discrete uniform flux heat sources, *ASME*, **HTD-121**, Dec, 55–60.

Kolte, J and Zirath, H (1989). A system comparison between water- and air-borne cooling systems for telephone exchanges, *Proc 11th Intl Telecom Energy Conf*, Firenze, Italy, Oct.

Komoto, M, Matsuo, Y, and Sano, T (1989). Liquid cooling dystem for integrated circuit chips, U.S. Patent 4,854,377, Aug.

Krause, W B, Klasi, M L, and Addala, S R (1989). Conjugate heat transfer from isothermal surface mounted heat sources embedded in a large substrate, *ASME*, **HTD-111**, Aug, 161–168.

LaBonte, M (1989a). Thermal analysis and reliability prediction tools, *Proc Natl Elec Packaging and Prod Conf — NEPCON East*, Boston, MA, Jun, 313–320.

LaBonte, M (1989b). Achieving accuracy in thermal analysis, *Electron Eng (UK)*, **61** (756), Dec, 35–42.

Lau, K S and Mahajan, R L (1989a). Convective heat transfer from longitudinal fin arrays in the entry region of turbulent flow, *ASME J Elec Packaging*, **111**, Sep, 213–219.

Lau, K S and Mahajan, R L (1989b). Effects of tip clearance and fin density on the performance of heat sinks for VLSI packages, *IEEE Trans Comp, Hybrids, Manuf Tech*, **CHMT-12** (4), Dec, 757–765.

Lau, J H (1989). Thermal stress analysis of SMT PQFP packages and interconnections, *ASME J Elec Packaging*, **111**, Mar, 2–8.

Lea, C (1989). Heat transfer fluids for vapour phase soldering — An appraisal, *Soldering & Surface Mount Tech (UK)*, **1**, Feb, 23–32.

Lee, C C, Min, J, and Palisoc, A (1989). An integration algorithm for the temperature solution of the four-layer infinite plate structure, *Proc 5th Semiconductor Thermal and Temp Symp*, San Diego, CA, Feb, 48–51.

Lee, S and Yovanovich, M M (1989). Conjugate heat transfer from a vertical plate with discrete heat sources under natural convection, *ASME J Elec Packaging*, **111** (4), Dec, 261–267.

Lee, S W (1989). Thermal evaluations of 144 pin plastic flat packs, *Proc 5th Semiconductor Thermal and Temp Symp*, San Diego, CA, Feb, 14–18.

Lee, T Y and Simon, T W (1989). High-heat-flux forced convection boiling from small regions, *ASME*, **HTD-111**, Aug, 7–16.

Lehmann, G L and Wirtz, R A (1989). The effect of variations in stream-wise spacing and length on convection from surface mounted rectangular components, *ASME J Elec Packaging*, **111**, Mar, 26–32.

Leung, C W and Probert, S D (1989). Steady-state heat transfer from vertical fins protruding upwards from horizontal bases, *J Inst of Energy (UK)*, **62** (451) Jun, 94–101.

Levine, M (1989). Solid state cooling with thermoelectronics, *Elec Packaging and Prod*, Nov, 74–77.

Linton, R L and Agonafer, D (1989). Thermal model of a PC, *ASME*, **HTD-121**, Dec, 69–72.

Lycans, R, Morrish, T, and Dyko, M (1989). Evaluation of thermal management techniques for surface mount SEM printed circuit boards, *Proc 9th Ann Intl Elec Packaging Soc (IEPS) Conf*, **2**, 961–975.

Maddox, D E and Mudawar, I (1989). Single- and two-phase convective heat transfer from smooth and enhanced microelectronic heat sources in a rectangular channel, *ASME J Heat Transfer*, **111**, Nov, 1045–1052.

Mahalingham, M (1989). Surface mount plastic packages — An assessment of their thermal performance, *Proc 5th Semiconductor Thermal and Temp Symp*, San Diego, CA, Feb, 5–13.

Mahaney, H V, Ramadhyani, S, and Incropera, F P (1989). Numerical simulation of three-dimensional mixed convection heat transfer from an array of discrete heat sources in a horizontal rectangular duct, *Num Heat Transfer*, Part A, **16**, 267–286.

Majumdar, A and Tien, C L (1989). Network analysis for contact conductance, *Proc 9th Ann Intl Elec Packaging Soc (IEPS) Conf*, **1**, San Diego, CA, Sep, 125–136.

Mazenq, S, Highgate, D, and Probert, S D (1989). Electronics cooling using a simple gas-fluidized bed, *Appl Energy (UK)*, **34**, 81–88.

McShane, M, Lin, P, and Wilson, H (1989). A unique low cost pin grid array package with heatspreader, *Proc 9th Ann Intl Elec Packaging Soc (IEPS) Conf*, **1**, San Diego, CA, Sep, 199–198.

Missaggia, L J, Walpole, J N, Liau, Z L, and Phillips, R J (1989). Microchannel heat sinks for two-dimensional high-power-density diode laser arrays, *IEEE J Quantum Elec*, **25** (9), Sep, 1988–1989.

Moffat, R J and Anderson, A M (1988). Applying heat transfer coefficient data to electronics cooling, *ASME*, **HTD-101**, Nov, 33–43.

Moore, L V (1989). Thermal evaluation of streaming tape and Winchester disk drives for use in digital cross-connect switches and other telecommunication equipment, *Proc 9th Ann Intl Elec Packaging Soc (IEPS) Conf*, **1**, San Diego, CA, Sep, 106–117.

Mudawar, I and Anderson, T M (1989a). High flux electronic cooling by means of pool boiling — Part I: Parametric investigation of the effects of coolant variation, pressurization, subcooling, and surface augmentation, *ASME*, **HTD-111**, Aug, 25–34.

Mudawar, I and Anderson, T M (1989b). High flux electronic cooling by means of pool boiling — Part II: Optimization of enhanced surface geometry, *ASME*, **HTD-111**, Aug, 35–49.

Mudawar, I and Maddox, D E (1989a). Critical heat flux in subcooled flow boiling of fluorocarbon liquid on a simulated electronic chip in a vertical rectangular channel, *Intl J Heat and Mass Transf*, **32** (2), Feb, 379–393.

Mudawar, I and Maddox, D E (1989b). Enhancement of critical heat flux from high power microelectronic heat sources in a flow channel, *ASME*, **HTD-111**, Aug, 51–58.

Nagurny, N J (1989). Thermal modeling and testing of surface mounted components on conduction cooled military standard electronic modules, *Proc 9th Ann Intl Elec Packaging Soc (IEPS) Conf*, **1**, San Diego, CA, Sep, 375–384.

Naik, S, Babus'Haq, R F, O'Callaghan, P W, and Probert, S D (1989). Heat losses via a bolted contact from an electronic component, *Appl Energy (UK)*, **34** (4), 317–22.

Nakanishi, K and Yamada, M (1989). Integrated circuit chips cooling module having coolant leakage prevention device, U.S. Patent 4,870,477, Sep.

Nakayama, W, Nakajima, T, Ohashi, S, and Kuwahara, H (1989). Modeling of temperature transient of microporous studs in boiling dielectric fluid after stepwise power application, *ASME*, **HTD-111**, Aug, 17–23.

Negus, K J, Franklin, R W, and Yovanovich, M M (1989). Thermal modeling and experimental techniques for microwave bipolar devices, *Proc 5th Semiconductor Thermal and Temp Symp*, San Diego, CA, Feb, 63–72.

Nelson, R D and Gupta, O R (1989). End fed liquid heat exchanger for an electronic component, U.S. Patent 4,884,630, Dec.

Nonn, T, Dagan, Z, and Jiji, L M (1989). Jet impingement flow boiling of a mixture of FC-72 and FC-87 liquids on a simulated electronic chip, *ASME*, **HTD-111**, Aug, 121–128.

Ocken, A G, Charlier, M L, and Kowieski, R J (1989). Heat sink assembly for densely packed transistors, U.S. Patent 4,872,089, Oct.

Ohbayashi, Y, Hashimoto, K, and Sakurai, H (1989). Cross flow cooling fan device, U.S. Patent 4,859,888, Aug.

Osterman, M D and Pecht, M (1989). Component placement for reliability on conductively cooled printed wiring boards, *ASME J Elec Packaging*, **111** (2), Jun, 149–156.

Parks, H L (1989). Three-dimensional microelectronic package for semiconductor chips, U.S. Patent 4,841,355, Jun.

Patankar, SV and Spalding, DB (1972). A calculation procedure for heat, mass, and momentum transfer in three-dimensional parabolic flows, *Intl J Heat and Mass Transfer*, **15**, 1787–1806.

Patankar, S V (1980). *Numerical Heat Transfer and Fluid Flow*, McGraw-Hill, New York, NY.

Phelps, D W and Ward, W C (1989). Heat dissipation package for integrated circuits, U.S. Patent 4,878,108, Oct.

Prudhoe, R K and Doukas, L (1989). Temperature limiting systems for environmental reliability of telecommunication equipment enclosures, *Proc 11th Intl Telecom Energy Conf*, Firenze, Italy, Oct.

Pshaenich, A (1989). 300A Δ_F tester, *Proc 5th Semiconductor Thermal and Temp Symp*, San Diego, CA, Feb, 140–145.

Rabinovich, E M (1989). Ceramic materials for electronic packaging, *ASME J Elec Packaging*, **111**, Sep, 183–191.

Rajakumar, C and Johnson, D (1989). Finite element predictions of free convection heat transfer coefficients of simulated electronic circuit boards, *ASME J Elec Packaging*, **111** (2), Jun, 129–134.

Ramadhyani, S, Moffatt, D F, and Incropera, F P (1985). Conjugate heat transfer from small isothermal sources embedded in a large substrate, *Intl J Heat and Mass Transf*, **28**, 1945-1952.

Rizk, T A and Kleinstreuer, C (1989). Forced-convection cooling of a linear array of blocks in open and porous matrix channels, *ASME*, **HTD-111**, Aug, 155–159.

Saaski, E, Hannemann, R J, and Fox, L R (1989). Integral heat pipe module, U.S. Patent 4,833,567, May.

Sammakia, B, Vadala, P, and Homa, T (1989). A transient technique to measure the temperature coefficient of resistance for thin film resistors, *ASME J Elec Packaging*, **111** (2), June, 143–148.

Schnipke, R J, Hayward, J D, and Rice, J G (1989). A fluid flow and heat transfer analysis for evaluating the effectiveness of an IC package heat sink, *Proc 5th Semiconductor Thermal and Temp Symp*, San Diego, CA, Feb, 81–87.

Schulenburg, N, Kvamme, E, Nelson, A, Roh, J, Phillips, J R, Mansingh, V, and Blackman, J (1989). Evaluation of air-cooled microchannel heat sinks, *Proc 9th Ann Intl Elec Packaging Soc (IEPS) Conf*, **2**, San Diego, CA, Sep, 913–923.

Sihlbom, A and Svärd, H (1989). New reliable methods of measuring temperature distribution on printed circuit boards during operations, *Proc 9th Ann Intl Elec Packaging Soc (IEPS) Conf*, **1**, San Diego, CA, Sep, 97–105.

Simon, B R, Yuan, Y, Umaretiya, J R, Prince, J L, and Staszak, Z J (1989). Parametric study of a VLSI plastic package using locally refined element models, *Proc 5th Semiconductor Thermal and Temp Symp*, San Diego, CA, Feb, 52–58.

Simons, R E (1988). Bibliography of heat transfer in electronic equipment — 1986, in *Advances in Thermal Modeling of Electronic Components and Systems — Volume 1*, A Bar-Cohen and A D Kraus, eds, Hemisphere Pub Co, New York, NY.

Simons, R E (1990). Bibliography of heat transfer in electronic equipment — 1987 and 1988, in *Advances in Thermal Modeling of Electronic Components and Systems — Volume 2*, A Bar-Cohen and A D Kraus, eds, ASME Press, New York, NY.

Smolley, R (1989). Liquid cooled high density packaging for high speed circuits, U.S. Patent 4,847,731, Jul.

Somos, I L, Piccone, D E, and Tobin, W H (1989). Two-dimensional heat flow in a GTO having a highly interdigitated emitter, *Proc 5th Semiconductor Thermal and Temp Symp*, San Diego, CA, Feb, 73–80.

Song, S and Yovanovich, M M (1989). Contact interface gas heat transfer: A method of measuring thermal accommodation coefficient, *Proc 9th Ann Intl Elec Packaging Soc (IEPS) Conf*, **2**, Sep, 925–936.

Spigarelli, D J (1989). Thermal separation in surface mount attachment and rework, *Proc Natl Elec Packaging and Prod Conf — NEPCON West*, **2**, Anaheim, CA, Mar, 1877–1888.

Spigarelli, D J (1989). Development of convector technology for FPD and TAB soldering, *Proc 1989 Intl Symp on Microelec*, Baltimore, MD, Oct, 635–643.

Spitz, S L (1989). Ceramics keep circuits cool, *Elec Packaging and Prod*, Jul, 36–41.

Strom, B D, Carey, V P, and McGillis, W R (1989). An experimental investigation of the critical heat flux conditions for subcooled convective boiling from an array of simulated microelectronic devices, *ASME*, **HTD-111**, Aug, 135–142.

Subramanyam, S and Rolph, W D (1989). Thermal analysis of direct chip mount electronic systems using ADINA-T, *Computers and Structures* (UK), **32** (3–4), 853–859.

Suryanarayana, D (1989). Thermally conductive ceramics for electronic packaging, *ASME J Elec Packaging*, **111**, Sept, 192–198.

Suzuki, M, Udagawa, Y, and Yamamoto, H (1989). Conductive liquid cooling for the FACOM VP2000 supercomputer, *Proc 9th Ann Intl Elec Packaging Soc (IEPS) Conf*, **1**, San Diego, CA, Sep, 118–124.

Swager, A W (1989). Thermostat packaging adds design flexibility, *EDN*, Sep, 63–68.

Tanabe, F, Asada, O, and Tsuji, K (1989). Thermo-history control simulation in thermal transfer printing, *Elec and Comm in Japan*, **72** (1), 76–88.

Tanzer, H J and Numa, C T (1989). Thermal performance of PWB heat pipe substrates during exposure to dynamic environments, *Proc 9th Ann Intl Elec Packaging Soc (IEPS) Conf*, **2**, 976–989.

Tarter, T S (1989). A novel circuit for the evaluation of thermal impedance characteristics of MOS integrated circuits, *Proc 5th Semiconductor Thermal and Temp Symp*, San Diego, CA, Feb, 131–135.

Thomson, M, Ruel, C, and Donato, M (1989). Characterization of a flat plate heat pipe for electronic cooling in a space environment, *ASME*, **HTD-111**, Aug, 59–65.

Tousignant, L and Maddox, D E (1989). An overview of electronic cooling for a range of applications, *Proc Natl Elec Packaging and Prod Conf — NEPCON West*, **1**, Anaheim, CA, Mar, 185–187.

Tseng, A A (1989). Thermal analysis of a new IMPATT semiconductor device, *ASME J Elec Packaging*, **111** (2), Jun, 135–142.

Tustaniwakyj, J I and Halkola, K G (1989). Liquid cooled multi-chip integrated circuit module incorporating a seamless compliant member for leakproof operation, U.S. Patent 4,879,629, Nov.

Vogelsong, R S and Brezezinski, C (1989). Extending SPICE for electro-thermal simulation, *Proc IEEE 1989 Custom Integrated Circuits Conf*, IEEE Cat. No. CH2671-6/89/0000-0131.

Wadsworth, D C and Mudawar, I (1989). Cooling of a multichip electronic module by means of confined two-dimensional jets of dielectric liquid, *ASME*, **HTD-111**, Aug, 79–87.

Wall, C (1989a). Liquid heat sink controls computer heat flux, *Elec Packaging and Prod*, Feb, 78–79.

Wall, C (1989b). Thermal management of a microcomputer using a liquid heat sink, *Elec Manuf*, Jun, 19–20.

Weiss, J (1989). Thermal analysis tools for electronic systems design, *Proc Natl Elec Packaging and Prod Conf — NEPCON East*, Boston, MA, Jun, 324–327.

Weiss, J, Fortner, P, Pearson, B, Watson, K, and Monroe, T (1989). Modeling air flow in electronic packages, *Mech Eng*, Sep, 56–58.

Wellins, A and Sullivan, B (1989). Using a PC-based thermal analyzer, *Printed Circuit Design*, Aug, 11–13.

Wilson, E A (1989). Thermal management design for a large mainframe, *Elec Packaging and Prod*, Feb, 142–144.

Wirtz, R A, Hollworth, B R, and Fuller, H A (1985). An infrared thermographic study of convection in an array of surface mounted components, *5th Ann Intl Elec Packaging Soc (IEPS) Conf*, 497–507.

Wirtz, R A and McAuliffe, W (1989). Experimental modeling of convection downstream from an electronic package row, *ASME J Elec Packaging*, **111**, Sep, 207–212.

Witzman, S, Newport, D, and Nicoletta, T (1989). Free convection air cooling of the electronic equipment: Still a miracle waiting to be explored, *Proc 5th Semiconductor Thermal and Temp Symp*, San Diego, CA, Feb, 98–103.

Yamamoto, H, Suzuki, M, Udagawa, Y, Nakata, M, Katsuyama, K, Ono, I, and Kikuchi, S (1989). Cooling system for an electronic circuit device, U.S. Patent 4,879,632, Nov.

Yao, S C (1989). Thermal modeling of printed circuit boards, *Elec Packaging and Prod*, Feb, 94–96.

Yao, S C, Deb, S, and Hammouda, N (1989). Impacting spray boiling for thermal control of electronic systems, *ASME*, **HTD-111**, Aug, 129–134.

Yokono, Y and Ishizuka, M (1989). Thermal studies on finned LSI packages under forced convection, *Proc 5th Semiconductor Thermal and Temp Symp*, San Diego, CA, Feb, 20-24.

Yokouchi, K, Kamehara, N, Niwa, K, Udagawa, Y (1989). High-capability coolant for cryogenic devices, *Proc NEPCON EAST*, 1023–1028.

You, H I (1989). Experimental determination of heat transfer coefficient between a microstructure and fluid, *Intl Comm Heat and Mass Transf*, **16**, 537–546.

Zebib, A and Wo, Y K (1989). A two-dimensional conjugate heat transfer model for forced air cooling of an electronic device, *ASME J Elec Packaging*, **111**, Mar, 41–45.

Zushi, S, Miyamoto, M, Gou, H, and Kobayashi, F (1989). Electronic apparatus cooling system, U.S. Patent 4,837,663, Jun.

NAME INDEX

C

D

E

F

G

H

I

J

K

L

M

N

O

P

R

S

T

U

V

W

Y

Z

SUBJECT INDEX

D

G

Q

R

T

V

CONTENTS OF VOLUME 1

CONTENTS OF VOLUME 2